GENERAL
ZOOLOGY

Tenth Edition · Complete Version

GENERAL ZOOLOGY

LABORATORY GUIDE

Charles F. Lytle
North Carolina State University

J. E. Wodsedalek
Late of the University of Minnesota

wcb
Wm. C. Brown Publishers
Dubuque, Iowa

Book Team

Editor
Kevin Kane

Designer
Jeanne M. Rhomberg

Production Editor
Michelle M. Kiefer

Photo Research Editor
Shirley Charley

Permissions Editor
Carla D. Arnold

Product Manager
Matt Shaughnessy

wcb group

Wm. C. Brown
Chairman of the Board

Mark C. Falb
President and Chief Executive Officer

wcb

Wm. C. Brown Publishers, College Division

G. Franklin Lewis
Executive Vice-President, General Manager

E. F. Jogerst
Vice-President, Cost Analyst

George Wm. Bergquist
Editor in Chief

Edward G. Jaffe
Executive Editor

Beverly Kolz
Director of Production

Chris C. Guzzardo
Vice-President, Director of Sales and Marketing

Bob McLaughlin
National Sales Manager

Marilyn A. Phelps
Manager of Design

Julie A. Kennedy
Production Editorial Manager

Faye M. Schilling
Photo Research Manager

Cover photo credit: Douglas Faulkner/Sally Faulkner Collection

Contents

Preface

Zoology, like many other branches of science, has made significant progress during recent years. Based on the accumulated knowledge of several generations of zoologists, recent investigators have found important new applications of zoology in the health sciences, agriculture, and environmental sciences and have expanded our understanding of basic life processes.

A solid foundation of basic zoology is essential in the preparation of students training for careers in biology, zoology, genetics, physiology, medicine, veterinary medicine, agriculture, environmental science, and many other fields. The laboratory experience is a vital part of this training and furthers an understanding of the structure and function of animals. For nearly five decades this laboratory guide has assisted students and teachers in their laboratory work with accurate information, clear directions, and quality illustrations. This tenth edition continues the tradition of providing a comprehensive guide to the common types of animals representing the major animal phyla. We have attempted to provide a helpful introduction to the major groups and representative forms, incorporating new data and interpretations while retaining important historical perspectives.

Since this book is intended to aid students and teachers in many diverse institutions with different schedules, resources, and preferences, we have chosen to be inclusive rather than exclusive in our coverage. Thus, we have included more material than most zoology instructors will be able to incorporate into their courses. We hope that you will select those chapters, sections, and animal types most appropriate for your course.

Changes in This Edition

The principal changes in this guide are the addition of *performance objectives at the beginning of each chapter, 39 new illustrations,* and the *elimination of the chapter on ecology.* We have followed the suggestion of several teachers and revised some chapters to place *greater emphasis on physiology, behavior,* and *comparisons among animals within and among phyla.* Where appropriate, we have introduced new data and presented new interpretations of animal structure, function, and relationships.

Other important additions include new exercises for the malaria parasite *Plasmodium,* the Portuguese man-of-war *Physalia,* the garden spider *Argiope,* and Phylum Onychophora, and new classifications for the Protozoa, Arthropoda, and Echinodermata. We have also presented suggestions on the handling and care of animals in the laboratory and some important safety precautions for the handling of preserved specimens.

The new illustrations include 15 new photographs and 24 new line drawings. Ten of these illustrations are in chapter 13, Arthropoda, and consist of four photographs (trilobite, spider fang, crayfish appendages, crayfish dissection) and six drawings (spider anatomy). Two new composite drawings illustrate the diversity of the classes of molluscs and echinoderms. Several drawings replace ones in the previous edition to better illustrate certain animal features.

Basic Features of This Manual

In this tenth edition we have retained the basic organization, approach, and educational features of the previous edition. These features are intended to assist students and teachers in completing the various exercises.

A *Correlation Chart* at the beginning of the manual will help teachers and students coordinate the laboratory exercises with one of the major zoology texts in current use. Some important features in the manual include *boldface headings* within each chapter, indicating the major divisions of the exercise, and *boldface terms* throughout the text, emphasizing terminology students should learn and understand. Many chapters provide *space for student drawings* of specified animals or structures. In some drawings labels have been omitted, and students are asked to supply them. The manual also includes *blank tables and graphs* for students to complete with data from their laboratory observations and/or experiments.

Every chapter begins with a list of *Objectives* that point out specific things students should accomplish in each exercise. I have used such objectives in my own teaching for many years, and students and other instructors have found that the objectives help focus attention on the important tasks in the laboratory exercises. I also suggest that instructors use the list of objectives when composing lab quizzes, to ensure that students are tested on important concepts, processes, and facts. The objectives also provide excellent guidance for students studying for a quiz.

Most chapters contain a brief *Introduction,* with background information on the topic, and a *Materials List* for the exercises. Many chapters have one or more lists of *Demonstrations,* which are suggested to supplement the main activities of each exercise. Toward the end of each chapter is a list of *Key Terms* introduced in that chapter; and, with the exception of chapter 9, all the chapters have blank pages at the end for additional *Notes and Sketches.* At the conclusion of a laboratory exercise, each student will have a complete, consolidated record of his or her lab work. Thus, all of the necessary study materials for laboratory quizzes and for review will be in one place.

Acknowledgments

I am fortunate to have had the constructive comments and suggestions of many teachers for the improvement of this laboratory guide. I continue to encourage such suggestions; they keep me in touch with the changing needs and preferences of teachers and students across the nation and in other countries.

I wish to thank also several colleagues who have provided information, materials, or reviewed portions of the manuscript, including Dr. Phyllis Bradbury, Dr. Edward Lyke, Dr. Robin Leech, Dr. John Mackenzie, Dr. Jane Westfall, Dr. G. C. Miller, Dr. Maurice Farrier and Mr. Charles M. Williams. Dr. Robin Leech provided most of the material and did the drawings for the new section on *Argiope.* Creseda Buchanan helped with several dissections. Carol Majors prepared most of the new drawings and did several of the photographs for this edition. Other new photographs were provided by Dr. Wendell McKenzie, Dr. Thomas Bouillon, Dr. Jane Westfall, North Carolina State University, Indiana University Audiovisual Center, and Philips Electronic Instruments, Inc. Carolina Biological Supply Company provided important information on the safe handling of chemical preservatives.

I would also like to thank Dr. Gwilym S. Jones of Northeastern University and Dr. Neal D. Mundahl of Miami University (Ohio) for their critical comments and special help with this revision.

I am grateful to all who assisted in the preparation of this edition, particularly my wife, Carol, for her encouragement, as well as for her editorial and secretarial assistance.

Special Note on the Care and Treatment of Laboratory Animals

Experience in observing, handling, caring for, experimenting with, and dissecting animals is essential for the training of zoology students. Prospective employers and graduate and professional school admissions committees often cite the importance of such experience. Studies from textbooks, photographs, charts, models, and computer simulations are not adequate substitutes for direct laboratory experience with living and preserved animals.

For this reason, it is important for every zoology student to learn the proper methods of handling and caring for animals. As a student of zoology and of nature, you should have a basic respect for all animal life and for the concerns of others about the humane treatment and conservation of animals.

Always handle living animals with care and respect. With both vertebrate and invertebrate animals, take adequate precautions to avoid causing unnecessary pain or discomfort from your handling or experimenting. Any animals kept in the laboratory should have a clean and appropriate environment, adequate food and water, and regular care. At the conclusion of your studies, animals should either be disposed of in a humane manner or returned to a permanent animal care or culture facility, as directed by your instructor.

Your instructor should be familiar with the Animal Welfare Act and other relevant federal and state laws governing the handling and use of animals.

Safety Precautions When Using Preserved Animals

The chemicals used to preserve animals and parts of animals can be toxic and dangerous if used improperly or under improper conditions. Ethanol, isopropanol, formaldehyde, phenol, and ethylene glycol are commonly used preservatives. The following information, supplied through the courtesy of the Carolina Biological Supply Company, provides some excellent safety guidelines to follow when handling and dissecting preserved animal specimens.

All teachers are responsible for implementing proper safety procedures when using potentially hazardous chemicals and for communicating appropriate information about these materials to their students in accordance with applicable federal, state, and local regulations.

To achieve the necessary level of safety in the laboratory, the instructor should be familiar with the chemicals present and the precautions to be taken when using these chemicals.

Carolina provides specimens preserved in alcohol, formaldehyde, and *Carosafe*® (contains ethylene glycol). We recommend you follow these safety tips whenever preserved specimens or chemicals are used:

1. Wear safety glasses at all times.
2. Do not wear contact lenses.
3. Wear appropriate gloves and lab coat.
4. Work only in a well-ventilated area.
5. Prohibit eating, drinking, and smoking in the work area.
6. In the event of contact, wash skin with soap and water; flush eyes with water.

Special chemicals may present hazards which require additional precautions to those listed above.

Formaldehyde should always be used in a well-ventilated area to prevent irritation to the eyes, skin, or respiratory tract. The use of goggles lessens eye irritation from formaldehyde vapors. If direct contact to eyes or skin occurs, wash thoroughly with water. Smoking should not be allowed.

Isopropanol is very flammable, so avoid sparks, open flames, and heat. Direct contact with isopropanol should be avoided through use of safety glasses, gloves, and lab coats. Wash thoroughly with water if direct contact to eyes or skin occurs.

Carosafe® is the safest of the three preservatives, but safety glasses should be worn to avoid direct contact with the eyes. Wash eyes thoroughly with water if contact occurs.

When working with preserved materials, be careful with sharp objects such as pins, scalpels, and the spines and teeth of specimens. When using a scalpel, we recommend cutting away from oneself and keeping fingers out of the cutting path.

Comparative Safety Data of Preservatives

	Formaldehyde	Isopropanol	Ethylene Glycol
Physical Data			
Hazardous Components (OSHA)	Methanol (TWA 200 ppm) Formaldehyde (TWA 3 ppm)	Isopropanol (TWA 400 ppm)	Ethylene Glycol (NR)
Flash Point	184° F (Combustible)	53° F (Flammable)	241° F
Explosion Limits (lower)	7%	2%	3.2%
Extinguishing Media	Alcohol foam, Water fog, Carbon dioxide, Dry chemical	Alcohol foam, Carbon dioxide, Dry chemical	Water fog, Carbon dioxide, Dry chemical
Unusual Fire Hazard	Vapor heavier than air; may travel along ground to ignition source	None	None
Threshold Limit Value (NIOSH)	2 ppm (air)	air: 400 ppm (skin)	100 ppm (vapor)
Effects of Overexposure			
Eyes	Vapor causes severe irritation, redness, tearing, blurred vision. Liquid may cause severe or permanent damage	Direct contact may cause irritation	Direct contact may cause irritation
Skin (Direct contact)	Mild irritation, dermatitis, possible sensitization	Mild irritation	Mild irritation (Reddening of skin after 24 hours of continuous exposure)
Inhalation	Irritation of respiratory tract, dyspnea, headache, bronchitis, pulmonary edema, gastroenteritis	Irritation of respiratory tract; headache, nausea; at high concentrations, narcosis	Reported irritant effects at extremely high (10,000 mg/m^3) concentrations of vapor
Oral	rat LD$_{50}$ 800 mg/kg	rat LD$_{50}$ 5840 mg/kg	rat LD$_{50}$ 5840 mg/kg

TWA:	Time Weighted Average
NR:	Not Restricted
ppm:	Parts per million; air (unless otherwise specified)

Special Note on the Care and Treatment of Laboratory Animals

Correlation Chart

Units of study correlated with chapters in some leading zoology textbooks

Units of study in Lytle/ Wodsedalek *General Zoology Laboratory Guide,* 10th edition, 1987	Hickman et al. *Integrated Principles of Zoology,* 7th edition, 1984	Hickman et al. *Biology of Animals,* 4th edition, 1986	Storer et al. *General Zoology,* 6th edition, 1979	Storer et al. *Elements of Zoology,* 6th edition, 1979	Villee et al. *General Zoology,* 6th edition, 1984	Villee et al. *Introduction to Animal Biology,* 1979
Chapters						
1. Microscopy		2				
2. Animal Cells and Tissues	4, 5, 6	2	3	2	2	2
3. Mitosis and Meiosis	4, 35	2, 3	3, 10	2, 10	2, 13	2, 13
4. Development	6, 35, 36	13	10	10	13, 14	13
5. Protozoa	8	16	15	15	20	20
6. Porifera	9	17	16	16	21	21
7. Cnidaria (Coelenterata)	10	18	17	16	22	22
8. Platyhelminthes	11	19	18	17	23	23
9. Hints for Dissection	6	6	14	14	4	4
10. Pseudocoelomate Animals	12	20	19	17, 18	24	24
11. Mollusca	13	21	21	19	25	25
12. Annelida	14	22	22	20	26	26
13. Arthropoda	15, 16, 17	23	23, 24, 25	21, 22	27	27
14. Echinodermata	20	25	26	18	29	29
15. Chordata	22	26	27	23	30	30
16. Shark Anatomy	23	27	29	24	31	31
17. Perch Anatomy	23	27	30	24	31	31
18. Frog Anatomy	24	28	31	25	32	32
19. Fetal Pig Anatomy	27	31	34	27	34	34
20. Animal Behavior	34	14	—	—	36	36

1
Microscopy

Objectives

After completing the laboratory work in this chapter, you should be able to perform the following tasks:

1. Identify the main parts of a compound microscope and explain their function.
2. Define and explain focus, working distance, resolving power, and magnification.
3. Describe the proper use and care of a compound microscope.
4. Explain the difference between compound and stereoscopic microscopes and give examples of appropriate uses of each type.
5. Use both compound and compound light microscopes in the correct way.
6. Explain the operating principles of a phase contrast microscope and give examples of its use.
7. Describe the two main types of electron microscopes and give examples of their use.
8. Compare the design principles of a transmission electron microscope, a scanning electron microscope, and a compound electron microscope.
9. Explain the importance of microscopes in biological studies.

The Compound Microscope

The **compound microscope** is one of the most important and useful tools of the zoologist. It is used to study cells and cell parts, the organization of tissues, the structure of bone, and the structure of developing embryos, among many other important applications. Since many of the exercises in this course will require the use of the compound microscope, it is important to review some aspects of its construction, use, and care.

A modern compound microscope is illustrated in figure 1.1. Since there are numerous makes and models of compound microscopes in use, the microscope assigned for your use may differ slightly from the one illustrated. The operating principles and procedures, however, will be similar to those outlined later. Your laboratory instructor will point out any important differences between your microscope and the one illustrated.

A microscope is an expensive precision instrument and must be handled with care. Always carry your microscope with both hands. Grasp the arm of the microscope firmly with one hand and support the base with the other hand. Place it gently on your desk with the arm facing you.

Materials List

Living Specimens
  Artemia larvae
Prepared Microscope Slides
  Sample tissue slide for compound microscope exercise
Audiovisual Materials
  Wall charts showing parts of compound and stereoscopic microscopes

---

Parts of the Microscope

Identify the principal parts and controls of the microscope with the aid of figure 1.1. At the top of the microscope is the **eyepiece,** or ocular lens, which is inserted in an inclined **body tube.** Microscopes with one eyepiece and body tube are monocular microscopes; those equipped with two eyepieces and body tubes for simultaneous viewing with both eyes are binocular microscopes. Below the eyepiece is the **arm** attached to the **base.** Also attached to the arm is the movable **stage,** which holds a microscope slide or other object for viewing. **Stage (slide) clips** aid in holding the slide in position on the stage. Above the object is the **revolving nosepiece** with two or more **objective lenses.**

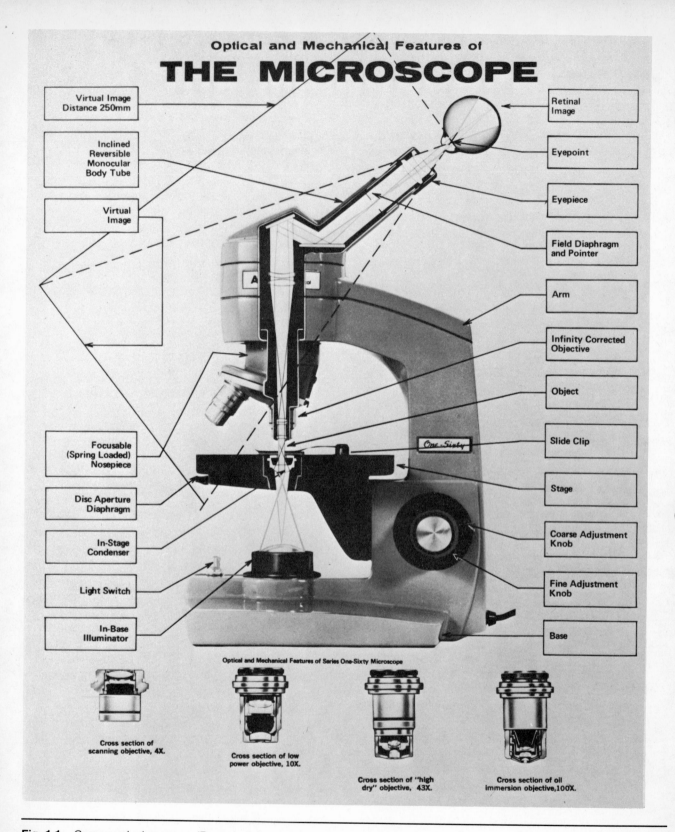

Fig. 1.1 Compound microscope. (Reproduced courtesy of Reichert Scientific Instruments.)

Within the stage is another lens system, the in-stage **condenser,** which serves to concentrate light rays from the in-base **illuminator.** Adjacent to the in-base illuminator is the **light switch.** Instead of an in-base illuminator, some microscopes have a **substage mirror** to reflect light from an auxiliary light source.

Beneath the condenser, locate the **disc aperture diaphragm;** rotating the disc diaphragm increases or decreases the amount of light on the specimen. More expensive microscopes often have an **iris diaphragm** with movable elements instead of a disc diaphragm. Raising and lowering the condenser also regulates the illumination of the specimen, although in most of your work in this course you will obtain satisfactory results by adjusting the condenser to the position that gives maximum illumination (usually near its uppermost position) and then making any further needed reductions in illumination with the disc diaphragm. This simplified method of controlling light does not produce the precise illumination required for advanced microscopy, but it produces results satisfactory for most routine purposes.

Accurate observation of a specimen requires positioning the objective lens at a specific distance from the specimen; this distance is determined by the specific construction of each objective lens and is called the **working distance** of the lens. When the objective lens is located at the proper working distance, the specimen will be in **focus.** The working distance of the lens varies **inversely** with the magnification of the objective; low-power objectives have longer working distances and high-power objectives have shorter working distances.

Focusing the lens system is accomplished by mechanically changing the distance between the specimen and the objective lens. Coarse and fine **adjustment knobs** for this purpose are provided on the side of the arm near the base. On some modern microscopes both coarse and fine adjustment are controlled by a single knob, but most models have separate controls. How many focusing control knobs are there on your microscope?

Magnification

The principal purpose of a microscope is to magnify the image of an object. The **magnification** of an object is determined by the construction of the ocular and objective lenses of the microscope, and the total magnification is the product of the separate magnification of these two lenses.

Example: 10X ocular × 10X objective = 100X total magnification

Student microscopes used in introductory biology and zoology courses are commonly equipped with 10X ocular lenses and both 10X and 43X objective lenses. These two lenses are mounted on a revolving nosepiece and are called the **low-power** and **high-power** objectives, respectively. Often, a 3.5X objective lens is also used on student microscopes; this is called a **scanning lens.** It is useful for viewing relatively large objects or for preliminary location of a specimen on the microscope slide. Occasionally a 90X or 100X **oil immersion lens** may be present for viewing very small objects like bacteria. Special instructions and precautions are needed for the use of oil immersion lenses.

Magnification is changed in microscopes with multiple objectives by rotating the revolving nosepiece until the desired objective is in position below the body tube of the microscope. A newer type of compound microscope (zoom lens type) uses a system of movable lenses to produce a variable magnification rather than a series of fixed magnifications as in microscopes equipped with a rotating nosepiece. What advantages and disadvantages of the continuously variable magnification obtained with a zoom-type microscope can you list?

Observe the 10X objective on your microscope and note that it is also marked 16 mm, the working distance of the lens. What is the significance of the working distance? What is the working distance of your 43X objective?

All the other parts of the microscope are accessory to the main purpose of magnification by the lenses. They consist mainly of mechanical devices to hold the specimen, to regulate the light necessary for clear vision, and to facilitate focusing.

Resolving Power

While magnification is the increase in size of the image of an object, the ability of a lens or a microscope to reveal the fine detail of a specimen is called its **resolving power.** In microscopy, an increase in the apparent size of a specimen is not always accompanied by an increase in the clarity of detail within the specimen. Beyond certain limits, further magnification simply makes the apparent image of a specimen become progressively fuzzy or indistinct as it becomes larger. The resolving power of a microscope depends largely upon the design and quality of its objective lenses, and it is really resolving power rather than magnification that limits the **useful** magnification of a compound microscope. Magnifications beyond approximately 2,000 diameters require an entirely different microscopic system.

In actual laboratory practice, the resolving power of a microscope can be determined by measuring the shortest distance between two points that can be visually distinguished as separate points.

Illumination

Proper illumination of the specimen is an extremely important matter, since improper lighting of the specimen can produce poor images, inaccurate observations, and/or unnecessary eyestrain.

If your microscope is equipped with a substage mirror, examine the two surfaces of the mirror—one surface is flat, and one surface is concave. The flat surface of the mirror is the one used most often in normal laboratory situations. When the mirror is properly adjusted, light is reflected by the mirror and passes through the condenser, the specimen, and the lenses to your eye. The concave (curved) surface of the mirror is used less frequently and only in certain situations where there is a need to concentrate the light rays. Adjust your mirror to reflect the maximum light on the specimen, and reduce the illumination as necessary by closing the iris diaphragm.

Illumination is also controlled by the substage condenser. For routine work in an introductory laboratory, the condenser should not be used to adjust illumination, but should be set at or near its uppermost position, and the light intensity adjusted by means of the iris diaphragm.

On microscopes equipped with in-base illuminators, the light level is adjusted by the disc (or iris) diaphragm and condenser only.

Focusing

To obtain a clear image of a specimen, you must carefully adjust the distance between the lenses and the specimen. This adjustment of the spatial relationships between the lenses and the specimen is called **focusing.** When a clear image of the specimen can be seen through the ocular lens, the specimen is said to be **in focus.** At this adjustment of the lens system, the specimen is located precisely at the working distance of the objective lens.

Locate once again the focusing control knob on your microscope (or the coarse adjustment knob if your microscope has separate coarse and fine focusing controls) and turn the control slightly in one direction. Observe that movement of the knob increases or decreases the distance between the stage and the objective lens. On your microscope, which part (the stage or the lens) moves, and which remains stationary? **Caution:** Never allow the objective lens to touch a slide or the condenser lens, which may protrude through the hole in the specimen stage. Anything coming in contact with the objective lens presents a danger of scratching, cracking, or otherwise damaging this critical and expensive part of the microscope. Always remember that this expensive scientific instrument is assigned to you for your use and safekeeping. Be certain that you use it properly and carefully, and that you keep it in good condition. Promptly notify your laboratory instructor of any malfunctions or if you have any difficulty in the use of your microscope.

Procedure for Use of the Microscope

Good microscopy requires the adoption of work habits that facilitate observation, minimize fatigue and eyestrain, and protect the equipment from damage. The steps outlined below should be followed each time you use the compound microscope.

1. Place the microscope directly in front of you on the laboratory table. Remember to carry the microscope by using both hands after removing it from the storage cabinet. Clean the objective and ocular lenses by wiping gently with a clean sheet of lens paper; never use anything but lens paper for cleaning the lenses.

2. Rotate the revolving nosepiece until the low-power (10X) objective lens clicks in place directly over the center of the condenser. While viewing the objective lens from one side of the microscope tube, **carefully** adjust the distance between the lens and the specimen stage, using the coarse adjustment control, to approximately one-half inch.

3. Select a microscope slide provided by your laboratory instructor and examine it to locate the position of the specimen on the slide. Then place the slide on the stage, with the specimen centered over the condenser lens. Make certain that the cover glass and specimen are on top of the slide.

4. Open the disc (or iris) diaphragm fully and turn on your microscope lamp. If your microscope is equipped with a substage mirror and an auxiliary lamp, adjust the positions of the lamp and the mirror until you obtain an evenly lighted, circular microscope field. Reduce the light on the specimen as necessary for clear observation by adjusting the diaphragm.

5. While observing the objective lens and the microscope slide from the side again, carefully readjust the distance between them to about one-eighth inch. DO NOT LOOK THROUGH THE OCULAR WHILE PERFORMING THIS STEP.

6. Now look through the ocular and slowly **increase the distance between the objective lens and the specimen** until the specimen comes into focus. This may involve either raising the objective lens or lowering the specimen stage, depending upon the design of your particular microscope. **Never** focus in the opposite direction—by decreasing the distance between the objective lens and the specimen—**while looking through the ocular.** Further, center the specimen in your field of view as necessary, and bring the specimen into sharper focus with the fine focusing control.

7. Rotate the revolving nosepiece until the high-power objective lens (43X) clicks into position directly over the condenser lens. The objective lenses on most modern microscopes are factory installed and adjusted so only minor changes in focusing are necessary when magnification is changed. Such lenses are **parfocal**—that is, their planes of focus are identical, or nearly so, although their working distances are different. You should find it necessary to make only a slight adjustment with the fine focusing control in order to achieve good focus after changing from the 10X objective to the 43X objective. **Use only the fine focus control** while looking through the

ocular with the high-power objective in position. You will also find it necessary to open the iris diaphragm slightly when you switch from low to high power. Why?

Returning the Microscope after Use

1. Rotate the revolving nosepiece to place the **low-power objective** (or scanning lens if your microscope has one) into position over the condenser lens.

2. Remove the microscope slide from the specimen stage and return it to its proper box or tray. If you have been using wet mounts, clean the specimen stage with a **clean cloth** or **cleaning tissues** as provided by your laboratory instructor.

3. Clean the objective and ocular lenses of the microscope with **lens paper.**

4. If your microscope has an in-base illuminator, disconnect the power cord, carefully wind the cord around the base of the microscope, and secure the end of the cord.

5. Return the microscope to its storage cabinet. Remember to use both hands in carrying the microscope. Check once again to make certain that the low-power objective lens (or scanning lens) is in position and that you have not left a slide on the stage.

Special Precautions

1. **Never** focus down (raise the stage or lower the objective, depending on the type of your microscope) while looking through the microscope.

2. **Always** locate the specimen under low power before switching to high power.

3. **Never** turn your microscope upside down or lay it on its side. The ocular might fall out and could be damaged.

4. **Always** keep the microscope clean and dry. Use only lens paper to clean the lenses.

5. **Never** use the coarse focusing control when the high-power lens is in position. Focus only with your fine focusing control when using high power.

6. **Always** try to relax and keep both eyes open when using the microscope. This helps to prevent undue eyestrain. With a little practice, you can learn to concentrate on the specimen and to disregard the image received by the other eye if you are using a monocular microscope.

The Stereoscopic Microscope

The **stereoscopic microscope,** also frequently called the dissecting microscope or stereomicroscope, is another common and extremely useful laboratory instrument (see figure 1.2). Stereoscopic microscopes are useful for viewing objects such as small insects, frog eggs, or large protozoa at low magnifications.

The image seen through a stereoscopic microscope is not inverted, as it is when seen through a compound microscope. Also, the internal configuration of lenses and prisms provides dual light paths and thus produces a stereoscopic or three-dimensional image. The effective magnification of such a lens system is more limited than is that of compound microscopes. Most stereomicroscopes provide magnifications in the range of 5–50X, although useful magnification of up to 100–200X can be obtained in the best quality research stereomicroscopes.

Older model stereomicroscopes frequently had a single pair of objective lenses and thus offered a single magnification; more recent and more expensive models are equipped with movable nosepieces with several pairs of matched objectives or an internal system of movable prisms.

Two recent designs of stereomicroscopes now in common use in teaching, as well as in research laboratories, utilize different mechanical and optical systems to achieve changes of magnification. One of these types of stereomicroscopes employs a fixed-position ocular, objective paired lenses, and an internal rotating drum on which several paired prisms are mounted. Two large knobs located on either side of the microscope head serve to rotate the internal drum and thus to change magnification. Focusing is accomplished by raising or lowering the objective, as in a compound microscope. The **focusing control knobs** are located on the sides of the microscope arm.

The other newer type of stereomicroscope utilizes a zoom-type lens system and provides a continuously variable magnification by the rotation of a dial located on top of the microscope head (see figure 1.2). Focusing is accomplished in this type of stereomicroscope by means of two lateral focusing control knobs as in the previously discussed types of stereomicroscopes.

Stereomicroscopes may be equipped with either an opaque or a transparent glass stage plate or disc. When equipped with a transparent stage plate and a substage beneath the regular stage, a stereomicroscope can be used to view objects in transmitted as well as in reflected light. This is often a very useful feature for biological studies.

Exercises in the Use of the Stereomicroscope

1. Remove your stereomicroscope from its storage cabinet and determine which type has been provided for your use in this course. Are there other types of stereomicroscopes present in the laboratory? Compare them with the type you have.

2. Place your stereoscopic microscope directly in front of you on the table and focus a concentrated light on the center of the stage. Select a microscope slide

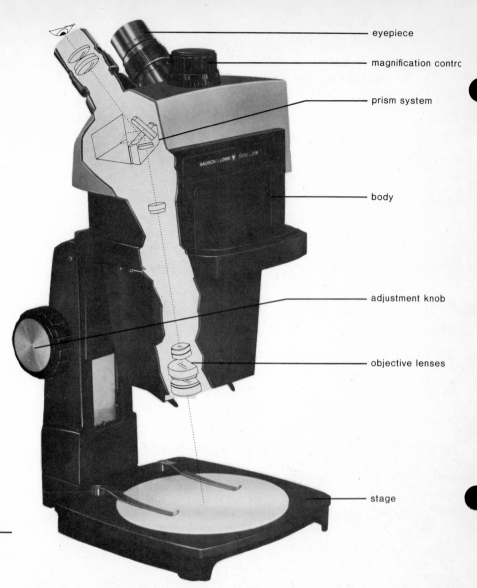

eyepiece

magnification contro

prism system

body

adjustment knob

objective lenses

stage

Fig. 1.2 Stereoscopic microscope. (Reproduced courtesy of Bausch and Lomb Optical Company.)

with a relatively large specimen, such as a fluke, a tapeworm, an insect wing, or other suitable specimen, and examine it through the microscope.

3. Experiment with several such slides as provided by your laboratory instructor until you become familiar with the use of your stereomicroscope. Try various adjustments of your light on the specimen, and try both transmitted and reflected light on specimens that are transparent or translucent if your microscope is provided with a substage and a substage mirror. If you do not have a substage, you may be able to achieve a similar effect by illuminating the specimen at right angles to your plane of viewing. Place your lamp close to the tabletop and focus the light beam from the side as sharply as possible on your specimen. Can you observe any structural details not discernible when the light comes from above or at a higher angle?

4. Place a few newly hatched larvae of the brine shrimp *Artemia,* or of a similar small living animal, in a

watch glass and examine them in both direct and transmitted light. Make a simple outline drawing showing the major structures that you see.

5. When you have completed your study with the stereoscopic microscope, wipe the stage clean of any spills, using a clean cloth or cleaning tissue, and return the microscope to its storage cabinet.

Other Types of Microscopes

Compound and stereoscopic microscopes are only two of the several different types of microscopes that have been developed by scientists in their continuing efforts to observe small objects more closely and in greater detail. Two other kinds of microscopes of particular importance in zoology are the phase contrast microscope and the electron microscope. In their design, both types of microscopes employ physical principles that differ significantly from those employed in ordinary light microscopes.

Phase Contrast Microscopy

Phase contrast microscopy permits the direct examination of transparent, unstained materials, including many kinds of living cells and tissues. Basically, this type of microscopy depends upon a special method of illumination which increases the contrast in an unstained specimen resulting from slight variations in the thickness and refractive index of its parts. The light passing through the specimen is manipulated in the lens system in such a way that these minor physical differences within the specimen are transformed into varying degrees of brightness and darkness. Thus, a living cell viewed through a **phase contrast microscope** has the appearance of being stained, although no chemicals that might alter its structure or kill it have been added to the cell. Figure 1.3 illustrates the type of image obtained in this kind of microscopy.

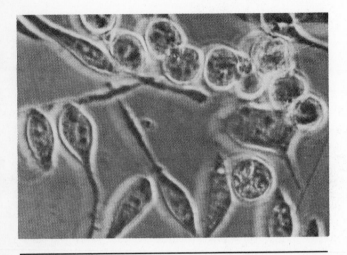

Fig. 1.3 Phase contrast microscopy. Photograph of several living macrophage cells (a type of large white blood cell). (Photograph by Kenneth E. Muse.)

Electron Microscopy

Electron microscopy has become a very valuable research technique in zoology in the last 30–40 years. In electron microscopy a specimen is irradiated with a concentrated **beam of electrons** rather than with rays of visible light, as in a light microscope. The two main electron microscopes commonly used by zoological researchers are **transmission electron microscopes** and **scanning electron microscopes.**

Transmission electron microscopes (figure 1.4) are capable of much greater resolving power and, therefore, much higher magnifications than are ordinary light microscopes. Direct magnifications up to 200,000 diameters

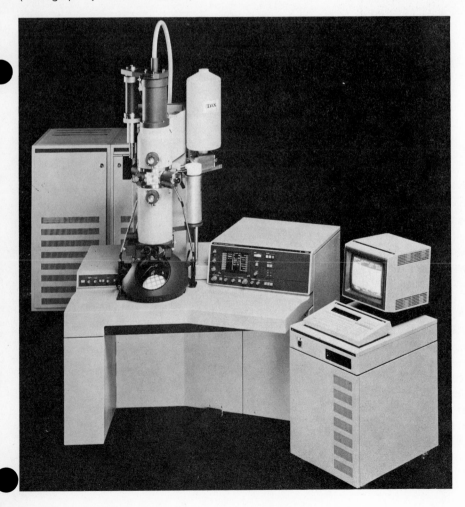

Fig. 1.4 Transmission electron microscope equipped with a special X-ray analyzer on the right. (Photograph courtesy of Philips Electronic Instruments, Inc.)

are commonly achieved with electron microscopes compared to a maximum of about 2,000 diameters with a compound light microscope. Photographic enlargements yield final magnifications greater than 1,000,000 diameters in many electron micrographs.

A transmission electron microscope works on principles generally similar to those of the light microscope. The two major differences between the light and electron microscope are (1) the nature of the illumination (visible light vs. electrons) and (2) the type of lenses for directing the illumination (glass lenses vs. electromagnetic lenses). Since you have already studied the light microscope, a brief explanation of the electron microscope should help you to understand better the basic working principles of both kinds of microscopes.

A transmission electron microscope is a large and complex instrument and resembles an inverted light microscope in its basic design (see figure 1.5). At the top of the microscope is the electron source, called an electron gun, in which a beam of electrons is emitted from an electrically heated tungsten filament. The electron beam passes downward from the electron gun, in a column and under a high vacuum, through the specimen and a series of electromagnetic lenses, to a viewing screen. Since the human eye is not sensitive to electrons, observations are made from a phosphorescent viewing screen or from photographic plates exposed to the electron beam after it passes through the specimen.

The image observed on the viewing screen consists of light and dark patterns produced by areas of varying electron density in the specimen. Changes in focus and magnification of the image are accomplished by adjustments of the electric current passing through the electromagnetic lenses along the path of the electron beam.

Since the wavelengths of electron beams are much shorter than those of visible light rays, the electron microscope has a much greater resolving power than does an ordinary light microscope. The resolving power of any optical system depends on the wavelength of the light (or other form of radiant energy, such as electrons) used in the system. Resolutions of 5–10 angstrom units (1 angstrom = 10^{-7} mm) are frequently achieved by biologists using transmission electron microscopes. Figures 1.6, 2.13, 2.19, and 2.20 are examples of photographs taken with a transmission electron microscope.

The **scanning electron microscope** (figure 1.7) is an even newer research tool and has had many important applications in biological research during the past 20 years. The scanning electron microscope differs in principle from both light and transmission electron microscopes. This kind of electron microscope has proved to be especially useful in providing three-dimensional images of small objects, information about chemical composition, electrical properties, and structural details of the surface of specimens. The scanning electron microscope has a great depth of field

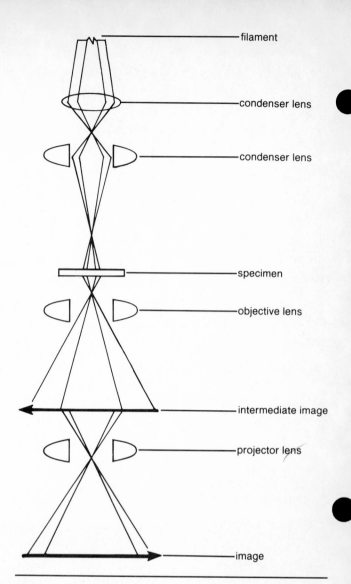

Fig. 1.5 Diagram illustrating the lens system and electron beam of a transmission electron microscope.

(7 to 10 times that of a light microscope at comparable magnifications), making possible photographs of excellent three-dimensional quality (see figure 1.8).

A specimen in the column of a scanning electron microscope is irradiated with a concentrated beam of electrons. This concentrated electron beam is moved over the surface of the specimen in a systematic pattern. Electrons from this concentrated beam cause the emission of secondary electrons from the surface of the specimen as the beam moves across the specimen. A cathode-ray tube (similar to the picture tube of a television receiver) displays the pattern of secondary electrons emitted from the surface of the specimen. Thus, a three-dimensional view showing minute details of the surface can be obtained. Photographs of the image on the cathode-ray tube are taken to make permanent records of the observations.

Chapter 1

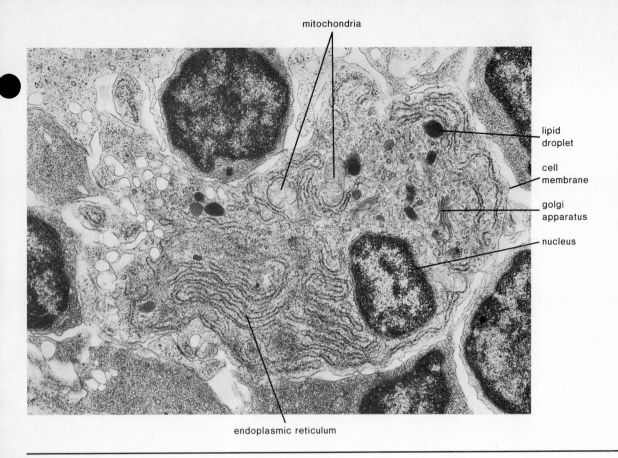

mitochondria

lipid droplet

cell membrane

golgi apparatus

nucleus

endoplasmic reticulum

Fig. 1.6 Transmission electron micrograph of a plasma cell.
(Photograph by Kenneth E. Muse.)

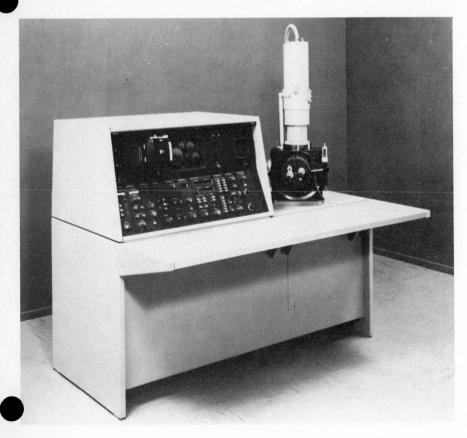

Fig. 1.7 Scanning electron microscope. (Photograph courtesy of Philips Electronic Instruments, Inc.)

Microscopy

 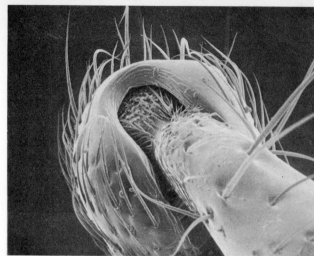

Fig. 1.8 *(left)* Scanning electron micrograph of an ant. Magnification 300×. *(right)* Enlarged view of the foreleg joint from the circled area of the illustration at left. Magnification 600×. (Photographs by Kenneth E. Muse.)

Key Terms

Compound microscope a type of light microscope with two separate lens systems, an eyepiece and an objective lens, which together serve to magnify the image of an object. Provides magnification to about 1,000–2,000 diameters.

Magnification the ratio of the apparent size to the actual size of an object when viewed through a microscope.

Phase contrast microscope a special type of light microscope that permits the observation of thin, unstained materials. Special lenses and illumination techniques increase the contrast in an unstained specimen due to variations in the refractive indices of its parts.

Resolving power the ability of a microscope to reveal fine detail in a specimen. More precisely, resolving power is defined as the shortest distance between two points that allows them to be distinguished as separate points.

Scanning electron microscope a type of electron microscope that provides three-dimensional images of very small objects. A concentrated beam of electrons is focused and moved along the surface of a specimen and induces the emission of secondary electrons from the specimen. These secondary electrons produce a magnified image of the specimen on a cathode-ray tube.

Stereoscopic microscope a type of microscope with two separate optical paths that provide a magnified three-dimensional image. Provides useful magnifications to about 100–200 diameters.

Transmission electron microscope a type of electron microscope used to view specially prepared thin specimens at magnifications to about 200,000 diameters. A concentrated beam of electrons passes through the specimen and produces a pattern of light and dark areas on a phosphorescent screen because of the differential passage of electrons through the specimen. Darker areas represent areas of greater electron density within the specimen.

Working distance the distance from the front of the objective lens of a compound microscope to the top of the specimen.

Notes and Sketches

2
Animal Cells
and Tissues

Objectives

After completing the laboratory work in this chapter, you should be able to perform the following tasks:

1. Describe the Cell Theory and explain its importance in zoology.
2. Describe the principal organelles of a typical animal, such as a starfish egg, that are visible in a light microscope.
3. List six organelles seen in a transmission electron micrograph of a generalized (typical) animal cell not generally visible with a compound microscope.
4. Distinguish between a cell, a tissue, an organ, and an organ system.
5. List five types of animal tissues and give examples of each. Identify typical examples of each in microscope slides.
6. Describe the histological structure of a compact bone and explain the role of a Haversian canal, lacuna, canaliculi, and lamellae.
7. Distinguish between smooth muscle, cardiac muscle, and striated muscle from microscopic slides or photographs.
8. Describe the structure of a vertebrate neuron and give the function of each main part.
9. Describe the composition of human blood and give the functions of erythrocytes, leucocytes, and platelets.
10. Identify erythrocytes, leucocytes, and blood platelets in microscopic preparations.

The Cell Theory

Cells are the fundamental structural and functional units of virtually all living organisms. The body of an animal typically is made up of many different kinds of cells that are organized into tissues, organs, and organ systems that carry out certain essential functions. Thus, a knowledge of cell structure and function is essential for the proper understanding of reproduction, growth, heredity, and all other normal and abnormal animal functions.

The importance of cells is summarized in a statement called the Cell Theory, which is one of the most important unifying concepts in biology. This theory was developed in the nineteenth century, when naturalists were experimenting with their latest technology—the compound microscope. The Cell Theory is generally attributed to two German scientists, botanist Jakob Schwann and zoologist Theodor Schwann, who published their ideas in 1838–1839; but other scientists also have contributed to the modern version of the theory.

The main points of the Cell Theory can be summarized as follows:

1. All organisms are composed of cells.
2. All cells come from other cells.
3. All vital functions of an organism occur within cells.
4. Cells contain the hereditary information necessary for regulating cell functions and for transmitting information to the next generation of cells.

Materials List

Prepared Microscope Slides
   Starfish eggs or sea urchin eggs
   Cuboidal epithelium (rabbit kidney)
   Columnar epithelium (rabbit kidney)
   Stratified epithelium (human skin)
   Loose connective tissue
   Hyaline cartilage
   Human bone, cross section
   Smooth muscle
   Striated muscle
   Neurons (smear from spinal cord of cow)
   Blood film (human), Wright stain

Ciliated epithelium (Demonstration)
Adipose tissue (Demonstration)
Cardiac muscle (Demonstration)
Amphibian blood (Demonstration)
Chemicals
Wright's blood stain
Distilled water
Miscellaneous Supplies
Clean microscope slides
Coverslips
Fresh cartilage (from frog sternum, ends of long bone, etc.)
Cross sections of long bones from pig, cow, or other mammal

Basic Cell Structure

Our knowledge of cell structure has increased dramatically in the last few decades because of the availability of electron microscopes as described in chapter 1. Figure 2.1 illustrates the appearance of a relatively simple cell seen in an ordinary compound light microscope of the type usually available in general zoology laboratories. Figure 2.2 is a diagram illustrating the parts of an animal cell seen in transmission electron micrographs. Note the obvious difference in structural detail. We now know that animal cells are composed of many kinds of organelles, the distinctive parts of cells that carry out specific functions. Several types of cellular organelles can be visualized only in electron micrographs because they are too small to be seen with a compound microscope.

Since electron microscopy is a very complex process, in this exercise we will concentrate on those aspects of cell

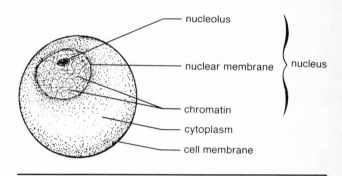

Fig. 2.1 Starfish egg.

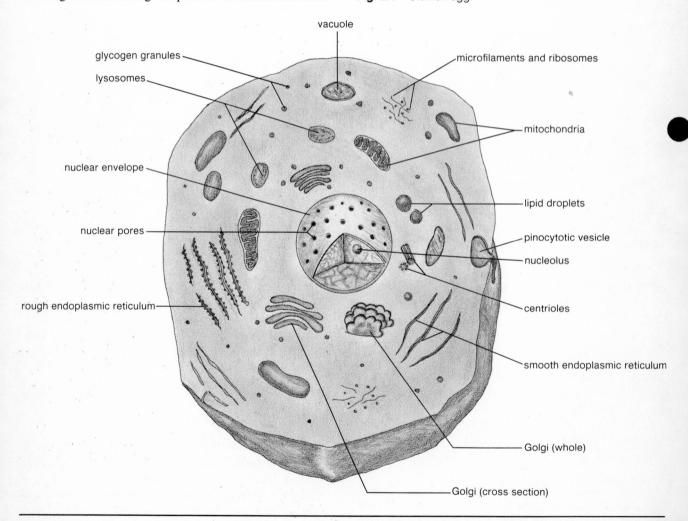

Fig. 2.2 Generalized animal cell, as seen with the aid of a transmission electron microscope.

Chapter 2

and tissue structure observable with a compound microscope and compare our findings with information available from more sophisticated techniques and instruments.

Good examples of undifferentiated animal cells are provided by the unfertilized eggs of many animals. For our introductory study of cells, we will use the unfertilized eggs of the starfish. Sea urchin eggs are very similar to starfish eggs and serve equally well. Prepared microscope slides containing many stained starfish eggs will be provided for your study.

Obtain a prepared microscope slide and, under low power on your compound microscope, observe the numerous starfish eggs. Select a well-stained cell similar to that illustrated in figure 2.1 and move the microscope slide to center the cell in your field of view. Rotate the high-power objective into position, regulate the light as needed, and readjust the focus. Observe the two well-differentiated parts of the **cell,** the central **nucleus** and the surrounding **cytoplasm.** Externally, the cytoplasm is bounded by a thin **cell membrane.** Although the cell membrane of animal cells is seen only as a thin outer boundary of the cell, it plays a very important role in the functions of the cell. The special properties of the cell membrane control the passage of materials into and out of the cell. Animal cells lack the thickened cellulose cell walls outside of the cell membrane that are usually found in plant cells.

In your specimen, identify the spherical **nucleus,** bounded by the **nuclear membrane** and containing numerous darkly staining masses of **chromatin** material. Also within the nucleus of the starfish egg, find the darkly staining **nucleolus.** On table 2.1 list the cell organelles you were able to observe with a compound microscope. In table 2.2 make a similar list of all the additional cellular organelles that you find in the diagram of an animal cell in figure 2.2 and in the transmission electron micrographs in figures 1.6, 2.19, and 2.20 that you were unable to observe with your light microscope. What conclusions can you make from a comparison of the two lists?

Table 2.1. Cellular Organelles Observed with Light Microscope.

1. _____

2. _____

3. _____

4. _____

5. _____

6. _____

Table 2.2. Cellular Organelles Commonly Observable in Transmission Electron Micrographs but NOT Seen with Light Microscope.

1. _____

2. _____

3. _____

4. _____

5. _____

6. _____

7. _____

8. _____

9. _____

10. _____

Animal Tissues

The bodies of most kinds of animals are multicellular, and in most multicellular organisms the cells are organized into definite tissues. A **tissue** can be defined simply as an aggregate of cells performing a similar function or functions. Tissues, in turn, are organized into **organs,** which consist of one or more tissues grouped into a structural and functional unit. Organs with related functions work together as **organ systems** in many types of animals. An **organism** consists of several integrated organ systems.

Animal tissues are commonly divided into four principal types: (1) **epithelial tissue,** (2) **connective tissue,** (3) **muscular tissue,** and (4) **nervous tissue.** Two other tissue types are also sometimes distinguished: (1) **vascular tissue** or blood, which is really a specialized type of connective tissue, and (2) **reproductive tissue,** which includes the specialized reproductive cells or gametes: eggs and sperm.

Epithelial Tissue

In epithelial tissues, the cells are closely spaced with small amounts of intercellular substance and are arranged in thin layers or sheets. They form the covering of the outer surface of the body, line the internal cavities and ducts, and form glands. There are several kinds of epithelial tissue, each distinguished by the shape of the cells and their arrangement.

Fig. 2.3 Drawing of squamous epithelial cells from human skin.

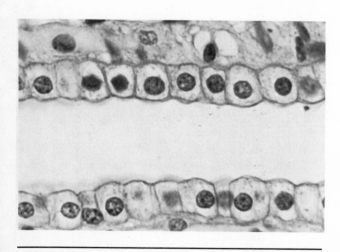

Fig. 2.4 Cuboidal epithelium. (Courtesy Carolina Biological Supply Company.)

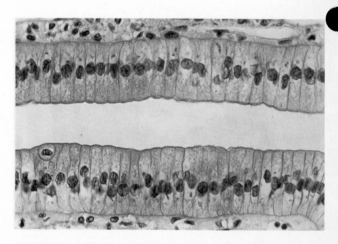

Fig. 2.5 Columnar epithelium. (Courtesy Carolina Biological Supply Company.)

Squamous epithelial tissue consists of thin, flattened cells arranged in one or several stratified layers. They line such cavities as the mouth, esophagus, vagina, and the coelom.

Take a clean toothpick and gently scrape the inside of your cheek to remove a few of the surface epithelial cells. When properly done, this operation is painless and bloodless. Touch the end of the toothpick to the center of a clean microscope slide and roll the toothpick carefully back and forth to transfer some of the epithelial cells to the slide. Add a drop of 1% methylene blue solution and a coverslip. Observe your preparation under a compound microscope. Draw some of the squamous epithelial cells from the lining of your mouth in the space provided in figure 2.3. Other types of epithelial tissues include **cuboidal epithelium, columnar epithelium,** and **stratified epithelium.**

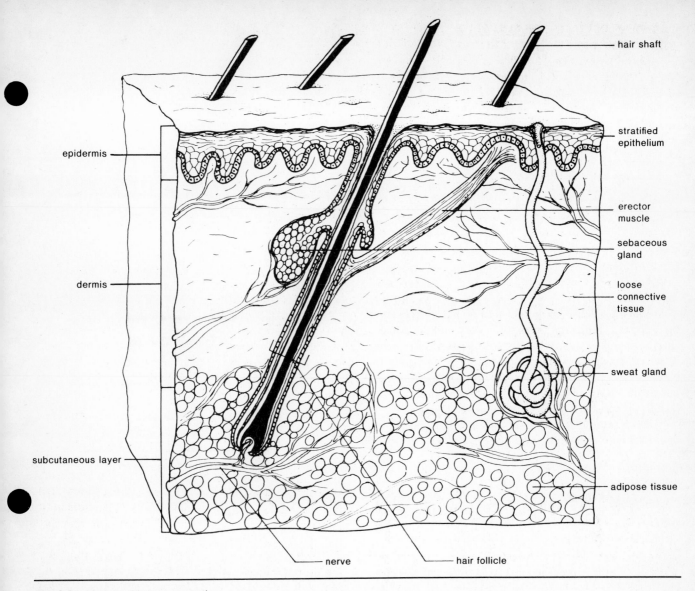

Labels on figure:
hair shaft
stratified epithelium
erector muscle
sebaceous gland
loose connective tissue
sweat gland
adipose tissue
epidermis
dermis
subcutaneous layer
nerve
hair follicle

Fig. 2.6 Human skin, cross section.

Examine a slide of **cuboidal epithelium** (figure 2.4) from the kidney tubules of a rabbit or other appropriate tissue. Note that the height and width of the single layer of cuboidal cells are about equal. A **basement membrane** lies at the base of the cells. Note the prominent nucleus.

Columnar epithelium (figure 2.5) consists of a single layer of tall, closely packed cells. Obtain a slide with columnar epithelium from the intestine of a frog, rabbit kidney, or other appropriate tissue, and identify the outer **brush border,** the **nucleus,** and the **basement membrane.** Columnar epithelium from the intestine also often exhibits secretory **goblet cells.** Some columnar epithelial tissues, such as those lining the oral cavity and esophagus of the frog, bear **cilia** on their distal (outer) margins.

Stratified epithelium (figure 2.6) makes up the outer part of the skin or integument of all vertebrates. Study a slide of human skin and observe the complex structure of the multilayered integument, or outer covering of the human body.

Connective Tissue

All connective tissues exhibit relatively large amounts of nonliving, intercellular substance produced by the living cells. This substance provides support and protection for the animal body.

You will study three examples of connective tissue in this exercise: **loose connective tissue, cartilage,** and **bone.**

Loose Connective Tissue

Loose connective tissue (figure 2.7) consists of scattered cells surrounded by a clear, jellylike **ground substance** and two types of fibers: **elastic fibers** and **nonelastic** (collagenous) **fibers.** Loose connective tissue is often called areolar tissue. It is found connecting the various tissues of the body in man and other vertebrate animals. Nerves, blood vessels, and cells lie embedded in loose connective tissue.

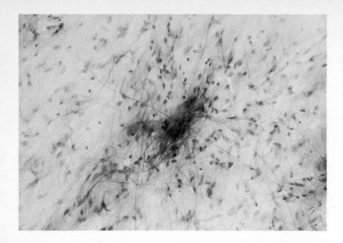

Fig. 2.7 Loose connective tissue. (Photograph by Kenneth E. Muse.)

Obtain a slide of loose connective tissue and observe the cells, the elastic and nonelastic fibers, and the apparently open spaces where the ground substance has been dissolved during the preparation of the slide. Identify the bundles of thin, wavy, **nonelastic fibers** and the thicker, but single, **elastic fibers.** Other types of connective tissue include **adipose** (fat) **tissue, cartilage,** and **bone.**

Draw examples of elastic and nonelastic fibers in figure 2.8 and label each type.

Cartilage

Cartilage (figure 2.9) consists of a firm but elastic **matrix** secreted by numerous cartilage cells embedded within the matrix. **Hyaline cartilage** is found at the ends of long bones. Study a microscopic section of hyaline cartilage and observe the scattered cartilage cells and the **homogeneous matrix** (chondrin). The cartilage cells are found in spaces in the matrix called **lacunae.** Do you find any examples of more than one cell in a single lacuna? How would you explain the thin layer of matrix occasionally found between two cells?

Bone

Bone plays an important role in the mechanical support of the body, and also in protecting vital parts from injury. The skull and the vertebral column, for example, play dual roles in the support and protection of the brain and the spinal cord. The heart and the lungs are also protected by the bony framework of the rib cage embedded in the thoracic wall.

The intercellular matrix of bone also plays an important physiological role in the storage of calcium, which can be withdrawn from the bone when the calcium level in the blood is lowered. The bone marrow contains tissue that plays an important role in the formation of blood elements.

Fig. 2.8 Drawing of elastic and nonelastic fibers.

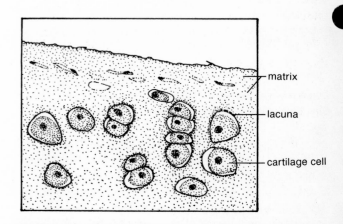

Fig. 2.9 Cartilage tissue.

Examine a thin section of **compact bone** (figure 2.10) prepared from a long bone, such as the humerus or femur of a human or other mammal, and identify: (1) the central **Haversian canal** through which passes many small blood vessels and nerves; (2) the **concentric layers** of bone (lamellae); (3) the **lacunae,** or spaces that house the bone cells or osteocytes; and (4) the numerous fine **canaliculi,** which serve to interconnect the lacunae and the Haversian canals.

Chapter 2

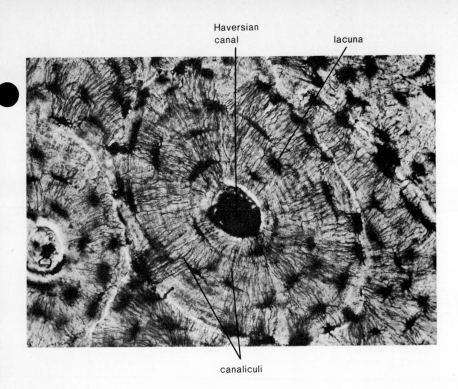

Haversian canal   lacuna

canaliculi

Fig. 2.10   Bone tissue. (Courtesy Carolina Biological Supply Company.)

Muscular Tissue

Muscular tissue is specialized for **contractility** and therefore has the capacity to perform mechanical work. Three different types of muscular tissue are distinguished: **smooth muscle, striated** (skeletal) **muscle,** and **cardiac** (heart) **muscle.**

Smooth Muscle

Smooth muscle consists of elongated, spindle-shaped cells with a single, central nucleus. This type of muscle is sometimes referred to as nonstriated muscle because it lacks the cross striations seen in both striated and cardiac muscle.

Smooth muscle forms the simplest type of muscle tissue and is generally found in parts of the body where rapid movement or contraction is not essential, such as in the walls of the digestive tract, in the walls of blood vessels, and in the walls of the urinary bladder and the uterus. Examine a slide of smooth muscle that has been teased apart to show the individual spindle-shaped cells or a section through some smooth muscle that shows the individual cells cut longitudinally. Locate the tapered smooth muscle cells and note the location of the nucleus. Identify the cell membrane and the fine longitudinal threads, the **myofibrils,** in the cytoplasm. Observe also the demonstration slides of smooth muscle from other types of material.

Draw several smooth muscle cells in figure 2.11 and label the **nucleus, cell membrane,** and **myofibrils.**

Striated Muscle

The large muscles attached to various parts of the skeleton are composed of striated, or voluntary, muscle (see figure 2.12). This type of muscular tissue is made up of long cylindrical fibers containing many nuclei. What is unusual about the location of the nuclei in striated muscle as compared to most other kinds of cells? The multinucleate (syncytial) condition in striated muscle arises during embryonic development of the muscle as a result of the fusion of several mononucleate cells.

Note the conspicuous cross striations in these fibers and the outer limiting membrane, called the **sarcolemma.** Observe also the fine longitudinal **myofibrils** running lengthwise through the striated muscle fibers.

The individual myofibrils of striated muscle also have a very distinctive banded or striated structure. Because of their small size, individual myofibrils can be studied only with an electron microscope. Figure 2.13 is an electron micrograph showing parts of four adjacent myofibrils.

The functional unit of the myofibril is the **sarcomere,** which extends between two adjacent Z-lines. During muscle contraction, the sarcomeres shorten (the distance between Z-lines decreases). Recent investigations have shown that the contraction of muscle is due to interactions between two contractile proteins, **actin** and **myosin.** These two proteins make up a substantial portion of each myofibril.

The large **mitochondria** provide energy in the form of ATP (adenosine triphosphate) to power the contraction.

Cardiac Muscle

A third type of muscle tissue, **cardiac muscle,** is found in the walls of the heart of vertebrate animals. Cardiac muscle (figure 2.14) consists of striated muscle fibers, which branch and unite (anastomose) with other fibers to

Animal Cells and Tissues

19

Fig. 2.11 Drawing of smooth muscle cells.

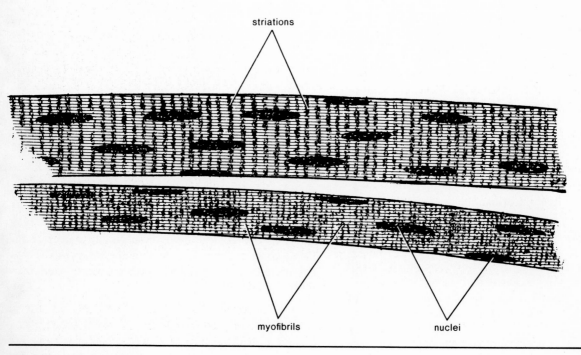

striations

myofibrils   nuclei

Fig. 2.12 Striated muscle.

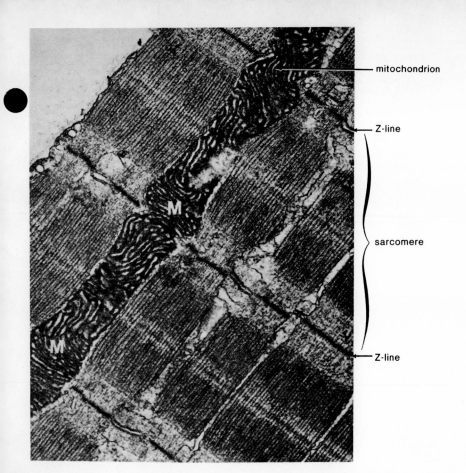

mitochondrion

Z-line

sarcomere

Z-line

Fig. 2.13 Striated muscle fibrils. (Electron micrograph by Kenneth E. Muse.)

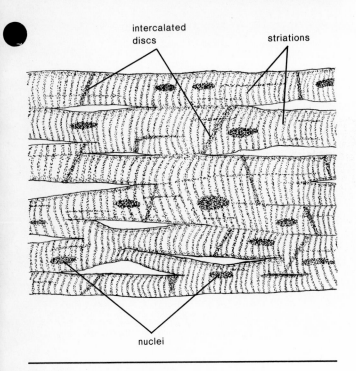

intercalated discs

striations

nuclei

Fig. 2.14 Cardiac muscle.

form a continuous network of muscle fibers. Cardiac muscle contains cross striations like striated muscle, but the fibers of cardiac muscle are divided into cell-like units by many **intercalated discs,** which partially divide the fibers. This type of muscle, like smooth muscle, is involuntary; and throughout the life of the organism, cardiac muscle contracts and relaxes rhythmically and automatically. In fact, the heart of some of the lower vertebrates, such as a frog or a turtle, can be removed from the body and placed in a physiological salt solution where it will continue to beat for many hours, or even days.

Observe the demonstration slides of human and other vertebrate cardiac muscle and observe: (1) the branching and anastomosing network of muscle fibers, (2) the cross striations, (3) the scattered nuclei, (4) the outer sarcolemma, and (5) the intercalated discs.

Nervous Tissue

The nervous system of vertebrate animals consists of the brain, spinal cord, and the nerves. Nervous tissue consists of highly specialized cells which carry impulses from one part of the body to another. A nerve cell, together with its branches or processes, which in some cases are several feet long, is called a **neuron.** A neuron consists of a **cell body** containing the **nucleus** and two or more elongated **nerve processes.** In a **motor neuron,** which transmits impulses

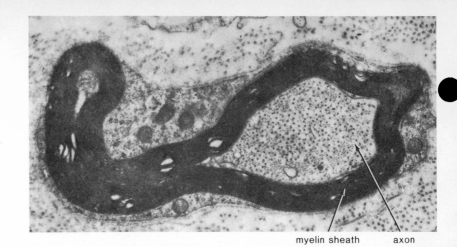

Fig. 2.15 Neuron in ganglion of cat. (Electron micrograph by Kenneth E. Muse.)

myelin sheath   axon

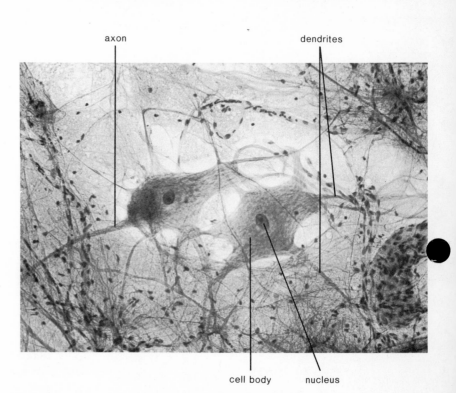

axon   dendrites

cell body   nucleus

Fig. 2.16 Neurons from spinal cord smear from a cow. (Courtesy Carolina Biological Supply Company.)

from the central nervous system to the muscles or other effectors, there are several short **dendrites** and one long **axon.** The dendrites carry nerve impulses toward the cell body, and the axon carries nervous impulses away from the cell body. A **myelin sheath** surrounds the axon.

Figure 2.15 is an electron micrograph showing a cross section of a neuron in a cat. Observe the thick, many-layered myelin sheath surrounding the axon. Note the many small tubules cut in cross section within the axon.

Examine also a slide with a stained smear preparation of the gray matter from the spinal cord of a cow. The details of the individual neurons are more readily seen in such a preparation than in sections. Observe the large neurons (easily seen under low power). Compare with figure 2.16. Select a typical cell and note the **cell body, nucleus, nucleolus,** numerous **dendrites,** and the longer **axon.** Observe the numerous fine fibrils, the **neurofibrils,** within the cytoplasm of the cell body and extending into the processes.

Vascular Tissue

The blood of living vertebrates is a red liquid that is constantly in motion as it circulates through a closed system of tubes—the blood vessels. In man, the blood comprises about 7 percent of the body weight. It consists of a colorless liquid, the **plasma,** and several types of **blood cells** suspended within it. In permanent, stained preparations the plasma is not seen, and many of the cells, particularly the red corpuscles, may be distorted by the reagents used in the preparation of the slides.

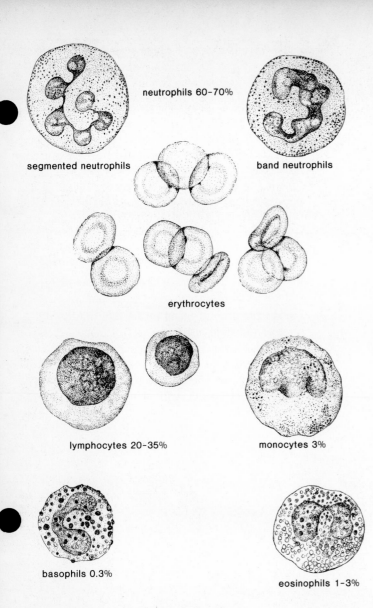

neutrophils 60-70%

segmented neutrophils

band neutrophils

erythrocytes

lymphocytes 20-35%

monocytes 3%

basophils 0.3%

eosinophils 1-3%

Fig. 2.17 Types of human blood cells.

Human Blood

Suspended in the plasma of human blood are the **red** and **white cells** and structures called **blood platelets.** Several types of human blood cells are shown in figure 2.17. Study the types of human blood cells in the prepared slides provided for you, or prepare your own blood smear slides by following the directions given at the end of this section.

1. **Erythrocytes.** Observe that human red blood cells, or erythrocytes, are small, circular, biconcave, and **lack nuclei.** Their chief function is to carry oxygen to the cells of the body. They contain large amounts of the iron-containing protein **hemoglobin,** which combines with oxygen in the lungs and releases it in the body tissues. The biconcave shape of human red blood cells is clearly shown in figure 2.18, which is a photograph taken with a scanning electron microscope. Figure 2.19 shows cross sections of two red blood cells within a capillary of a mouse in a transmission electron micrograph. Note the outer membrane and the uniform dark, granular cytoplasm of

the erythrocytes. Observe the lack of nuclear material in both cells. The absence of a nucleus in mature red blood cells is characteristic of all mammals.

2. **Leucocytes.** Human blood contains several types of leucocytes or white blood cells. The types and relative frequencies of each type are shown in figure 2.17. Study your blood smear and identify as many types of leucocytes as you can find.

A cross section of a leucocyte of a mouse is shown in figure 2.20. Observe the large nucleus and the complex structure of its cytoplasm. Compare it with the relatively simple structure of the erythrocytes in figure 2.19. Why do you think that mammalian leucocytes would have a more complex structure than mammalian erythrocytes? How is their structure related to their function?

Figure 2.21 is a scanning electron micrograph of a **macrophage** from the human lung. Note the many irregular extensions from the surface of the

Animal Cells and Tissues

Fig. 2.18 Human erythrocytes. (Scanning electron micrograph by Kenneth E. Muse.)

cell. Macrophages are large phagocytic cells formed from monocytes and lymphocytes which engulf and break down bacteria and cellular debris.

3. **Blood platelets.** Platelets are small, nonnucleated bits of cytoplasm that bud off from large cells (megakaryocytes) in the bone marrow. They are about one-third the size of erythrocytes and are not usually preserved in prepared microscope slides. Platelets function in **blood clotting** by temporarily plugging punctures that may occur in blood vessels and also by releasing substances that trigger later chemical reactions necessary for clotting.

Directions for Preparing and Staining a Blood Smear

1. Place a small drop of fresh blood about one-half inch from one end of a clean microscope slide.

2. Spread the drop of blood into a thin film covering most of the slide, using the end of a second clean microscope slide. Hold the second slide at about a 45° angle with the first slide and spread the drop of blood thinly over the surface of the first slide. Your laboratory instructor will demonstrate the technique for you.

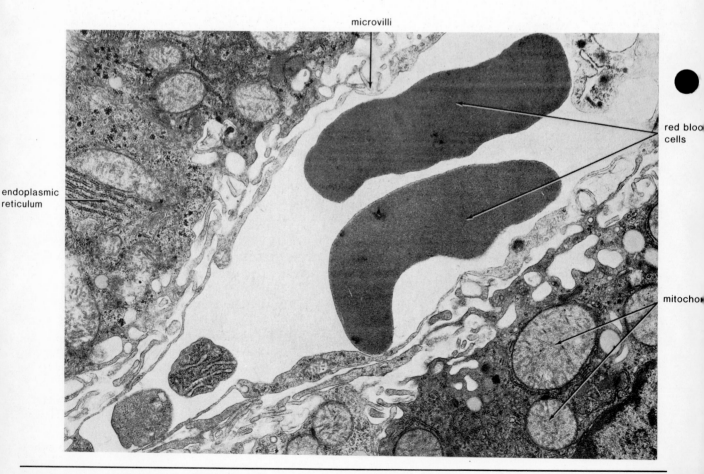

microvilli

endoplasmic reticulum

red blood cells

mitocho

Fig. 2.19 Erythrocytes in capillary of mouse liver. Magnification 13,750×. (Electron micrograph by Kenneth E. Muse.)

Chapter 2

3. Allow the blood smear to dry.

4. Lay the slide face up on a staining rack (or two glass rods) and cover the blood smear with a few drops of Wright's blood stain (a methylene blue-eosin mixture).

5. Stain for about 1–2 minutes.

6. Dilute the stain with about twice the volume of distilled water.

7. Drain the slide and rinse with distilled water until the thinner portions of the slide are pink.

8. Wipe the reverse side of the slide clean and let the slide air dry.

9. When dry, the slide may be examined under a compound microscope. A coverslip is not necessary, but one may be added for protection if desired. Consult your laboratory instructor for directions.

Demonstrations

1. Ciliated epithelium (microscope slide).
2. Adipose (fat) tissue (microscope slide).
3. Fresh cartilage (from the sternum of a frog, ends of a long bone, etc.).
4. Cross sections (1–2 inches thick) of long bone to illustrate gross structure of bone.
5. Cardiac muscle (microscope slide).
6. Amphibian blood (microscope slide).

Key Terms

Cardiac muscle   type of muscle found in the heart wall of vertebrate animals. Consists of a branching and anastomosing network of striated, multinucleate muscle fibers.

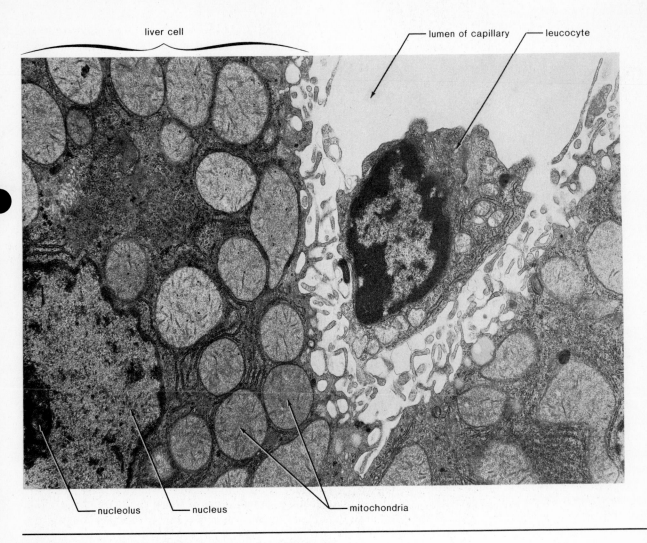

Fig. 2.20   Leucocyte in capillary of a mouse. Magnification 21,000×. (Electron micrograph by Kenneth E. Muse.)

Animal Cells and Tissues

pseudopodia

Fig. 2.21 Macrophage from human lung. Magnification 20,000×. (Scanning electron micrograph by Kenneth E. Muse.)

Cell membrane outer limiting membrane of an animal cell; composed of phospholipids, proteins, and cholesterol.

Connective tissue type of animal tissue that binds, supports, and protects other body parts. Includes several kinds of tissue, such as loose connective (areolar) tissue, adipose (fat) tissue, cartilage, and bone.

Epithelial tissue type of animal tissue that lines the outer surface and inner cavities and ducts of the body.

Erythrocytes red blood cells whose chief function is the transport of oxygen; contain large amounts of hemoglobin. May be nucleated (as in frogs and salamanders) or without a nucleus in the mature stage (as in humans).

Leucocytes white blood cells that are usually colorless in life and which perform many essential functions, including engulfment of foreign particles, production of antibodies, and wound healing. Several distinct types of leucocytes are usually present in the blood of an animal.

Muscular tissue type of tissue specialized for contraction; contains fibrils constructed of contractile proteins. Three types are distinguished—smooth muscle, striated muscle, and cardiac muscle.

Nervous tissue type of tissue specialized for the conduction of electrical impulses; important in the coordination of body activities. The basic functional unit is the neuron, or nerve cell.

Neuron the basic, functional unit of the nervous system of animals. The neuron is a single nerve cell, which consists of a cell body with a nucleus and two or more long extensions or processes (axons and dendrites).

Nucleus the central organelle of an animal cell, which contains the genetic material (chromosomes) and which controls the metabolism of the cell.

Platelet a small, nonnucleated body in the blood of humans and other mammals. Formed by megakaryocytes in the bone marrow and serves mainly to plug leaks in blood vessels and to release chemical substances that initiate clotting.

Reproductive tissue specialized cells (or tissue) that function in sexual reproduction—eggs and sperm.

Smooth muscle type of muscle tissue found associated with internal organs in higher animals. Consists of spindle-shaped cells with a single central nucleus. Also known as involuntary muscle.

Striated muscle type of muscle tissue found attached to parts of the skeleton. Consists of long, cylindrical fibers with a cross-banded or striated appearance in microscopic preparations; each fiber contains many nuclei. Striated muscle is also known as skeletal or voluntary muscle since many striated muscles are under voluntary control.

Vascular tissue made up of the blood cells and their suspending fluid, the plasma. Human blood consists of the red blood cells (erythrocytes), white blood cells (leucocytes), and platelets suspended in the plasma. Other animals may have additional types of blood cells.

3
Mitosis and Meiosis

Objectives

After completing the laboratory work in this chapter, you should be able to perform the following tasks:

1. Explain the basic difference between mitosis and meiosis.
2. Describe the cell cycle and explain the function of its principal stages.
3. Describe the mitotic apparatus of animal cells and list its principal parts.
4. Briefly describe the structure and function of chromosomes, centromeres, spindle fibers, aster rays, and centrioles.
5. Identify the principal mitotic stages of an animal cell in microscope slides or photographs.
6. Explain the chief events that occur in prophase, metaphase, anaphase, and telophase.
7. Define random sample and explain its significance in biological research.
8. Describe a method for estimating the relative duration of the various stages of mitosis in a population of animal cells.
9. Explain the terms haploid and diploid and how they relate to the process of reproduction in animals.
10. Distinguish between a chromosome and a chromatid.
11. List the principal stages of meiosis and identify each stage in microscopic preparations or illustrations of animal cells.

Introduction

All animals depend on cell division for their growth and repair processes. Each cell has a precise set of genetic information built into its chromosomes. This information is essential for the proper functioning of each cell and for prescribing the characteristics of the next generation of cells. The division of the nucleus of the cell and the precise distribution of the duplicated chromosomes between the two new cells in this type of cell division is called **mitosis.**

Meiosis is a specialized type of cell division that usually occurs during the formation of the gametes or sex cells of multicellular animals. During meiosis, the normal diploid (2N) chromosome number of the somatic (body) cells is reduced by half to the typical (1N) chromosome number of the gametes. Meiosis is extremely important for the survival and evolution of animals because it provides for recombinations of genes during each generation. Thus, variations occur among the offspring of each generation and natural selection can operate to select the better adapted individuals.

Materials List

Prepared Microscope Slides
   Whitefish blastula (mitosis) (or *Ascaris* embryo)
   Ascaris ova (meiosis)
Audiovisual Materials
   Color transparencies of mitosis in whitefish blastula
   Color transparencies of meiosis in *Ascaris* ova
   Wall charts illustrating mitosis and meiosis

Mitosis

The division of nuclei by mitosis is exhibited by the somatic cells in most plants and animals. You should already have some idea of the complexity of cell division from your study of cell structure in an earlier laboratory session. During its life span, a cell passes through a regular sequence of physiological events called the **cell cycle** (figure 3.1). This sequence includes several distinct stages, each characterized by certain metabolic activities of the cell. In actively dividing cells, the cycle may last only a

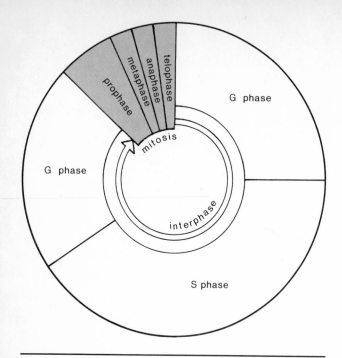

Fig. 3.1 The cell cycle.

few hours; in other cells, the cycle may last for days or weeks. At the completion of the cell cycle a new generation of cells is produced.

The cell cycle is made up of four phases: G_1, S, G_2, and M. Together G_1, S, and G_2 constitute **interphase.** The G_1 **(gap) phase** is a period of active protein synthesis and the formation of new cell organelles, like mitochondria, Golgi, ribosomes, and endoplasmic reticulum. It is a phase of rapid cell growth. The **S (synthesis) phase** is the time when DNA and other molecules making up the chromosomes, like histones, are synthesized. Duplication of the DNA strands and of the chromosomes also occurs during S phase. In the G_2 **(gap) phase** the cell synthesizes the proteins, actin and tubulin, plus the enzymes and other materials necessary for the mitotic spindle. **Mitosis (M phase)** follows the G_2 phase and involves the division of chromosomes and the formation of two new nuclei. Mitosis consists of four distinct stages but actually takes up only about 5–10 percent of the complete cell cycle in most cells.

Abundant evidence now demonstrates that the duplication of the genetic material (DNA synthesis) in the nucleus occurs in the **nondividing cell prior to the initiation of mitosis.** The chromosomes are **already doubled** when they become visible during prophase, the first stage of mitosis. The term **mitosis** refers specifically to the process of nuclear division, the orderly distribution of the chromosomes between two daughter nuclei, starting with prophase and continuing through metaphase, anaphase, and telophase. Technically, therefore, mitosis occurs **after** duplication has been completed.

Nuclear division usually occurs in close association with division of the cytoplasm (**cytokinesis**). The fact that these two processes do not always occur together, along with evidence from numerous experiments that have demonstrated that various chemical and physical treatments of dividing cells have different effects on mitosis and cytokinesis, clearly shows that different chemical and **physical** processes are involved in nuclear division (mitosis) and in cytoplasmic division (cytokinesis).

An important point to remember is that cell division is a **dynamic** series of events during which the cell undergoes dramatic and often rapid physiological and morphological changes. The so-called stages of mitosis merely represent a few morphologically identifiable points in this continuum.

The Mitotic Apparatus

During mitosis, a new structure, called the **mitotic apparatus,** is formed within the dividing cell (see figure 3.2). The mitotic apparatus plays an essential role in mitosis and provides the mechanism by which the duplicated chromosomes of the dividing cell are distributed between the daughter cells. A knowledge of the structure of the mitotic apparatus and its principal parts will help you to understand the process of mitosis and to learn how genetic continuity is maintained throughout many generations of cells. The mitotic apparatus is of vital importance to you; without it, a corn plant would not grow, a cut finger would not heal, a broken leg would not mend, and you would be unable to have children.

There are five main components of the mitotic apparatus: the **aster rays,** the **chromosomes,** the **centromeres,** the **centrioles,** and the **spindle fibers.** Most dividing animal cells exhibit all five of these components. Most plant cells and certain invertebrate animal cells, however, lack centrioles and aster rays. Four of these five parts of the mitotic apparatus appear to play important roles in the process of mitosis, but the function of one of these structures, the aster rays, is still unknown. The functions of these various components will be considered later in this exercise.

Several important events take place in a living cell during the early stages of mitosis. These events include:

1. The breakdown of the **nuclear membrane** and the mingling of the nuclear contents with the cytoplasm of the cell.

2. The condensation of the chromosomal material within the nucleus to form discrete, visible chromosomes.

3. The separation of the centrioles and their migration to opposite sides of the cell.

4. The formation of spindle fibers and aster rays.

The spindle fibers and aster rays are formed by the coalescence or condensation of relatively small protein molecules already existing within the cytoplasm of the cell. Thus, the formation of the components of the mitotic apparatus represents a recombination of molecules preexisting in the parent cell rather than the synthesis of new

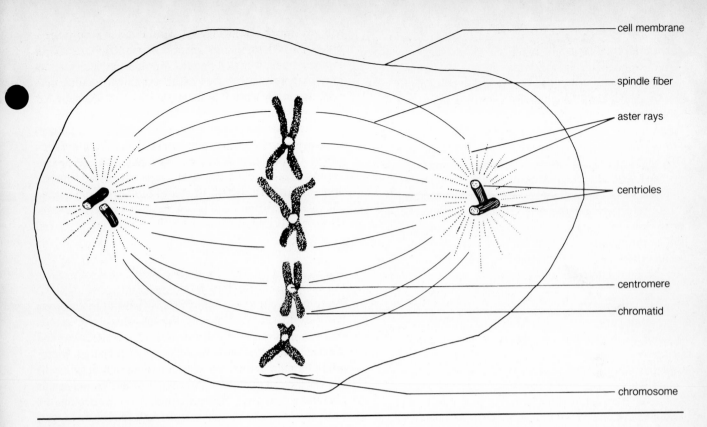

— cell membrane

— spindle fiber

— aster rays

— centrioles

— centromere

— chromatid

— chromosome

Fig. 3.2 Mitotic apparatus.

protein molecules. There is relatively little synthesis of new molecules during the process of mitosis; instead, most of the new molecules formed within the cell are synthesized during the interphase.

Chromosomes

Chromosomes are more or less elongate structures present throughout the life cycle of a cell. Their structure, however, appears quite different when observed at various times during the life cycle of a cell. During mitosis, the chromosomes can be seen as short, rodlike structures that are formed by condensation or contraction of very fine, threadlike structures present in the nucleus prior to mitosis. Thus, the appearance of chromosomes, like the appearance of the other components of the mitotic apparatus during the early portions of mitosis, results mainly from the reorganization of preexisting materials rather than from the actual synthesis of new materials.

Figure 3.3 is a photograph of the metaphase chromosomes in a male human cell with the normal 46 chromosomes (23 pairs). Observe the different sizes and shapes of the chromosomes. Each species of animal (and plant) has a characteristic chromosome number. In humans this number is 46; in the fruit fly *Drosophila* it is 8; in the dog it is 78; and in the cat it is 60.

The chromosomes contain the hereditary units, or genes, which specify each of the characteristics of a cell and determine all of its capabilities. Thus, the chromosomes are responsible for the control of cell metabolism and enable cells to differentiate and to organize into specialized tissues, organs, and organ systems.

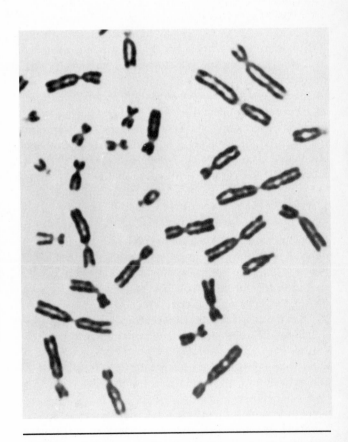

Fig. 3.3 Chromosomes from a male human lymphocyte in metaphase. (Photograph by Wendell Mackenzie.)

Mitosis and Meiosis

Centromeres

The centromeres firmly bind the chromosome to the spindle fiber. During the latter portion of mitosis, the chromosome pairs move toward opposite **poles** of the mitotic spindle. Despite numerous scientific studies, it is still not clear whether the chromosomes are *pulled* toward the poles by contraction of the spindle fibers or whether they are *pushed* toward the poles by growth or lengthening of the spindle fibers, or some combination of these processes. We do know, however, that damaged chromosomes in cells injured by X rays or by chemical agents sometimes separate into two or more fragments. Those chromosomal fragments that lack centromeres do not become attached to the chromosome and do not move toward the poles. Such observations clearly demonstrate the functional role played by the centromeres in mitosis.

Spindle Fibers and Asters

Special research techniques have been devised that allow scientists to isolate and to study the mitotic apparatus from living cells. Biochemical studies on such preparations of isolated mitotic apparatuses have demonstrated that the spindle fibers and **asters** are constructed mainly of two common contractile proteins, actin and tubulin. These proteins are involved in the movements of all eukaryotic cells.

Aster rays are constructed of the same proteins and thus might be expected to play some related role in mitosis; however, aster rays have no chromosomes attached to them, and scientists have not yet found evidence that the asters actually do play a significant role in mitosis. In fact, although asters are typical components of the mitotic apparatus in animal cells, they are absent from most types of higher plant cells. Thus, aster rays are apparently not essential for the process of mitosis since this process commonly occurs in plant cells even though asters are lacking.

Centrioles

Centrioles are tiny structures usually found in pairs near the nucleus in nondividing animal cells with a well-formed nucleus. The centrioles migrate toward opposite sides of the cell during the early stages of mitosis and seem to form centers or foci for the spindle fibers and aster rays at each end of the mitotic apparatus. Another name sometimes given to the centrioles is **division centers,** a name that suggests the role attributed to these structures by some scientists.

Recent laboratory studies have indicated that centrioles have the ability to initiate the formation of asters and spindle fibers. Scientists have found that laboratory preparations of isolated tubulin do not form into microtubules unless intact centrioles are added. This evidence clearly points to the centrioles as the organizational centers (or "inducers") of the **mitotic spindle** and asters in animal cells.

It appears that clearly identifiable centrioles generally are lacking in most plant cells, although plant cells typically exhibit a well-defined mitotic spindle. This fact has frequently been cited as an argument against centrioles being essential as the organizer of the spindle apparatus. New research, however, has shown that even plant cells have something that serves as a spindle organizer. The spindle organizers in dividing cells of flowering plants, however, appear to be rather amorphous structures that are difficult to see under either light or electron microscopes. Further research is needed to help us understand the role of centrioles and the processes by which the elaborate spindle apparatus is organized in dividing plant and animal cells.

Stages of Mitosis

An important point to remember is that mitosis involves a dynamic series of events during which the cell is undergoing dramatic, and often rapid, physiological and morphological changes. When studying prepared microscope slides of cells in mitosis, you are merely seeing cells that have been killed and stained at specific points in this continuous process. For convenience in describing the process, mitosis is usually divided into four main stages: **prophase, metaphase, anaphase,** and **telophase** (figure 3.4). The stage between successive mitoses is called **interphase.**

Many rapidly growing tissues provide good material for the study of mitotic cell division. Obtain a microscope slide prepared from some appropriate tissue such as the whitefish blastula, early embryos in the uterus of the roundworm *Ascaris,* or the skin of an amphibian tadpole. Study first one or more cells in **interphase.** Observe the structure of the nucleus and the arrangement of the chromatin material. Can you identify definite chromosomes? How many nucleoli do you find? Is the number of nucleoli the same in all interphase cells?

Observe other cells on the slide and select cells in each of the main stages of mitosis for further study. During your study, attempt to follow the sequence of stages as described below and try to visualize the changes that occur during the transition from one "stage" to the next. Draw on figure 3.5 each of the stages of mitosis listed below.

1. **Early prophase stage** during which the **chromatin** material shortens to form long, coiled, threadlike chromosomes.
2. **Middle prophase stage** with relatively thick chromosomes.
3. **Late prophase stage** in which the chromosomes are further shortened and thickened. Under high magnification, late prophase chromosomes can be seen to consist of two separate strands, the chromatids, joined by a single centromere.
4. **Metaphase stage** showing the chromosomes arranged in a ringlike fashion on the equatorial plane and attached to fibers of the mitotic spindle.

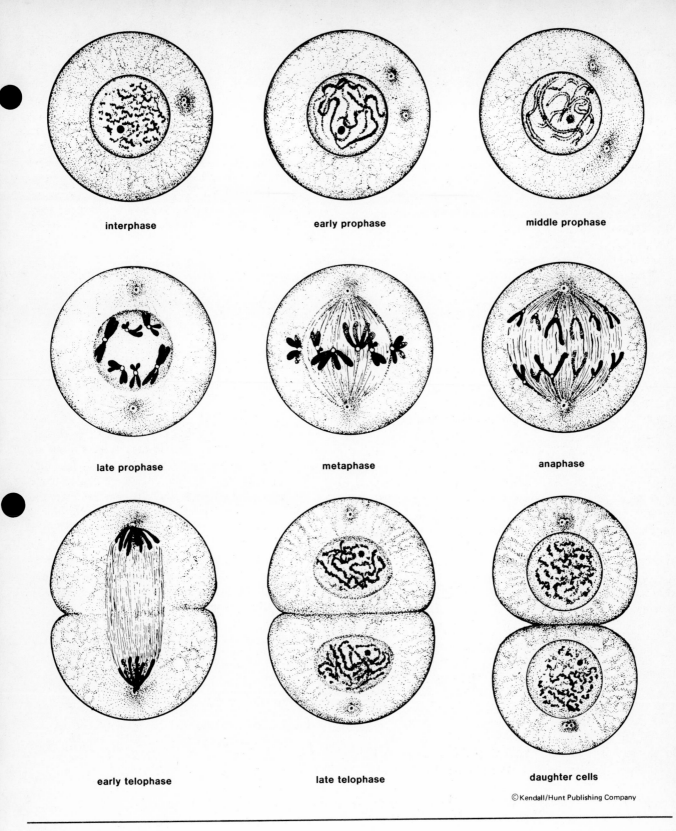

interphase

early prophase

middle prophase

late prophase

metaphase

anaphase

early telophase

late telophase

daughter cells

Fig. 3.4 Stages of mitosis.

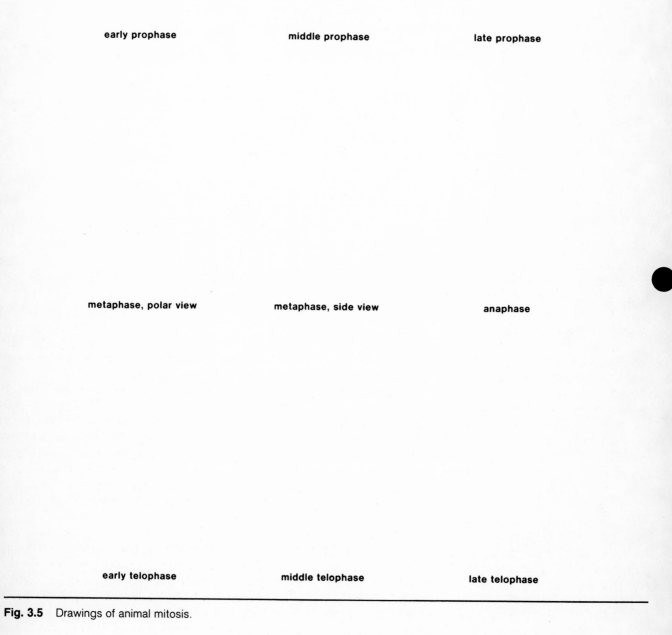

early prophase middle prophase late prophase

metaphase, polar view metaphase, side view anaphase

early telophase middle telophase late telophase

Fig. 3.5 Drawings of animal mitosis.

5. **Anaphase stage** during which the centromeres divide and the two chromatids of each chromosome move apart to opposite poles. The anaphase stage is relatively brief.

6. **Early telophase stage** showing the full set of chromosomes at each end of the elongated cell and the beginning of a **cleavage furrow** around the middle of the cell.

7. **Middle telophase stage** in which the individual chromosomes start to uncoil and lengthen and begin to appear less distinct.

8. **Late telophase stage** during which the nuclei in the two daughter cells reorganize. The chromosomes disappear, nuclear membranes and the nucleoli reappear, and the separation of the daughter cells is completed.

Timing in the Cell Cycle

The different stages in the cell cycle are not of equal duration; some stages are relatively long while others are very brief. The actual duration of mitosis and the relative duration of individual stages of mitosis vary widely from one cell type to another. Cells of a particular type also vary in their rates of division depending upon numerous physiological and environmental conditions. Nevertheless, you can obtain an estimate of the relative duration of the various stages in a population of cells and learn something about the kinetics of cell division by employing some relatively simple experimental techniques. You will need a whitefish blastula slide for this experiment.

Consider the whitefish embryo from which your mitosis slide was prepared as a **population of dividing cells** and see what can be determined about the kinetics of cell division, the relative duration of various stages of mitosis, and the relationship between mitosis and interphase in the life cycle of a cell. The whitefish embryo represents a population of relatively homogeneous cells; during the earliest stages of development the embryo consists of few cells which divide more or less synchronously. As development continues and the number of cells in the embryo increases as a result of cell division, the degree of synchrony decreases and division becomes progressively more randomized.

Your whitefish embryo slide represents a slice through a fish embryo containing several thousand cells. It therefore represents a sample taken from a larger population of cells. Still, the number of cells on your slide is too great for you to count readily in the limited time available in the lab. Select a still smaller sample of cells from the embryo to yield some numbers that you can use to estimate some characteristics of the population of cells that made up the original whitefish embryo.

Take your whitefish embryo slide and select a random sample of fifty cells. How can you be sure that you obtain a **random sample** of cells? Record the number of cells in each of the four mitotic stages and those in interphase and carefully record the results of your count in your notebook.

Random sample

A *random sample* can be defined most simply as a sample from a population in which *every member* of the population has an *equal chance* to be included. Thus, a sample in which the individuals selected are determined by use of a table of random numbers (or the random number generator in a computer) would be a random sample. A sample in which the individuals selected are determined by taking every tenth individual would not necessarily be a random sample. Why?

Note: Count only cells in which the nucleus appears in the section and enough nuclear material (chromosomes, mitotic spindle, etc.) can be seen to allow accurate identification of the stage.

Since the cells in your slide were all killed (fixed) at approximately the same time, they represent a sample of the fish embryo cells stopped in action at a particular point in time. Thus, the frequency of cells in the various stages is proportional to the relative duration of the stages.

Construct a bar graph on figure 3.6 to illustrate the results of your count. Show the number of cells in your sample in each stage (prophase, metaphase, anaphase, telophase, and interphase). In which stage did you find the most cells? The fewest cells? Which stage, therefore, would you conclude is the longest in duration? The shortest? What are the relative lengths of the other stages as estimated from your sample?

Compare the data from your slide with that obtained by other students from their slides. How do their counts compare with yours? How much variation does there appear to be in the counts on different slides? What are some of the sources of **variation** that may account for the differences in your counts? Are your conclusions regarding the relative lengths of the various stages the same as those of other students in the class or do they differ? Why?

Now, combine the data from your count with that of the other students in your lab section to obtain estimates of the characteristics of another population of cells. The population represented by these pooled data consists of the cells in several (20–24) different whitefish embryos, and the fifty cells counted by each student represent a series of subsamples taken from that larger population of cells.

From these pooled data, construct another bar graph on figure 3.7 to show the number of cells in each stage. Estimate the relative lengths of the various stages in the

Fig. 3.6 Graph of the distribution of cells in various stages of division. Data from your slide.

Chapter 3

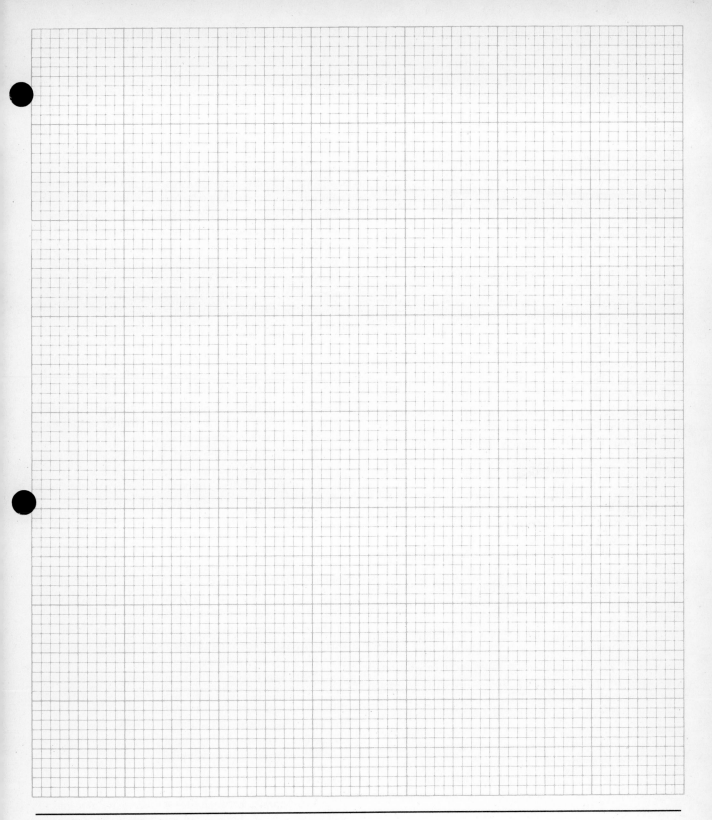

Fig. 3.7 Graph of the distribution of cells in various stages
of division. Pooled data from the whole class.

new cell population from these pooled data and compare the results with those from your count of a single slide. How do your results compare with and how do your conclusions differ from those based on data from a single whitefish blastula slide? Explain.

Meiosis

Meiosis is a special kind of nuclear division which ensures the constancy of chromosome number in the cells of succeeding generations of organisms. Sexually reproducing animals usually form male and female **gametes** at some point in their life history. **Fertilization** normally occurs at a later time in the life cycle and involves the fusion of male and female gamete nuclei. Thus, in order to maintain a constant number of chromosomes in successive generations (and to avoid doubling the chromosome number each time), some mechanism is necessary to provide a reduction (halving) of chromosome number between successive fertilizations. The process that results in the reduction in chromosome number is called **meiosis.**

The somatic cells of every species of animal have a definite and characteristic number of chromosomes. In man, this number is forty-six; in the fruit fly *Drosophila,* the number is eight; in the roundworm *Ascaris,* the number is four. This is referred to as the **diploid** (2N) number of chromosomes because the chromosomes are arranged in pairs. One member of each chromosome pair came from the father, and one chromosome came from the mother.

During the formation of gametes in animals, the number of chromosomes is reduced by half, and the resulting gametes have the **haploid** (1N) chromosome number. The subsequent fusion of the two haploid gametes (egg and sperm) during fertilization results in a return to the diploid chromosome number.

Meiosis generally consists of two successive nuclear divisions called the **first** and **second meiotic divisions** (figure 3.8). Meiosis differs in two important respects from ordinary mitosis. First, the final number of chromosomes in a gamete resulting from meiosis is **only half** that of the parent cell, and each gamete or spore receives only **one chromatid** from each homologous pair of chromosomes that was present in the original parent cell. Second, during the reduction in number, the chromosomes are **assorted at random** so that each gamete or spore receives a chromatid from one or the other member of each homologous pair. This random assortment of genetic material during meiosis plays a very important role in heredity. **Homologous chromosomes** are the paired chromosomes found in diploid cells that are very similar in size and shape but differ both in origin (one comes from the father and one from the mother) and in genetic composition. (The father and mother usually contribute different sets of alleles to the offspring.)

Meiosis, like mitosis, is a dynamic process during which the cells are undergoing continuous changes. Nonetheless, a good understanding of the process can be achieved by describing it as a sequence of two nuclear divisions, each with four distinct stages and with an intervening interkinesis stage between the first and second meiotic divisions.

Principal Stages of Meiosis

First Meiotic Division (Reduction Division)

Prophase I. Many of the events in the first meiotic prophase are similar to prophase in mitosis. The chromosomes become visible as the very thin chromatin strands coil and condense. The nucleoli and the nuclear membrane disappear, and the mitotic spindle appears.

The key difference between the first meiotic prophase and prophase in mitosis is that, in meiosis, each pair of homologous chromosomes comes closely together in an intimate pairing process called **synapsis.** By the end of the first meiotic prophase, each homologous chromosome pair is seen as two double-stranded chromosomes closely held together. This structure, formed as a result of synapsis and consisting of two chromosomes with a total of four chromatin strands, is called a **tetrad.** Tetrads are found only during prophase I of meiosis. Note the six chromosomes at the end of prophase I as illustrated in figure 3.8. Each chromosome consists of two **distinct strands,** or chromatids, attached to **a single centromere.**

Metaphase I. At the onset of metaphase I, the synapsed chromosome pairs move together as a unit to the equatorial plane. Each chromosome pair is attached to a single spindle fiber by its **two adjacent centromeres.**

Anaphase I. The important difference between this stage and the corresponding stage in ordinary mitosis is that there is **no division of centromeres** in anaphase I of meiosis. The centromeres of the homologous chromosomes simply move apart, and the two double-stranded chromosomes of each pair migrate toward opposite poles. Thus, **half** of the chromosomes move to one pole, and **half** of the chromosomes move to the opposite pole. Note the three chromosomes migrating toward each pole in anaphase I, as shown in figure 3.8.

Telophase I. The chromosomes reach the poles of the mitotic spindle in each of the daughter cells, the spindle disappears, and new nuclear membranes appear around the reforming nuclei in the daughter cells. The chromosomes begin to elongate, gradually fade from view, and nucleoli reappear within the nuclei.

Interkinesis

Interkinesis. Between the two successive divisions in meiosis is a brief stage called **interkinesis.** This stage is generally similar to an interphase between mitotic divisions, but there is **no duplication of genetic material** (no DNA synthesis) in interkinesis.

Chapter 3

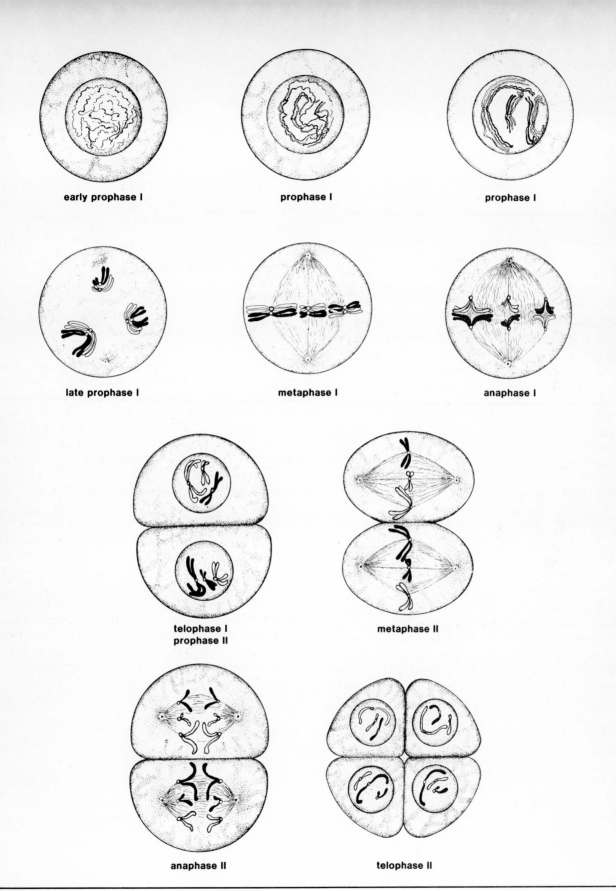

early prophase I

prophase I

prophase I

late prophase I

metaphase I

anaphase I

telophase I
prophase II

metaphase II

anaphase II

telophase II

Fig. 3.8   Meiosis.

Fig. 3.9 Drawings of selected stages of meiosis.

Chapter 3

Second Meiotic Division (Nonreduction Division)

Prophase II. The second meiotic division is essentially an ordinary mitotic division. There is no **synapsis** in prophase II; the double-stranded chromosomes reappear and move independently toward the equatorial plane.

Metaphase II. Each double-stranded chromosome attaches **separately** to a spindle fiber.

Anaphase II. The **centromeres divide** at the end of metaphase II, and during anaphase II the newly separated, single-stranded chromosomes move toward opposite poles.

Telophase II. The new chromosomes in the nuclei that appear in the daughter cells are **single-stranded** and contain only **half** the number of chromosomes as in prophase I.

The actual demonstration of meiosis to an inexperienced observer is difficult because of the small size of the chromosomes in most kinds of cells and the problems inherent in obtaining good microscopic preparations showing cells in clearly recognizable stages of meiosis. For these reasons, meiosis is often presented in introductory biology and zoology courses by means of a series of carefully selected demonstration slides. If possible, a time-lapse motion picture illustrating meiosis should also be seen to help you gain an understanding of the dynamic sequence of events in meiosis.

Study the demonstration materials on meiosis provided in the laboratory and **draw** selected stages of meiosis in the space provided in figure 3.9 as directed by your laboratory instructor.

Key Terms

Aster includes all of the aster rays surrounding one pole of the mitotic apparatus in an animal cell. Absent in plant cells.

Aster ray one of the fibrils, or rays, making up an aster.

Centriole self-replicating organelles usually found in pairs adjacent to the nucleus of an interphase cell. Also found centered in the asters of the mitotic apparatus of most animal cells.

Centromere (Kinetochore) a narrow region of a chromosome that binds the two chromatids together prior to separation and to which the spindle fibers attach during mitosis.

Chromatid one strand of a duplicated chromosome.

Chromatin genetic material in the interphase nucleus, representing the chromosomes in a long, thin, thread-like form.

Chromosome filamentous structure that carries the genetic material of the cell (DNA). Chromosomes can be very long and uncoiled in interphase, shorter with double strands (chromatids) in prophase, and still shorter with a single strand in anaphase.

Cleavage furrow indentation of the cell membrane around the equator of an animal cell at the beginning of cytokinesis.

Cytokinesis division of the cytoplasm of a cell.

Diploid or 2N cells containing both members of each homologous pair of chromosomes.

Equator by analogy with the earth, a line that goes around the middle of the cell, equidistant from the poles.

Haploid or N cells containing only one member of each homologous pair of chromosomes.

Homologous chromosomes chromosomes with the same size and shape and carrying the genetic material for the same characteristics. One member of each homologous pair comes from each parent.

Interkinesis the period intervening between the first and second meiotic divisions. No chromosomes are duplicated and no DNA is synthesized during interkinesis.

Interphase the stage between successive nuclear (mitotic) divisions consisting of G_1, S, and G_2 phases. It is the stage during which the cells are metabolically active.

Meiosis a special type of nuclear division in which the chromosome number is reduced from 2N to 1N by separating the members of the homologous pairs of chromosomes.

Mitosis nuclear division resulting in two new nuclei with the same genetic material as the original nucleus.

Mitotic apparatus a special structure formed during mitosis consisting of the spindle fibers, asters, and centrioles.

Mitotic spindle all of the spindle fibers collectively.

Nonreduction division The second meiotic division. This division follows interkinesis and resembles a mitotic division. There is no reduction in chromosome number.

Nuclear membrane (Nuclear envelope) the double membrane surrounding the nucleus in an interphase cell.

Nucleolus a dense organelle within the interphase nucleus of eukaryotic cells; site of synthesis of ribosomal RNA and ribosome assembly.

Nucleus a membrane-bound organelle containing the genetic material of a cell which controls cell metabolism.

Poles opposite ends of a cell where spindle fibers converge during mitosis and meiosis.

Reduction division the first meiotic division, during which the number of chromosomes in a cell is reduced by half.

Spindle fiber one of the microtubular filaments extending between the poles of the cell in mitosis and meiosis made up of contractile protein.

Synapsis the pairing of homologous chromosomes in prophase I of meiosis, the centromeres touching each other.

Tetrad a group of four chromatids from a pair of homologous chromosomes formed by a synapsis during prophase I of meiosis.

4
Development

Objectives

After completing the laboratory work in this chapter, you should be able to perform the following tasks:

1. Describe the structure of a spermatozoan of a frog or other representative animal and identify its principal structures visible in a light microscope.

2. Explain the processes of differentiation, growth, and morphogenesis and discuss their importance in animal development.

3. Briefly describe the major events in starfish or sea urchin development from fertilization to the gastrula stage and identify representative stages in a microscopic preparation.

4. Identify the principal structures in the blastula and gastrula stages of a starfish or sea urchin.

5. Discuss the organization of a frog egg and tell how it differs from a starfish egg.

6. Describe the major events in frog development from fertilization to the tadpole stage and identify representative stages from living or preserved specimens.

7. Discuss the organization of a chick egg and its adaptations for development on land.

8. Describe the four extraembryonic membranes surrounding a chick embryo and explain the function of each.

9. Identify the principal structures seen in whole mounts of 24-, 33-, 56-, 72-, and 96-hour chick embryos.

10. Identify the major structures in representative cross sections of a 33-hour chick embryo.

Introduction

Animal development usually begins with the fertilization of an egg by a sperm. The nuclei of the egg and sperm fuse, and the male and female parents' genes share in determining the characteristics of the offspring. Compared to other biological processes, embryonic development is relatively slow. New cells, tissues, and organs make their appearance in the embryo over a period of hours, days, or weeks.

Embryonic development can be divided into five major phases: (1) **gametogenesis,** the formation of the haploid male and female gametes (sperm and eggs); (2) **fertilization,** activation of the egg and fusion of the sperm and egg nuclei to form the diploid zygote; (3) **cleavage,** the subdivision of the zygote into many cells by mitosis; (4) **gastrulation,** the formation of germ layers; and (5) **organogenesis,** the initiation and differentiation of specific organs.

Materials List

Living Specimens
    Frog embryos at various stages
    Frog sperm
    Chick embryos, 33, 56, 72, and 96 hours old
Prepared Microscope Slides
    Starfish embryos
    Chick embryo, 24 hour, whole mount
    Chick embryo, 33 hour, whole mount
    Chick embryo, 33 hour, cross sections
    Chick embryo, 56 hour, whole mount
    Chick embryo, 72 hour, whole mount (Demonstration)
    Chick embryo, 96 hour, whole mount (Demonstration)
    Frog ovary with ova, cross section (Demonstration)
    Frog testis, cross section (Demonstration)
    Selected slides of early frog development (Demonstration)
Chemicals
    Amphibian Ringer's solution
    Chick Ringer's solution

In this exercise we shall study examples of development from three different animals to illustrate different portions of embryonic development and some variations in the development of different kinds of animals. The **starfish** illustrates the earliest stages of development, the **chick** serves to illustrate the later stages of development and adaptation for terrestrial life, and the **frog** is well suited for the study of development from fertilization to hatching.

Gametes

Gametes are the mature germ cells, **eggs** and **spermatozoa.** Both living sperm and stained microscope slides should be available for your study of male gametes. Observe the rapid movements of the living sperm. On a stained microscope slide identify the anterior **head,** the narrower **middle piece,** and the long posterior **tail.** The sperm of different animal species vary considerably in size and shape, particularly in the shape of the head. Draw a sperm cell in figure 4.1 and label each of these parts.

Eggs, or ova, are much larger in size than sperm and contain varying amounts of stored food materials for the nourishment of the developing embryo. Since the eggs of the starfish, frog, and chick differ substantially in this regard, they will be described separately as we study the development of each animal.

Influence of Yolk

The pattern of development in animals is strongly influenced by the relative amount of **yolk** present in the egg. Based upon the amount and distribution of yolk, eggs can be classified into four main types: (1) **isolecithal eggs,** as found in starfish and man, which have relatively little yolk and in which the yolk is uniformly distributed throughout the egg; (2) **mesolecithal eggs,** as found in the frogs, toads, and salamanders, which have a moderate amount of yolk and which have a concentration of yolk in the vegetal (lower) hemisphere; (3) **telolecithal eggs,** as found in birds and reptiles, which have a large amount of yolk and the cleaving portion of the embryo is restricted to a small disc at one end of the egg; and (4) **centrolecithal eggs,** as found in insects, which have much yolk and in which the actively developing portion of the embryo forms a thin layer around the outside of the large central yolk mass.

Among the animals chosen for this exercise, therefore, we have examples of three egg types:

Starfish egg—isolecithal
Frog egg—mesolecithal (although some textbooks classify frog eggs as "moderately telolecithal")
Chick egg—telolecithal

Fig. 4.1 Drawing of sperm.

Differentiation, Growth, and Morphogenesis

Differentiation

Development is the complex series of processes by which a new organism arises from an egg (or directly from a parent organism in the case of asexual reproduction). Different kinds of organisms exhibit many differences in the details of their development, but there are some important basic similarities in the development of all organisms. One of these similarities is a progressive increase in the complexity of organization. A fertilized egg of a frog, for example, undergoes not only a large increase in size during its period of development, but, more importantly, new kinds of cells, tissues, and organs are formed within the frog embryo during its development. Such a progressive increase in complexity of structure and function is called **differentiation.**

Growth

Development also involves an increase in mass of the organism through the addition of new cells and/or an increase in size of existing cells. In addition to becoming more complex in structure and function, a developing frog (or human) embryo also becomes larger in size. This increase in mass of the embryo and of its constituent parts is called **growth.**

Morphogenesis

A third important process of development is **morphogenesis.** A living organism is not merely a bag of assorted parts heaped together in a random fashion. The parts of every living organism are arranged in a specific pattern and bear definite relationships to one another. Morphogenesis includes those processes through which the characteristic form (both external and internal) of an organism becomes established. For example, the movements of cells and masses of cells in an embryo to form a wing or a limb are important processes of morphogenesis.

Starfish Embryology

Starfish embryos are often used for introductory studies of development because they illustrate clearly the essential principles of early development of multicellular animals. Sea urchin embryos are also frequently used, and their development is very similar to that of starfish in these early stages. Sea urchins also have the added advantage that living eggs and embryos are relatively easy to obtain for laboratory study. With living sea urchin embryos, it is possible to observe most of the early developmental events described later with an ordinary compound microscope.

Obtain a whole-mount microscope slide with stained starfish embryos at various stages of development. Examine the slide first under low power of your microscope and select good representatives of the main stages described in the following section.

Color transparencies are excellent aids for the study of early embryology. They can be viewed easily in a small slide viewer or with the aid of a projector. Be sure to study the projection slides and any other demonstration materials available to assist in your understanding of starfish development.

Summary of Early Starfish Development

1. **Unfertilized egg.** Observe the large **nucleus** and the darkly staining **nucleolus.** A large food reserve, which provides energy for early development of the starfish embryo, is present as yolk in the cytoplasm. Surrounding the egg, or ovum, is a **cell membrane.**

    Prior to fertilization, the nuclear membrane breaks down, and the contents of the nucleus mingle with the cytoplasm. After penetration of the cell membrane by a spermatozoan, the haploid sperm nucleus migrates through the cytoplasm and fuses with the haploid egg nucleus. Also following fertilization, a new membrane is formed around the egg. This new membrane rises slightly above the surface of the egg and can often be observed in microscopic preparations. It is called the **fertilization membrane.**

2. **Two-cell stage.** Shortly after fertilization, the fertilized egg divides into two cells. This is the first of a number of rapid cell divisions that take place during the next several hours. This series of cell divisions is called **embryonic cleavage.**

3. **Four-cell stage.** Observe the membrane enclosing the four **blastomeres** (embryonic cells).

4. **Eight-cell stage.** What is the orientation of the cleavage plane that produced this stage relative to the last cleavage plane?

5. **Sixteen-cell stage.**

6. **Thirty-two-cell stage.** Note the rapidly diminishing size of the blastomeres. What has happened to the overall size of the embryo?

7. **Morula.** Solid mass of cells formed after many cleavages. No central cavity.

8. **Blastula.** Several more cleavage divisions occur, leading to the formation of a hollow ball of ciliated cells. This stage is the blastula, and after escaping from its enveloping membrane, the ciliated blastula swims about freely. The central cavity is the **blastocoel.**

9. **Early gastrula** (figure 4.2 *left*). A few hours later, the cells at one end of the blastula begin to push (or are pulled) into the blastocoel. This enfolding of the embryo is called **invagination** and results in the formation of a two-layered structure. These two layers are **primary germ layers,** and similar layers are formed by virtually all multicellular animals. (Sponges are one of the few peculiar exceptions to this rule.) The layers are an outer **ectoderm** layer, which forms the skin and the nervous system of the starfish, and an inner **endoderm** layer, which primarily forms the lining of the digestive tract of the adult starfish (and a few accessory organs of the digestive tract).

10. **Midgastrula** (figure 4.2 *center* and *right*). Invagination continues, and a hollow tube of endoderm extends into the blastocoel. The opening in the center of the endodermal tube is the **archenteron** (primitive gut) of the starfish embryo. The external opening of the archenteron is the **blastopore** (future anus of the starfish).

11. **Late gastrula.** The endodermal tube later expands near its inner end and forms two lateral pouches. These lateral pouches become part of the internal body cavity or **coelom** of the starfish. The walls of these lateral pouches become differentiated and form the third primary germ layer, the **mesoderm.**

Draw representative stages of starfish development as indicated on figure 4.3.

Frog Development

Ripe, unfertilized eggs of the grass frog, *Rana pipiens,* are relatively large (averaging about 1.75 mm in diameter) and have a moderate amount of yolk (mesolecithal type of egg). Remember that the frog is a semiaquatic

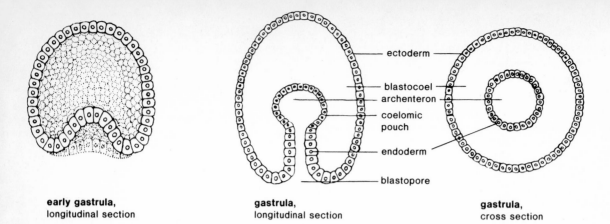

early gastrula, longitudinal section	gastrula, longitudinal section	gastrula, cross section

Fig. 4.2 Starfish gastrula.

animal; its eggs must be deposited in water and development of the embryos and tadpole larvae can take place only in an aquatic medium.

Obtain a few unfertilized frog eggs in water and study them under your stereomicroscope. Observe that the eggs are darkly pigmented. How is the pigment distributed on the surface of the eggs? Is the distribution of pigment uniform or nonuniform? The center of the dark portion of the surface of the egg is called the **animal pole.** The center of the lighter surface (opposite the animal pole) is called the **vegetal pole.** The line connecting these two poles is the **animal-vegetal axis.** Shortly after fertilization the egg rotates so that the animal pole is uppermost. How does this differ from the orientation of the animal-vegetal axis of the unfertilized eggs in your sample?

Closely applied to the surface of the unfertilized egg is a transparent **vitelline membrane,** which lifts from the surface of the egg following fertilization and thereafter is known as the **fertilization membrane.** Surrounding the vitelline membrane are three **jelly coats** made up largely of albumen, a protein secreted by the oviduct during the passage of the egg from the ovary. The jelly coats serve to protect the egg from injury and infection.

Fertilization is external in the frog. During mating the male frog grasps the female with his forelimbs, and as the female discharges her ripe eggs into the water the male releases sperm over them.

Shortly after fertilization the vitelline membrane rises to form the fertilization membrane, and later a **gray crescent** area appears on one side of the egg along the margin of the heavily pigmented zone. The gray crescent is a more lightly pigmented zone of the egg cortex and appears on the side opposite the point of sperm entrance. The center of the gray crescent area marks the future **posterior end** of the embryo; thus, at the time of sperm entry into the egg, the future **anterior-posterior axis** of the frog embryo is determined.

Development is fairly rapid following fertilization (see figure 4.4), and within a few hours at normal temperatures the fertilized egg becomes divided into many

cells, a hollow blastula is formed, and gastrulation is completed. Cleavage is **holoblastic** (all of the egg cytoplasm becomes divided) and **unequal** (the cells nearest the animal pole are smallest in size and those nearest the vegetal pole are largest in size). How is this related to the distribution of yolk in the frog egg?

Table 4.1 illustrates the schedule of development in *Rana pipiens* at 18° C. Other common species of frogs, toads, and salamanders follow a generally similar pattern of development but differ in the times required to reach the various stages. Several representative stages of development in *Rana pipiens* are shown in figure 4.4.

Table 4.1 Schedule of Development in *Rana pipiens* at 18° C (after Shumway and Rugh)

Stage	Hours at 18° C
Fertilization	0
Formation of gray crescent	1
Rotation (animal pole now uppermost)	1.5
First cleavage (2 cells)	3.5
Second cleavage (4 cells)	4.5
Third cleavage (8 cells)	5.5
Blastula	18
Early gastrula (dorsal lip stage)	26
Midgastrula	34
Late gastrula (yolk plug stage)	42
Neural plate	50
Neural folds	62
Ciliary movement	67
Neural tube formation	72
Tail bud stage	84
Muscular contractions	96
Heartbeat	5 days
Gill circulation, hatching (ruptures fertilization membrane)	6 days
Circulation in tail fin	8 days
Internal gills formed, opercular fold present	9 days
Operculum closed, covers internal gills	12 days
Metamorphosis into frog	3 months

Chapter 4

fertilized egg   two-cell stage   four-cell stage

eight-cell stage   morula   **blastula**, section

early gastrula, section   **gastrula**, longitudinal section   **gastrula**, cross section

Fig. 4.3   Drawings of starfish embryos.

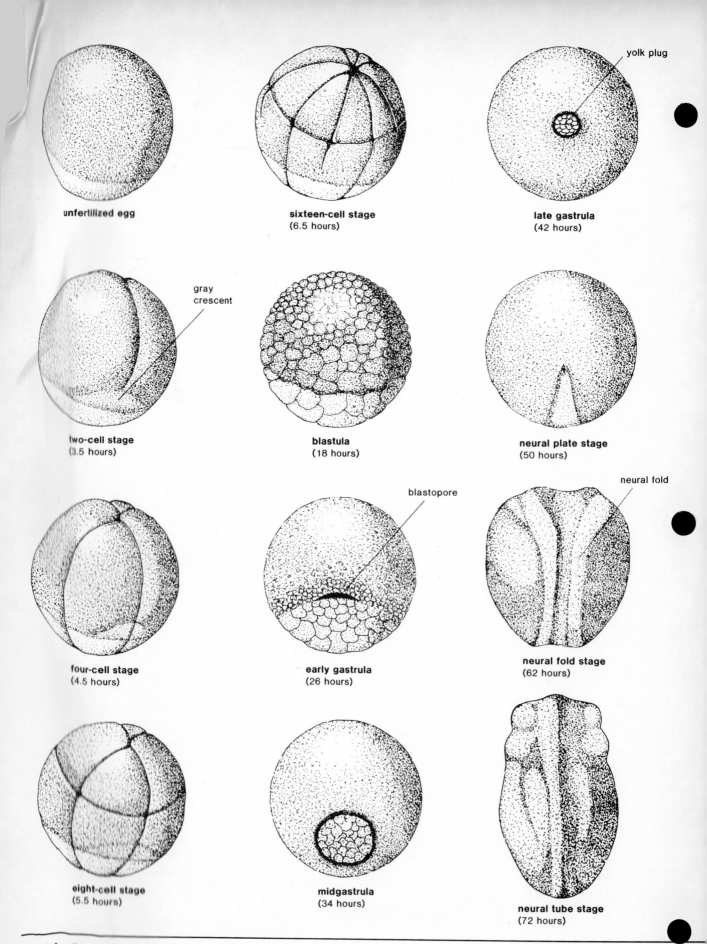

Fig. 4.4 Frog development, unfertilized egg through tadpole stages. (After Shumway and Huettner.)

Chapter 4

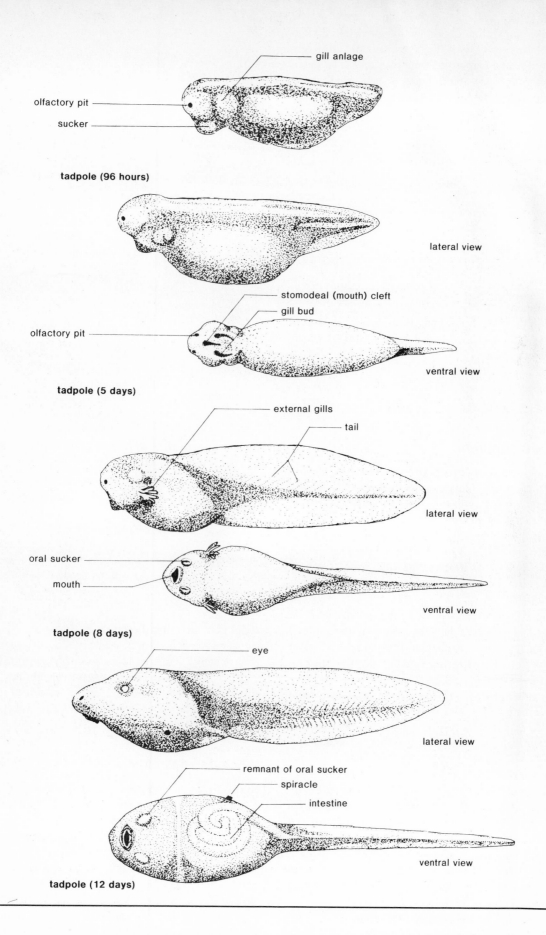

gill anlage

olfactory pit

sucker

tadpole (96 hours)

lateral view

stomodeal (mouth) cleft

gill bud

olfactory pit

ventral view

tadpole (5 days)

external gills

tail

lateral view

oral sucker

mouth

ventral view

tadpole (8 days)

eye

lateral view

remnant of oral sucker

spiracle

intestine

ventral view

tadpole (12 days)

Development

Study the frog embryos provided in your laboratory and compare the various stages with figure 4.4. Living embryos are best for study if they are available, although plastic-embedded and preserved embryos can also be used. Supplement your study of the living or preserved whole embryos by observing the microscopic demonstrations provided to illustrate the internal structure of the embryos at different stages. When you have completed your study, you should be able to describe and explain the major events in the development of a frog from an unfertilized egg to an adult.

Demonstrations

1. Microscope slide showing developing eggs in the uterus.
2. Microscope slide showing frog testis with developing sperm.
3. Wet mount of living frog sperm in 10% amphibian Ringer's solution.
4. Microscopic slides with cross sections and/or sagittal sections of selected early developmental stages of the frog through hatching.
5. Procedures for inducing ovulation in the frog by injection of pituitary extract and artificial insemination of frog eggs in vitro.

Chick Development

The eggs and embryos of birds exhibit several major differences from those of amphibians. Birds have a **cleidoic** (terrestrial) type of egg with several adaptations that permit development on land. Among the most important of these adaptations are: (1) **a hard outer shell,** (2) **a large food supply,** and (3) **extraembryonic membranes,** which envelop the embryo as it grows, provide protection, and ensure an optimal environment during embryonic development.

Four extraembryonic membranes are formed by a developing chick embryo: **yolk sac, chorion, amnion,** and **allantois.** Each of these membranes forms in a specific way and each performs a distinctive role in protecting the embryo (figure 4.5).

The **yolk sac** forms as a pouchlike outgrowth from the developing gut. It grows around the yolk, releases enzymes to digest the yolk, and transports the digested yolk products through its blood vessels to the developing embryo.

The **chorion** and **amnion** are sheets of living tissue that grow out of and around the embryo. These sheets ultimately join above the embryo and enclose the embryo in a double sac. Both the chorion and the amnion consist of two tissue layers, ectoderm and mesoderm. After closure of the amnionic folds, the amnion becomes filled with a watery fluid. Thus, the amnion maintains an aqueous environment to protect the growing embryo from dessication and serves as a physical protection for it.

The **allantois,** like the yolk sac, arises as a saclike outgrowth from the ventral surface of the gut. It has quite

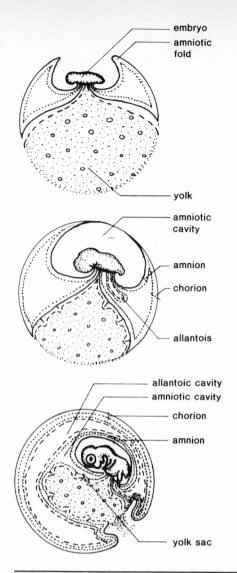

Fig. 4.5 Chick development, formation of extraembryonic membranes.

a different function, however. In birds, the allantois collects and stores metabolic waste products (largely crystals of uric acid). The allantois also grows and fuses with the chorion to form the **chorioallantoic membrane.** The chorioallantoic membrane is highly vascularized and facilitates the exchange of gases between the embryo and the external environment.

Whole Mounts of 24- and 33-Hour Chicks

Study first a whole mount of a 24-hour chick embryo. Consult figure 4.6 and study the whole-mount side under your stereomicroscope. **Caution:** Whole mounts of embryonic stages are relatively thick and should never be viewed under high power on your compound microscope. Note the approximate size and shape of the embryo. Observe the transparent area immediately surrounding the embryo and an opaque outer area. The inner portion of this opaque area contains many blood vessels that bring nutrients to the developing embryo.

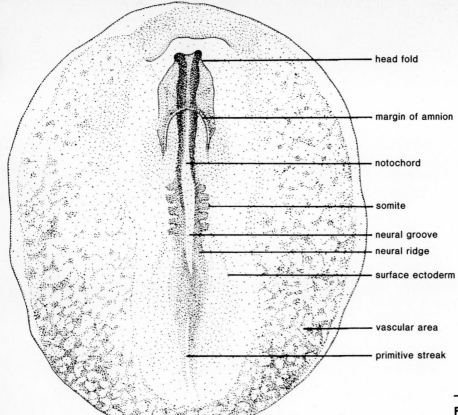

head fold

margin of amnion

notochord

somite

neural groove

neural ridge

surface ectoderm

vascular area

primitive streak

Fig. 4.6 Chick development, whole mount of 24-hour embryo, dorsal view.

Locate the **head** of the chick at the anterior end of the embryo, lying free and slightly elevated above the underlying membrane. Within the head, observe that the brain at this stage consists of a **neural tube,** which is continued posteriorly with the open **neural plate** consisting of a median **neural groove** and two lateral **neural folds.** Note also that the tissue of the neural folds is continuous laterally with the surface ectoderm, thus demonstrating the ectodermal origin of the nervous tissue.

Observe that the neural plate is continued posteriorly as the **primitive streak.** Find the **notochord,** which lies beneath the neural tube and the neural groove. Note that anteriorly the notochord appears as a well-defined rod, but more posteriorly it appears as a wide band of less dense tissue. Differentiation of the notochord, as of the neural structures, proceeds from anterior to posterior. Likewise, the **somites** differentiate from anterior to posterior, and new somites are formed by the coalescence and differentiation of mesoderm cells behind previously formed somites.

In a 33-hour embryo you can also observe the **heart** on the right side of the embryo, lying slightly ventral to the neural tube and notochord. Attached to the heart are two large **vitelline veins** (posterior) and a single **ventral aorta** (anterior). Posterior to the heart you can find the **crescentic fold,** which marks the posterior border of the forming **digestive tube.**

Make a drawing on figure 4.7 of a 33-hour chick embryo and include these structures mentioned.

Fig. 4.7 Drawing of a 33-hour chick embryo.

Development

Representative Cross Sections of 33-Hour Chick

Observe demonstration slides with selected cross sections through several different regions of the chick embryo similar to those indicated in figure 4.8. Each region reveals a distinctive pattern and illustrates the development of the organs located in that part of the embryo. Figures 4.9, 4.10, 4.11, and 4.12 illustrate four distinct regions of the embryo at about 33 hours.

Identify the principal structures of the embryo shown in these cross sections, including the **neural tube** with the **neural canal,** the **notochord,** the **mesodermal somites,** the dorsal **ectoderm** and the ventral **endoderm.** Lateral to the somites, note that the mesoderm is separated into two layers, a dorsal layer that in association with the ectoderm forms the **somatopleure,** and a ventral layer that in association with the endoderm forms the **splanchnopleure.** Observe the **coelom** between the somatopleure and the splanchnopleure, the paired **dorsal aortae** (which later fuse into a single dorsal aorta) alongside the **notochord,** and many **smaller blood vessels** within the mesoderm of the splanchnopleure. The following additional descriptions should help in your study of these cross sections.

Figure 4.9. Section through the brain (optic vesicles). Note that the blastoderm immediately under the free head region consists only of the ectoderm and the endoderm; laterally, the mesoderm is present also. The optic vesicles later separate from the brain to form the eyes.

Figure 4.10. Section through the region of the heart. Note the **epimyocardium,** the outer thick wall of the heart; the **endocardium,** the thin lining inside the heart; the **pharynx;** and the **dorsal aorta.**

Figure 4.11. Section through the somite region. Note the absence of the digestive tube, the relation of the yolk to the endoderm, and the connection of the somites with the lateral mesoderm. Identify the **neural tube** and the **notochord.**

Figure 4.12. Section through the primitive streak. Note the three germ layers, which meet and fuse in a thick band along the longitudinal axis. The depression of the ectoderm is the **primitive groove.** Label the three germ layers on figure 4.12.

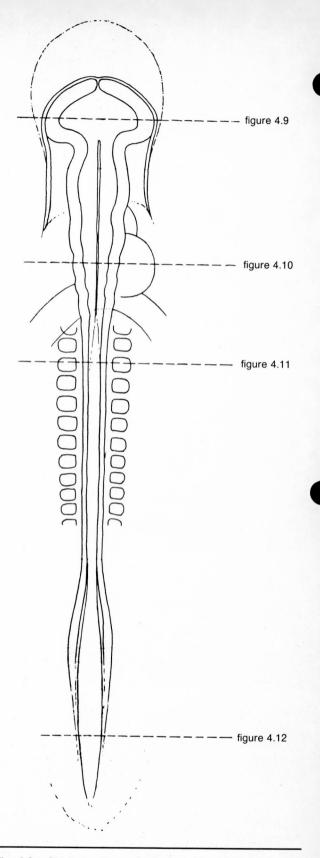

Fig. 4.8 Chick development, dorsal view of 33-hour embryo.

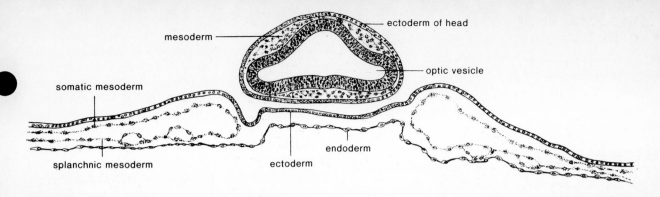

Fig. 4.9 Chick development, cross section of 33-hour embryo, through brain region.

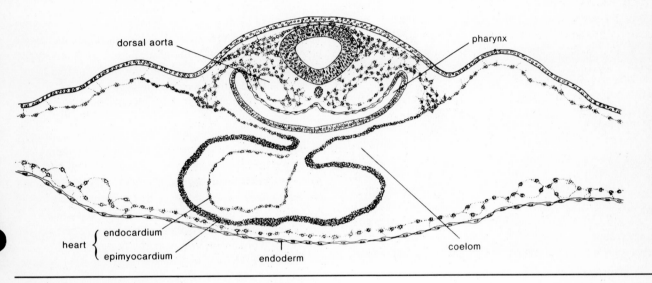

Fig. 4.10 Chick development, cross section of 33-hour embryo, through heart region.

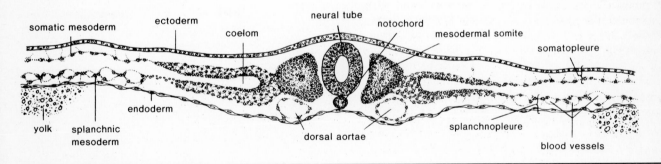

Fig. 4.11 Chick development, cross section of 33-hour embryo, through region of third somite.

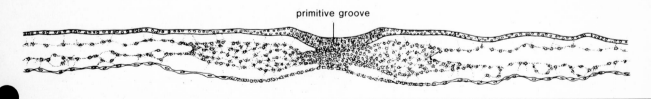

Fig. 4.12 Chick development, cross section of 33-hour embryo, through primitive streak region.

Development

Whole Mounts of 56-Hour Chick

Obtain a whole mount of a 56-hour chick embryo and compare it with that of the 33-hour chick. Figure 4.13 will aid you in the identification of the principal structures.

Note that the anterior part of the embryo is now flexed to the right. The **amnion** has grown dorsally and posteriorly to cover the anterior portion of the embryo. Locate the posterior margin of the amnion on your slide. Locate also the caudal fold of the amnion near the posterior end of the embryo. Later these portions of the amnion will join and completely enclose the embryo within the **amniotic cavity.**

Locate the three pairs of **gill slits** visible at this stage. What becomes of these gill slits in the adult? Note also the **aortic arches** which pass through the gill arches.

Nervous System and Sense Organs

Observe that several basic components of the nervous system are already well differentiated at 56 hours. Study the parts of the **brain** in your whole mount with the aid of figure 4.13. Note the anterior **prosencephalon,** the dorsal **epiphysis** (pineal gland), and the **mesencephalon.** Find also the **metencephalon** (which forms the cerebellum) and the **myelencephalon** (which forms the medulla oblongata).

Note the great enlargement of the **optic vesicles** to form the **optic cups.** Concurrent with the differentiation of the optic cup, the surface ectoderm overlying the optic cup thickens and invaginates to form a **lens vesicle.** Dorsal to the gill clefts, find the thickened ectoderm of the **auditory vesicle** (figure 4.13). This vesicle will later form the entire sensory part of the ear (inner ear).

Circulatory System

Observe that the heart, which was originally tubular and consisted of a single chamber in the 33-hour chick, has now become twisted and transformed into a **two-chambered structure.** Identify the single **atrium** and single **ventricle** present at this stage. Later the atrium divides into two atria to form a **three-chambered heart,** and at a still later stage of development the ventricle becomes subdivided to form a **four-chambered heart.** What is the phylogenetic significance of this progression from a single hollow tube to a four-chambered heart?

Locate in your whole mount of the 56-hour chick the ventral aorta, which arises from the ventricle and branches to form the **first, second,** and **third aortic arches.** Note that the aortic arches pass through the gill arches to form the still paired dorsal aortae. Find the **vitelline arteries,** a pair of large arteries leading from the embryo and carrying blood out into the yolk sac where nutrients are absorbed by the blood prior to return to the embryo via the **vitelline veins.** Locate the vitelline veins and trace them to the atrium. Observe the demonstration of a living chick embryo, see the beating of the heart and observe the circulation of the blood. Trace the flow of blood from the heart through the gill arches and work out the main course of circulation back to the heart.

Whole Mounts of Older Stages

The study of some older chick embryos will increase your understanding of the pattern of chick development. Some useful stages for study include 72- and 96-hour stages. Compare the orientation of embryos at 72 and 96 hours. How does the amnion compare in these two stages? What is the orientation of the 96-hour embryo in relation to the underlying yolk mass?

Among the major features of these stages also note the **four pairs of gill slits,** the **first gill arch** developing into the upper and lower jaws, the two pairs of **limb buds,** the paired **cerebral hemispheres,** the **nasal pits,** the **tail,** and the **allantois.** The allantois is small and concealed by the posterior limb buds in the 72-hour chick but appears in the 96-hour chick as a spherical, stalked vesicle on the ventral side of the embryo near the posterior end.

Demonstrations

1. Living chick embryos after 33, 56, 72, and 96 hours of incubation (in chick Ringer's solution).
2. Microscope slides with selected cross sections of 33-hour chick embryos.
3. Microscope slides with whole mounts of 72- and 96-hour chick embryos.

Chapter 4

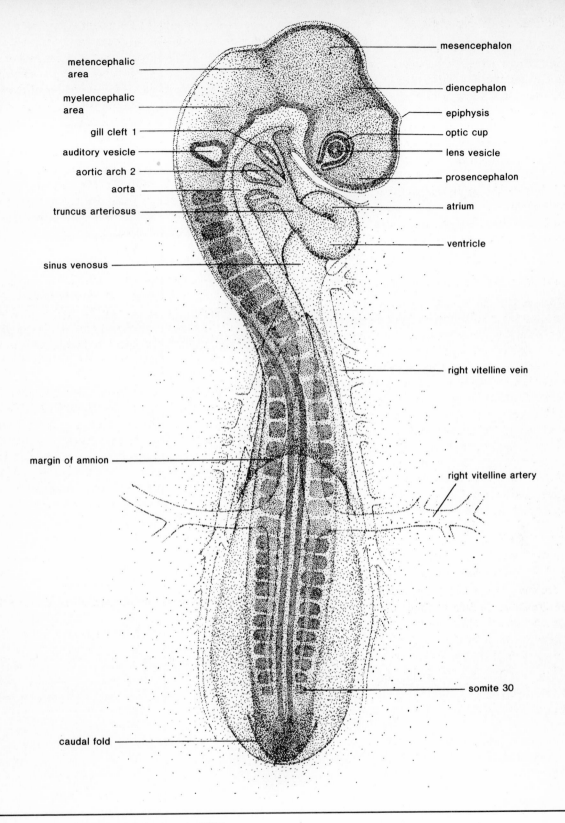

metencephalic area

myelencephalic area

gill cleft 1

auditory vesicle

aortic arch 2

aorta

truncus arteriosus

sinus venosus

margin of amnion

caudal fold

mesencephalon

diencephalon

epiphysis

optic cup

lens vesicle

prosencephalon

atrium

ventricle

right vitelline vein

right vitelline artery

somite 30

Fig. 4.13   Chick development, whole mount of 56-hour embryo.

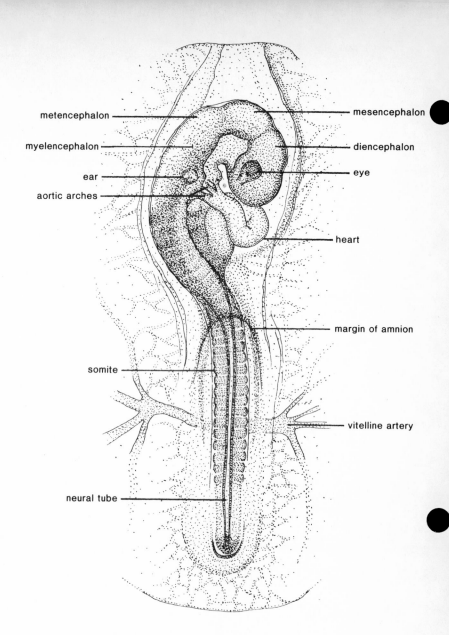

metencephalon

myelencephalon

ear

aortic arches

mesencephalon

diencephalon

eye

heart

margin of amnion

somite

vitelline artery

neural tube

Fig. 4.14 Chick development, whole mount of 48-hour embryo.

Living Chick Embryos

Living chick embryos are easily obtained and make excellent material for laboratory study of development. During its early stages of development, the chick embryo occupies a relatively small portion of the inside of the eggshell. Most of the space is taken up by the stored food materials, the yolk and the "white" of the egg. The portion of the egg that will become the chicken is comprised at first of a small disc of rapidly dividing cells (called the **blastoderm**) which later undergoes a process of gastrulation, develops a head, nervous system, circulatory system, and other organ systems.

Stages from 2–4 days (24–96 hours) are most satisfactory for study in an introductory zoology laboratory (see figures 4.14, 4.15, and 4.16). Obtain some fertilized eggs that have been incubated for 2–4 days and study the principal features of the embryo at each of these stages.

Directions for Opening Eggs

1. Put a penciled X on the egg to mark the side that was on top in the incubator tray. This step is important because the blastoderm rotates to the top of the egg during incubation. Otherwise, you may have difficulty finding the blastoderm after you crack the egg.

2. Fill a culture dish half full of warm (37° C) chick Ringer's solution. Ringer's solution contains 9gm NaCl, 0.4gm KCl, 0.24gm CaCl, and 0.2gm NaHCO₃ in a liter of distilled water.

3. Gently crack the eggshell on the side opposite the location of the embryo (i.e., opposite the X) by striking it carefully against the edge of the finger bowl.

4. With your thumbs placed over the X, lower the egg and your fingers into the solution and carefully pry open the cracked surface with your fingers, using your thumbs as a pivot.

Chapter 4

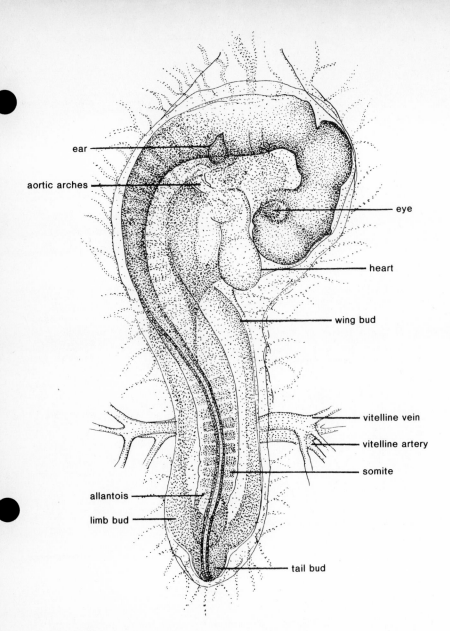

ear

aortic arches

eye

heart

wing bud

vitelline vein

vitelline artery

somite

allantois

limb bud

tail bud

Fig. 4.15 Chick development, whole mount of 72-hour embryo.

5. **Caution:** If you open the shell too slowly, the sharp edges of the shell may sever the delicate membranes of the embryo as it slides out of the shell. Likewise, if you open the shell too quickly, you can also damage these membranes.

6. If there is no embryo on the surface of the yolk, try another egg. Be sure to look on the sides and bottom of the egg also, in case the egg rotated as you opened the shell.

Study the embryo and identify as many structures of the embryo as possible using figures 4.14, 4.15, and 4.16 as a guide.

Repeat your study with other stages of development as available. Make notes on your observations on the pages provided for "Notes and Sketches" at the end of the chapter. Make a list of the most obvious features of each stage that you study.

Caution: Be sure to dispose of all eggs, eggshells, and other waste at the end of your study as directed by your instructor. Also remember to wash out your glassware when you have finished with it.

Key Terms

Allantois one of the extraembryonic membranes found in the chick; forms as an outgrowth of the gut. Functions in gas exchange and for the storage of metabolic wastes.

Amnion an extraembryonic membrane found in the chick; consists of layers of ectoderm and mesoderm. Encloses the developing embryo and provides a fluid environment to protect the embryo.

Archenteron the primitive gut of an embryo; formed by endoderm tissue.

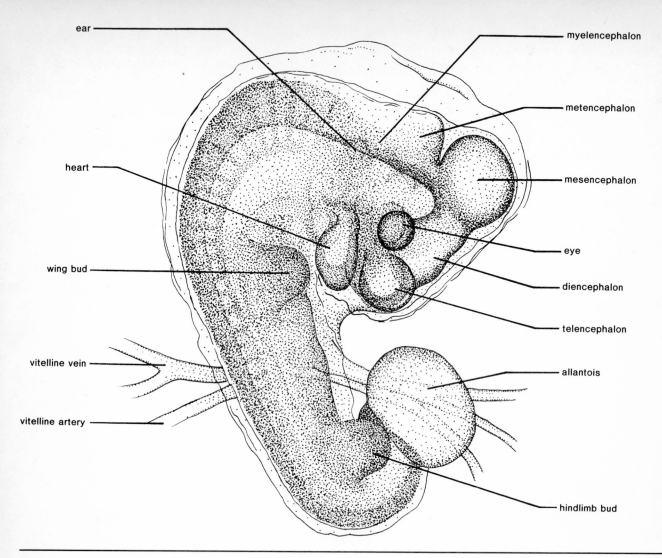

ear — myelencephalon

heart — metencephalon

mesencephalon

wing bud — eye

diencephalon

telencephalon

vitelline vein — allantois

vitelline artery

hindlimb bud

Fig. 4.16 Chick development, whole mount of 96-hour embryo.

Blastopore the opening in the gastrula through which the ectodermal cells invaginate. Often becomes the mouth or the anus of the adult, depending on the type of animal.

Blastula hollow-ball stage in early embryonic development; follows the morula stage and precedes the gastrula stage.

Centrolecithal egg type of egg with abundant, centrally located yolk, such as an insect egg.

Chorion an extraembryonic membrane comprised of ectoderm and mesoderm layers; protects the developing embryo.

Cleavage the period of rapid cell division in an embryo following fertilization; leads to the formation of the morula and blastula stages.

Differentiation formation of a specialized tissue or cell type from a simpler tissue or cell type.

Ectoderm primary germ layer found on the exterior of an early embryo; forms the nervous system, integument, and certain other tissues in the adult.

Endoderm primary germ layer found lining the gut of an early embryo; forms the lining of the digestive tract, the liver, pancreas, thyroid, and certain other organs in the adult.

Extraembryonic membranes membranes that form external to the embryo of the chick and most other terrestrial vertebrate animals; serve to enclose and protect the developing embryo. Includes the chorion, amnion, allantois, and yolk sac in the chick.

Fertilization the fusion of the male and female nuclei, which initiates embryonic development. The sperm cell must first penetrate the cell membrane of the egg and migrate to the egg nucleus.

Gametes specialized sex cells, eggs and sperm, necessary for sexual reproduction.

Gametogenesis the formation of eggs and sperm; involves meiosis and cellular differentiation of the male and female sex cells to form gametes.

Gastrula   developmental stage following the blastula. Invagination or infolding of cells at this stage leads to differentiation of the three primary germ layers: endoderm, ectoderm, and mesoderm.

Gastrulation   formation of the gastrula.

Growth   increase in size or mass of an embryo or individual.

Invagination   infolding of cells through the blastopore. Leads to the differentiation of endoderm and mesoderm.

Isolecithal egg   type of egg with a small amount of yolk, as a starfish egg.

Mesoderm   one of the primary germ layers; formed from invaginated ectoderm cells or as outpockets of the gut. Develops into the muscles, bone, connective tissues, circulatory system, and many other structures in the adult.

Mesolecithal egg   type of egg with a moderate amount of yolk, as in the frog egg.

Morphogenesis   the molding of a structure during embryonic development by cell and tissue movements.

Morula   stage formed during the latter part of embryonic cleavage; consists of a solid ball of cells.

Notochord   cartilaginous supporting rod parallel to the dorsal nerve cord of all chordates. May be replaced by a vertebral column.

Organogenesis   the formation of a specific organ, such as the heart, during embryonic development.

Primary germ layer   one of the three tissue layers differentiated early in development—ectoderm, mesoderm, and endoderm.

Somite   a block of mesodermal tissue along the nerve cord in an embryo; differentiates into segmental muscles and other tissues in the adult.

Telolecithal egg   type of egg with a large amount of yolk and the embryo restricted to one end, as a chick egg.

Yolk   stored food reserves in the egg and embryo; rich in lipids and proteins.

Yolk sac   one of the extraembryonic membranes of the chick and many other vertebrates; forms as an outgrowth of the gut and encloses the yolk.

5
Protozoa

Objectives

After completing the laboratory work in this chapter, you should be able to perform the following tasks:

1. Identify the major structures in a specimen of *Euglena* and tell the function of each structure.
2. Describe the structure of a *Volvox* spheroid and of an individual *Volvox* cell.
3. Describe the reproductive processes and life cycle of *Volvox*.
4. Explain the possible evolutionary significance of *Volvox* and the Volvocine Series.
5. Identify the principal structures in a specimen of *Amoeba proteus*.
6. Describe the techniques for preparing a wet mount and a hanging drop for microscopy and explain the uses of each.
7. Identify the main structures in a specimen of *Paramecium* and give the function of each structure.
8. Describe the functions of cilia in the feeding and locomotion of *Paramecium*.
9. Compare the locomotion of *Euglena, Amoeba,* and *Paramecium*.
10. Explain the life cycle of *Plasmodium* and identify the principal stages in microscope slides or photographs.
11. Explain the relationship of *Plasmodium* to the disease malaria.

Introduction

Protozoa are eukaryotic unicellular organisms, generally microscopic in size, that live as single individuals or in simple colonies. Within the unicellular body of a protozoan are many organelles that are analogous to the organs and organ systems of higher animals. Thus, protozoans exhibit a great deal of intracellular complexity. Most scientists believe that multicellular animals (metazoa) evolved from some group or groups of protozoans. Historically, there has been some disagreement among biologists over which group of protozoans may have been the real ancestors of the metazoa, but the flagellates are often cited as the most probable ancestral group.

For many years biologists considered the Protozoa to be the simplest group of animals and traditionally included them as the most primitive phylum in the animal kingdom. Recently, however, most biologists have come to recognize that the Protozoa have more in common with other unicellular eukaryotic organisms than with multicellular animals. Therefore, Protozoa are often placed in a separate kingdom, Kingdom Protista, with certain organisms formerly considered to be unicellular algae and fungi. Classification of the Protozoa is further complicated by recent evidence suggesting that the differences among various groups are sufficiently important that they should be divided into separate phyla. A revised classification of the Protozoa includes seven phyla.

Regardless of their taxonomic status, the Protozoa are an important assemblage of organisms with many animal-like characteristics. For this reason, and because of their possible phylogenetic significance as possible ancestors of multicellular animals, they should be included in any course in general zoology.

More than 70,000 species of Protozoa have been described, including species widely distributed in many different kinds of moist or wet habitats: in fresh, marine, and brackish waters; in sewage; in moist soil; in or on the bodies of many species of animals; and in or on some plants.

In this chapter we shall consider representatives of four common and well-differentiated groups of Protozoa: the Mastigophora, the Sarcodina, the Ciliophora (formerly called Ciliata), and the Apicomplexa (formerly called Sporozoa).

Classification

Phylum Sarcomastigophora

Protozoa with locomotion by means of flagella and/or pseudopodia; with a single type of nucleus.

Subphylum Mastigophora (Flagellata)

Locomotion by one or more flagella. Examples: *Euglena*, *Volvox*, *Trypanosoma* (blood parasites of man and other animals), and *Gonyaulax* and *Gymnodinium* (dinoflagellates, implicated in the red tides of coastal waters).

Subphylum Sarcodina

Locomotion by pseudopodia. Examples: *Amoeba proteus*, *Arcella* and *Difflugia* (testate amoebae), *Entamoeba histolytica* (a human parasite), *Globigerina* (a foraminiferan), and *Actinosphaerium* (a heliozoan).

Phylum Ciliophora (Ciliata)

Protozoa with cilia or ciliary organelles present in at least one stage of the life cycle; with two distinct types of nuclei (macronucleus and micronucleus). Examples: *Paramecium*, *Tetrahymena*, *Euplotes*, *Vorticella*, *Stentor*, *Blepharisma*, and *Trichodina*.

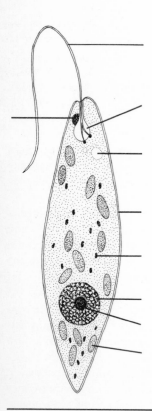

Fig. 5.1 *Euglena*. Label each structure where indicated.

Phylum Apicomplexa

Protozoa typically lacking locomotory organelles (except for gametes in some groups); with a characteristic set of anterior organelles called the apical complex (visible only with the electron microscope); microspores present at some stage in the life cycle; all species parasitic. Formerly included in the Sporozoa.

Materials List

Living Specimens
  Amoeba proteus
  Euglena
  Volvox
  Paramecium caudatum
Prepared Microscope Slides
  Arcella (Demonstration)
  Difflugia (Demonstration)
  Entamoeba histolytica (Demonstration)
  Actinosphaerium (Demonstration)
  Globigerina (Demonstration)
  Peranema (Demonstration)
  Symbiotic flagellates from termite or wood roach (Demonstration)
  Dinoflagellates (Demonstration)
  Volvox, cell walls (Demonstration)
  Flagellates illustrating Volvocine Series (Demonstration)
  Paramecium, pellicle (Demonstration)
  Paramecium, trichocysts (Demonstration)
  Representative ciliates (Demonstration)
  Plasmodium (Demonstration)
  Eimeria (Demonstration)
Chemicals
  Lugol's solution
  Protoslo (or methyl cellulose)
Miscellaneous Supplies
  Congo red stained yeast cells
Audiovisual Materials
  Chart of *Amoeba proteus*
  Chart of *Volvox* life cycle
  Chart of *Paramecium*
  Chart of *Plasmodium* life cycle
  Chart of *Eimeria* life cycle

A Solitary Flagellate: *Euglena*

Subphylum Mastigophora

Euglena (figure 5.1) is a common green flagellate often found in the greenish surface scum of standing or slowly moving water. *Euglena* is an enigmatic organism with a curious mixture of plant and animal characteristics and, therefore, sometimes is considered to represent a borderline case between the plant and animal kingdoms. *Euglena* is smaller than *Amoeba* and *Paramecium* and, therefore, the details of its internal structure are more difficult to observe.

Prepare a wet mount from a culture of living *Euglena* and observe the locomotion of an active specimen under your compound microscope. The active swimming movements result from the beating of the long flagellum, which pulls the organism through the water. A second, shorter, flagellum is present within the flagellar pocket but does not aid in the swimming movements. At certain times *Euglena* also exhibits another type of wormlike locomotion during which waves of contraction pass along the body in a characteristic fashion. This type of locomotion is peculiar to *Euglena* and related organisms and is appropriately termed **euglenoid movement** or **metaboly.** It appears to result in part from the elasticity of the thick outer covering of the body, the pellicle.

After the wet mount begins to dry out, temporarily immobilizing some of your specimens, study the anatomy of a stationary *Euglena*. You will also find it useful to supplement your observations with the study of a prepared microscope slide.

Identify the following structures under high power on your compound microscope: (1) **pellicle,** the thick outer covering of the body; (2) **chloroplasts** with green chlorophyll; (3) **nucleus,** exhibiting a large central **endosome** in stained preparations; (4) **flagellar pocket;** (5) **contractile vacuole;** (6) a red **stigma,** or eye spot; (7) the long anterior **flagellum;** and (8) **paramylum grains,** a type of starch that represents stored food materials. Label each of these structures in figure 5.1.

Euglena is quite sensitive to light, and changing the light intensity tends to cause the *Euglena* to move away. Bright light tends to make this protozoan remain stationary.

After you have completed your observations of the living specimen, add a drop of Lugol's solution (iodine and potassium iodide). This solution will kill the specimen and stain the flagellum to make it more readily visible.

As suggested by the presence of chloroplasts, the nutrition of *Euglena* is normally autotrophic; organic molecules (sugars) are synthesized from inorganic nutrients absorbed from the medium. Light from the sun provides the energy necessary for this process.

Biochemical tests have shown the paramylum granules to be a form of starch similar to that found in plants. Thus, both the presence of the chloroplasts and the storage of a plantlike form of starch indicate a close relationship of *Euglena* and its relatives to the plant kingdom.

Some species of *Euglena* are also able to survive, grow, and reproduce in the dark with no visible evidence of chloroplasts, chlorophyll, or stored food materials. How might such organisms obtain their food? Consult your textbook for further information.

A Colonial (?) Flagellate: *Volvox*

Subphylum Mastigophora

Volvox (figure 5.2) is a common green alga that often occurs in great numbers in freshwater ponds and lakes. Although *Volvox* is now clearly recognized as an alga, it is often studied in zoology courses because it illustrates the organization of a simple colonial (or multicellular) organism well and also because of its usefulness in illustrating one popular theory for the evolution of multicellular organisms from unicellular ancestors. It also demonstrates some basic similarities between plants and animals.

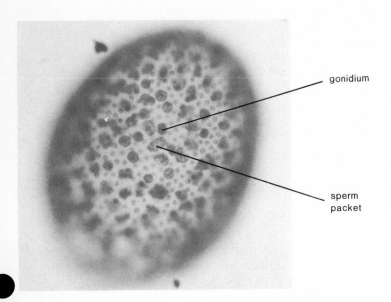

gonidium

sperm packet

Fig. 5.2 *Volvox,* male spheroid. (Photograph by Barbara Grimes.)

The spherical green *Volvox* are large enough to be seen swimming near the surface of a pond or in a laboratory culture, even without a microscope.

Obtain some living *Volvox* from a culture and prepare a wet mount. Be sure to add a few bits of broken coverslip or sand grains to protect the spherical *Volvox* bodies from being crushed by the weight of the coverslip. Observe the spheroid shape of *Volvox,* its swimming movements, and its prominent green color.

The spherical *Volvox* bodies are usually called colonies in textbooks, but recent studies have suggested that they are more similar to **multicellular individuals.** The coordination of cells and cellular function within the spheroid body is much greater than in most other colonial algae and protozoans.

The hollow *Volvox* spheroids average about 0.5mm in diameter and have many small green cells embedded in their outer walls (see figure 5.3). Each of the tiny body cells of *Volvox* contains a **nucleus,** a **contractile vacuole,** a green **chloroplast,** and two whiplike **flagella.** Some or all of the cells (depending on the species) may also have a red **stigma** (light-sensitive spot). The flagella project outward from the surface, and their beating keeps the spheroids in a constant spinning motion. Although the somatic cells of *Volvox* are very small and difficult to observe except with special microscopic preparations, the adjacent cells in some species of *Volvox* are connected by thin **cytoplasmic bridges** or strands. Other species of *Volvox,* however, lack these intercellular connections. Add a drop of 0.1% methylene blue solution to a wet mount of *Volvox* and study under high power on your compound microscope. Can you observe any cytoplasmic bridges?

The green color of the spheroid individuals results from the presence of a chloroplast in each cell. What can you therefore infer about the nutrition of *Volvox?*

Within the spheroid you should be able to observe one or more large reproductive cells or **gonidia.** Reproduction in *Volvox* involves both sexual and asexual processes. In asexual development, embryos are formed from the gonidia. Locate several gonidia within the interior of an asexual parent individual (see figure 5.4).

During asexual development the gonidia undergo a series of cell divisions remarkably like those seen in the embryonic development of many animal species. See if you can locate several different developmental stages of gonidia in the living specimens provided.

The beginning of sexual reproduction can be recognized when the gonidia form male and/or female spheroids. Most species of *Volvox* have separate male and female individuals (i.e., are **dioecious,** "two houses"), but some species produce both eggs and sperm in the same individual and are thus **monoecious** ("one house"). Figure 5.4 also shows both male and female individuals.

The living cultures available for study in the laboratory are usually all asexual. You should study the demonstration chart to learn about the life cycle of *Volvox.*

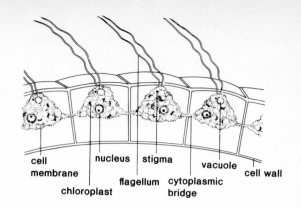

Fig. 5.3   *Volvox* cell structures, longitudinal section of cells at surface of spheroid.

Study the structure of *Volvox* using the living specimens provided in the laboratory. Observe the swimming of *Volvox* spheroids in a drop of water on a clean microscope slide first under low power, and then add a coverslip over the water drop and study under higher magnification to observe more details of the structure of the colony.

Observe: somatic cells, gonidia, eggs, sperm, and **zygotes,** using both the living specimens and the demonstration materials as necessary.

Evolution of Multicellularity

Volvox and several related green flagellates are often studied as models to illustrate one popular theory for the evolution of multicellular organisms. Most scientists believe that multicellular organisms arose from some unicellular form. The particular kind of unicellular organism is not known because this major evolutionary step took place more than 600 million years ago in the Precambrian Era. No well-preserved fossils have been found that actually document this transition from one to many cells, so biologists have searched among living plants and animals to seek **models** that might help their understanding of early evolution.

Volvox and several related green algae comprise the most popular model discussed by scientists. These related forms exhibit a graded series of solitary and colonial forms of increasing complexity and are called the **Volvocine Series** (figure 5.5). Among the important genera of algae comprising the series are: *Chlamydomonas, Gonium, Pandorina, Eudorina, Pleodorina,* and *Volvox.*

Other Mastigophora

Other important mastigophorans include the dinoflagellates, symbiotic flagellates that inhabit the digestive tracts of termites and wood roaches, some peculiar flagellates that may be related to sponges, and several important parasites of humans.

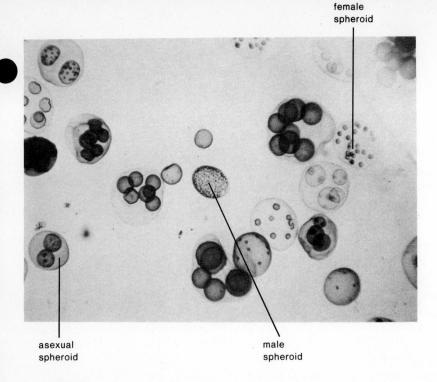

female
spheroid

asexual
spheroid

male
spheroid

Fig. 5.4 *Volvox,* male, female, and asexual spheroids. (Photograph by Barbara Grimes.)

Dinoflagellates are found in both fresh and marine waters; many species form a characteristic outer covering of cellulose. Certain freshwater dinoflagellates may cause an unpleasant odor or taste in human water supplies. *Gonyaulax* and *Gymnodinium* are two marine dinoflagellates often associated with the red tides of coastal waters of North America, Europe, and Africa, sometimes resulting in massive fish kills.

Some flagellates live as **symbionts** in the digestive tracts of wood roaches and termites (figure 5.6). Experiments have shown that the termites lack the digestive enzymes necessary to digest the cellulose in the wood they eat. The flagellates produce these enzymes (cellulases) and release them in the gut of the termites, thus making the digested products from the cellulose breakdown available for the nutrition of both the flagellates and the host. Termites from which the flagellates are experimentally removed soon die of starvation, no matter how much wood they ingest. The flagellates benefit from the continuous supply of cellulose and from the suitable anaerobic environment of the host hindgut. Such a mutually beneficial symbiotic relationship is called **mutualism.**

Proterospongia is a colonial flagellate with species that closely resemble the flagellated collar cells, or choanocytes, characteristic of sponges (see chapter 6). Some biologists have suggested that the sponges may have evolved from some ancient protozoan similar to *Proterospongia.*

Still other mastigophorans are **parasites.** *Trypanosoma* and *Leishmania* are two important genera that include several serious human parasites. *Trypanosoma* has a thin, undulating membrane connecting its long, whip-like flagellum with its body (see figure 5.7). Several species of *Trypanosoma* cause sleeping sickness and other diseases in humans. *Leishmania* includes species that cause severe diseases in tropical areas of Africa, Asia, and South America.

Demonstrations

1. Large flagella in other Mastigophora, such as *Peranema.*
2. Microscope slide of trypanosomes.
3. Microscope slide of symbiotic flagellates from digestive tract of termite or wood roach.
4. Microscope slide of dinoflagellates.
5. Microscope slide showing cell walls in *Volvox.*
6. Chart illustrating the life cycle of *Volvox.*
7. Microscope slides with related volvocine flagellates, such as: *Gonium, Pandorina, Eudorina,* and *Pleodorina.*

An Amoeba: *Amoeba proteus*

Phylum Sarcomastigophora

Subphylum Sarcodina

Amoeba proteus (figure 5.8) is a common protozoan found in ponds and streams. It often occurs on the undersides of plant leaves and among diatoms and desmids. The transparent amoeba constantly changes shape by extending pseudopodia, footlike extensions of the cytoplasm, which serve for locomotion and in food capture. *A. proteus* feeds on bacteria, small algae, and small protozoans.

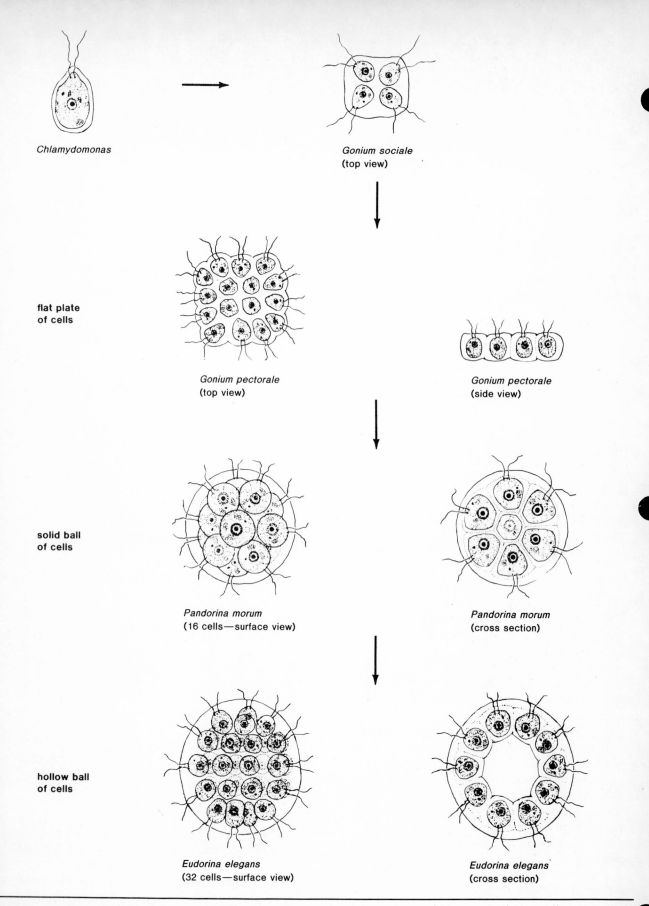

Chlamydomonas

Gonium sociale
(top view)

flat plate
of cells

Gonium pectorale
(top view)

Gonium pectorale
(side view)

solid ball
of cells

Pandorina morum
(16 cells—surface view)

Pandorina morum
(cross section)

hollow ball
of cells

Eudorina elegans
(32 cells—surface view)

Eudorina elegans
(cross section)

Fig. 5.5 Volvocine flagellates illustrating possible evolution
from unicellular to multicellular form.

Chapter 5

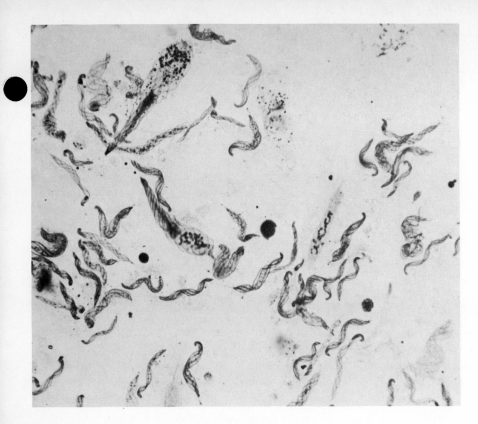

Fig. 5.6 Symbiotic flagellates from termite gut. (Photograph by Barbara Grimes.)

white blood cell

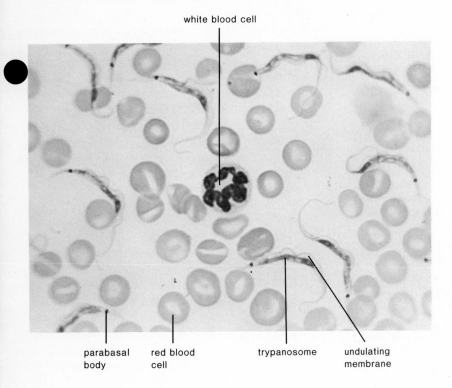

parabasal body   red blood cell   trypanosome   undulating membrane

Fig. 5.7 *Trypanosoma*, blood smear. (Courtesy Carolina Biological Supply Company.)

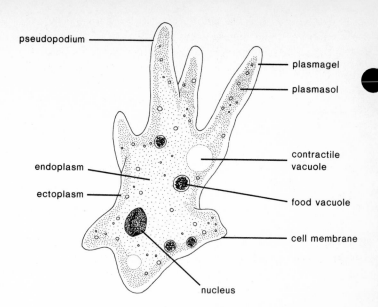

pseudopodium

plasmagel

plasmasol

endoplasm

ectoplasm

contractile vacuole

food vacuole

cell membrane

nucleus

Fig. 5.8 *Amoeba proteus.*

In feeding, an advancing pseudopodium flows over one or more food organisms to trap the food in a water-filled cup. The opening of the food cup then narrows until the food is completely enclosed in a food vacuole.

Prepare a wet mount to study living amoebae under your compound microscope.

Preparing a Wet Mount and a Hanging Drop. Obtain a clean microscope slide and add a drop of amoeba culture solution to the center of the slide. Take care to withdraw the drop of culture solution from the bottom of the culture dish or jar with a clean eyedropper or pipette. The amoebae are slightly heavier than the culture solution and are usually concentrated on the **bottom** of the vessel. Add a few bits of broken coverslip or grains of sand around the periphery of the drop to protect your specimens from being crushed, then carefully cover your preparation with a coverslip.

Another method of studying living Protozoa is to prepare a **hanging drop.** With this method you place a drop of the amoeba culture in a coverslip and invert the coverslip over the cavity of a depression slide. Both types of preparations can be observed for long periods of time if the outside edge of the coverslip is coated with petroleum jelly before it is placed in position on the microscope slide. Be sparing in your use of the petroleum jelly; avoid getting it into the culture droplet and on your microscope lens.

Study your preparation under the low power of the compound microscope to observe the general appearance of the amoeba. For best observation of a living amoeba, reduce the illumination to a minimum since living specimens are nearly transparent and almost invisible in bright light. Search your slide carefully before discarding it or asking your instructor for a new preparation.

Locate an actively moving amoeba and note its constantly changing shape. The long, fingerlike projections

are **pseudopodia** ("false feet"). Observe the lack of permanent orientation of the body of an amoeba; any portion may temporarily be anterior, posterior, right, or left.

With the aid of figure 5.8, identify and study the following structures found in the amoeba.

1. **Endoplasm** the inner granular region, which forms the bulk of the cytoplasm.
2. **Ectoplasm** the thin layer of clear cytoplasm which surrounds the endoplasm.
3. **Cell membrane** the outer membrane surrounding the amoeba. Frequently also called the plasmalemma.
4. **Plasmagel** the stiff, jellylike, granular outer layer of colloidal endoplasm in the **gel state.**
5. **Plasmasol** the central mass of colloidal endoplasm in a fluid, or **sol state.** Note the streaming movements within the plasmasol.
6. **Nucleus** a transparent structure with no fixed position in the cell. It has the shape of a biconcave disc and often exhibits a folded or wrinkled appearance. Examine also the nucleus in a stained microscope slide of *Amoeba proteus.* Observe the darkly staining granular chromatin material within the nucleus.
7. **Contractile vacuole** a clear vacuole found in the endoplasm which collects excess water from the surrounding cytoplasm and discharges it outside the body. Shortly after one contractile vacuole discharges its contents at the cell surface, a new contractile vacuole forms. Where are the new contractile vacuoles formed? Formerly, it was believed that the contractile vacuole also played an important role in the excretion of waste products from protein metabolism, but recent evidence has not supported this belief, and most specialists now agree that the contractile vacuole functions primarily in maintaining water balance in the cell (osmoregulation).

Chapter 5

8. **Food vacuoles** vacuoles containing bits of ingested food and the digestive enzymes that act to break down these food materials into soluble materials that can be utilized by the amoeba. How are the food vacuoles formed? How are the undigested contents of a food vacuole disposed of after digestion has taken place?

Amoeboid Movement. *Amoeba* moves about by extending pseudopodia into which some of the innermost cell contents flow. Various kinds of amoebae form pseudopodia of different size and form. These pseudopodia are important in feeding, support, and locomotion. The mechanism of amoeboid movement has been studied by many scientists because of its intriguing nature and because similar movements occur in many other kinds of cells, including human leucocytes. Also, scientists now believe that amoeboid movement may be closely related to the phenomenon of cytoplasmic streaming, which occurs in virtually all kinds of living cells.

The movement of an amoeba is accomplished by the forward flow of the relatively liquid **plasmasol** from the center of the amoeba toward and into an expanding pseudopodium. Around the periphery of the pseudopodium, the plasmasol changes into a stiff **plasmagel**. Thus, the plasmasol moves the pseudopodium forward, and the plasmagel serves to fix it in position.

Recent biochemical and biophysical studies have demonstrated that the mechanism of amoeboid movement is similar to that in muscle contraction. **Contractile proteins** similar to the actin and myosin found in vertebrate muscles are present in the cytoplasm of an amoeba. Amoeboid movement results from folding, unfolding, polymerization, and depolymerization of these proteins.

Study the locomotion of an *Amoeba* on your microscope slide and also the pattern of its internal protoplasmic movements. In an active specimen, locate and carefully follow the movement of some granules in the plasmagel at the temporary posterior end. Observe how the plasmagel of the endoplasm changes into plasmasol, which flows forward and then changes into the gel state again, just back of the tip of the forming pseudopodium.

Reproduction. The reproduction of *Amoeba proteus* occurs only through the asexual process of binary fission. The nucleus and cytoplasm of a parent cell divide to form two daughter cells approximately equal in size. Thus, each of the daughter cells is genetically identical to the parent cell—excluding the rare occurrence of a mutation in one of the daughter cells.

Other Sarcodina

Many members of this group of Protozoa are more specialized than Amoeba. *Pelomyxa carolinensis* is a large multinucleate amoeba often studied in zoology classes. Numerous species of amoebae live in shells or tests which

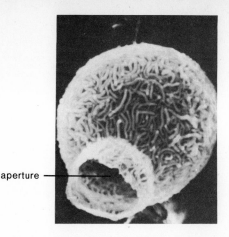
aperture

Fig. 5.9 Test of a freshwater amoeba. (Scanning electron micrograph by F. W. Harrison.)

they secrete, or which they form from sand grains or other materials (figure 5.9). *Difflugia* and *Arcella* are two common testate amoebae found in freshwater ponds and streams. Other species of amoebae are parasites or symbionts in the digestive tracts of various animals. *Entamoeba histolytica,* an important parasite of humans, is the cause of amoebic dysentery, a disease often spread by drinking water or by eating raw vegetables contaminated by human wastes.

The Foraminifera (figure 5.10) are an ancient and important group of marine sarcodines, which form tests of calcium carbonate or other materials. The shells of foraminiferans accumulate on the sea bottom and contribute to the formation of chalk and limestone. The White Cliffs of Dover in England are made up largely of foraminiferan tests, as is much of the Bedford limestone found in Indiana and Illinois and some of the limestone that was used to build the Egyptian pyramids. The distribution of certain species of Foraminifera is also very important to petroleum geologists as an indication of ancient environmental conditions that may have been favorable for the formation of petroleum.

Observe several of these other types of Sarcodina among the demonstrations.

Demonstrations

1. Models and charts of *Amoeba proteus.*
2. Culture of *Amoeba* under a stereomicroscope.
3. Microscopic slides with testate and parasitic representatives of the Subphylum Sarcodina, such as *Arcella, Difflugia, Entamoeba histolytica, Actinosphaerium* (a heliozoan), and *Globigerina* (a foraminiferan).

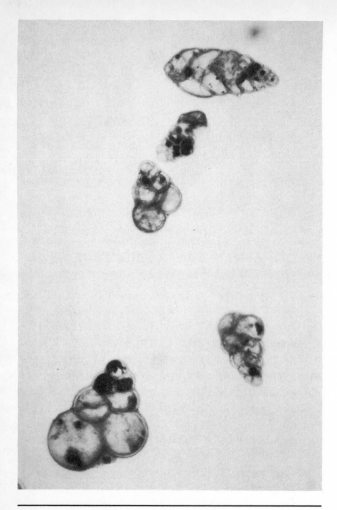

Fig. 5.10 Foraminiferan tests. (Photograph by Barbara Grimes.)

A Ciliate: *Paramecium caudatum*

Phylum Ciliata

Paramecium (figure 5.11) is a large, common, ciliated protozoan often found in water containing bacteria and decaying organic matter. There are several species of *Paramecium* that differ in various details of structure and that range in length from about 120–300 microns. The laboratory directions provided here are based upon *Paramecium caudatum,* a species frequently used for laboratory study and experimentation, but they will also apply, with minor modification (such as body size and number of micronuclei), to the study of other species of *Paramecium.*

Obtain a drop of *Paramecium* culture in a clean pipette and place it on a clean microscope slide with a similar-sized drop of methyl cellulose solution (or other similar agent) to slow movement. Methyl cellulose is a viscous material and serves mechanically to slow the swimming of the fast-moving *Paramecium.* Add a coverslip and observe your preparation under low power with

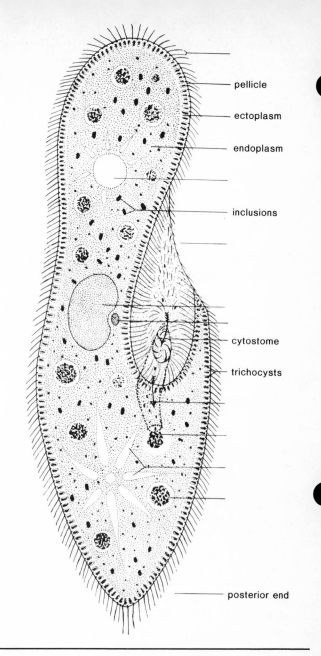

pellicle

ectoplasm

endoplasm

inclusions

cytostome

trichocysts

posterior end

Fig. 5.11 *Paramecium caudatum.* (Supply the missing labels.)

your compound microscope. Note the form, color, and behavior of the animals in your preparation. Observe the slipper-shaped body with an **oral groove** beginning at the anterior end and running diagonally across the anterior portion of the animal. At the posterior end of this groove is the **cytostome,** or "cell mouth," through which food particles are passed as a result of the action of the specialized oral cilia lining the oral groove.

Paramecium is much more complex in its structure than *Amoeba.* Select a large, immobile, or slowly moving specimen, and with the aid of figure 5.11, identify and study the following structures. Provide the missing labels in the figure.

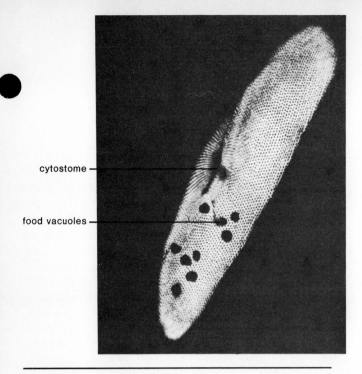

cytostome

food vacuoles

Fig. 5.12 *Paramecium,* nigrosin stain to illustrate sculpturing of bilayered pellicle. (Photograph by Barbara Grimes.)

1. **Cilia** the numerous cylindrical protoplasmic extensions that cover the surface of the *Paramecium* and that function in locomotion and in food gathering.
2. **Pellicle** the thick outer covering of the body through which the cilia project. The pellicle has a complex structure, but its details are difficult to observe without special techniques. Figure 5.12 shows some of the surface depressions in the bilayered pellicle.
3. **Trichocysts** tiny, rodlike structures embedded in the cortical (outer) cytoplasm beneath the pellicle. When properly stimulated, the trichocysts discharge their contents and form long threads. There is some evidence that the trichocysts may serve as a defense against predators, and they also serve to anchor the animal during feeding. In other types of ciliated Protozoa, trichocysts have been found to have still other functions. Special microscopic preparations are also necessary to observe trichocysts (see demonstration).
4. **Macronucleus** the large nucleus located near the center of the cell. Since it is transparent in a living animal, the structure of the macronucleus is best studied in a stained preparation. Experiments have demonstrated that the macronucleus controls most metabolic functions of the cell.
5. **Micronucleus** a smaller nucleus located close to and lying partly within a depression on the oral side of the macronucleus. The micronucleus is involved primarily in the reproductive and hereditary functions of the animal. This presence of two distinct types of nuclei is called **nuclear dimorphism** and is a condition found only in the Phylum Ciliophora. *Paramecium caudatum* has only a single micronucleus, but other species of *Paramecium* have from two to many micronuclei. As with the macronucleus, the structure of the micronucleus is best studied in a prepared microscope slide.

6. **Contractile vacuoles** two clear, slowly pulsating vesicles located near each end of the body. Each contractile vacuole is surrounded by several **radiating canals** (not often seen in ordinary student preparations) which collect water from the surrounding cytoplasm. Observe the behavior of the contractile vacuoles. Are they fixed in position? Do they contract alternately or simultaneously? The function of the contractile vacuoles in *Paramecium* is the same as in *Amoeba,* the collection and discharge of excess water from the cell. Freshwater Protozoa often have contractile vacuoles; marine Protozoa generally lack them. How would you explain this difference?
7. **Cytostome** (cell mouth) a permanent opening near the posterior end of the oral groove through which food is passed.
8. **Cytopharynx** a short tube extending from the cytostome posteriorly and downward into the cytoplasm where food vacuoles are formed.
9. **Food vacuoles** vacuoles located within the cytoplasm where they are carried by the streaming movements of the cytoplasm. Undigested materials are discharged through the **cytopyge,** or anal pore, located posterior to the oral groove.

Feeding

Paramecium is a filter-feeding organism and normally feeds on bacteria and yeast cells collected by a specialized food-collecting apparatus. An **oral groove** extends diagonally back along the body to a funnel-shaped **cytopharynx.** Food is swept along the oral groove by the action of specialized cilia lining the groove, is passed through the circular **cytostome** ("cell mouth") at the opening of the cytopharynx, and is passed through the cytopharynx into a newly forming **food vacuole.**

Prepare a wet mount with a drop of *Paramecium* culture to study the feeding process. Add a small amount of congo red stained yeast with the tip of a toothpick or clean dissecting needle. Try to pick up the smallest amount of yeast possible on the toothpick; too much yeast will cloud your preparation and obscure the *Paramecium*. (Note: To prepare the stained yeast, boil the yeast in a small amount of water in a test tube, cool to room temperature, add 1% congo red solution, and after a few minutes decant most of the supernatant dye solution to concentrate the yeast cells. This method works better than several others we have tried.)

With this preparation you can study the movement of the food particles, the formation of food vacuoles, and the subsequent movement of the food vacuoles within the cytoplasm. After the food vacuoles are formed, digestive enzymes are released into them and chemical digestion of the food particles begins. Note the color change in the vacuoles as the enzymes work. The color change is due to a change of pH in the vacuoles. Where do the digestive enzymes come from? Why don't they digest the other materials in the cell such as mitochondria and ribosomes? The diffusible products of digestion are released into the cytoplasm, and the undigestible remains are discharged at a specific site on the surface of the animal. This site is the **cytopyge,** or cell anus.

Cilia and Flagella

Most of the surface of *Paramecium* is covered by thin, hairlike projections called **cilia** (singular: cilium). **Cilia** are extensions of the cortical (outer) cytoplasm of the cell and play important roles in feeding and locomotion.

A great deal has been learned in recent years about the structure and function of cilia. These studies have revealed that cilia are closely related to the flagella (singular: flagellum) found on the surface of other kinds of Protozoa. The structural differences between cilia and flagella are minor. When the projections are short and numerous, they are called cilia. When they are long and few, they are flagella. Cilia generally exhibit a relatively simple back and forth movement. The movements of flagella are often more complex and may involve a series of helical waves propagated along the flagellum.

Both cilia and flagella have a common basic structure. A cross section reveals an outer membrane enclosing **nine pairs of microtubules and two single microtubules** in the center of the cilium or flagellum. This basic pattern is found in all cilia and flagella, not only among the Protozoa but also on the gills of molluscs, the ciliated epithelium lining the trachea of vertebrates, and the tail of spermatozoa.

Recent biochemical studies have also demonstrated that the movements of cilia and flagella involve **contractile proteins** similar to those found in striated muscle. This is another important illustration of the basic similarity of all living organisms.

Reproduction

Paramecium reproduces by a simple type of asexual reproduction in which the parent divides into two equal daughter cells. This type of asexual reproduction is termed **transverse fission** and is found in many kinds of Protozoa.

Living specimens are occasionally seen in the process of fission, but the details of fission are best studied in a stained microscope slide. Obtain a prepared slide of *Paramecium* in fission and observe the nuclei. During fission, the micronucleus first divides by **mitosis,** and the macronucleus later divides by **amitosis.** No visible chromosomes are formed in the macronucleus; the macronucleus simply constricts, and the two portions separate. Macronuclear division is followed by cytoplasmic division (cytokinesis). The process of fission may be completed rapidly, and under optimal conditions, *Paramecium* can reproduce asexually two or more times per day.

Draw the representative stages of fission in *Paramecium* as indicated on figure 5.13.

Unlike *Amoeba, Paramecium* can also reproduce sexually. The specialized type of sexual process exhibited by *Paramecium* is called **conjugation.** During this process, two individuals come together and adhere by their oral surfaces, undergo a complex series of changes in both the macronuclei and the micronuclei, exchange a single pair of micronuclei (one from each cell), separate, and resume asexual reproduction. Following the exchange of micronuclei in each *Paramecium,* the newly introduced micronucleus fuses with another (nonmigrating) micronucleus. Thus, there is an exchange of hereditary material and a subsequent fusion of hereditary material from the two parents, analogous to the situation in ordinary sexual reproduction studied previously.

Examine a prepared slide or a demonstration of *Paramecium* in conjugation. Draw a conjugating pair of paramecia in figure 5.14.

Other Ciliata

The Phylum Ciliata is a large and diverse group of protozoans. One important form is *Stentor* (figure 5.15), a large, trumpet-shaped ciliate, common in many lakes, ponds, and streams, which has been used in many experimental studies. *Stentor* has a spiral array of complex ciliary organelles leading to its cytostome and a beaded macronucleus. *Didinium* (figure 5.16) is a barrel-shaped predaceous ciliate with a voracious appetite. It feeds on other ciliates including *Paramecium.* A hungry *Didinium* can eat a *Paramecium* every two hours.

Blepharisma (figure 5.17) is a large pink ciliate with a terminal (posterior) contractile vacuole and cytopyge. *Tetrahymena* is a small ovoid ciliate that has been used in many experimental studies of biochemistry and genetics.

Vorticella (figure 5.18) is a sessile form with a long contractile stalk which attaches to submerged stones, shells, rocks, and other objects. Some relatives of *Vorticella,* like *Carchesium* and *Zoothamnion,* form stalked colonies. *Trichodina* (figure 5.19) is a related mobile genus which includes several parasites found on the gills of fishes and one occurring on freshwater *Hydra.* Various species of *Trichodina* are cylindrical or sometimes shaped like a derby hat and have a **flattened aboral disc** often equipped with **denticles** ("teeth") which aid in attachment to the host.

Micronucleus just divided into
two, macronucleus elongated.

Two micronuclei, macronucleus
dividing or constricted.

Macronucleus divided into two,
cell body dividing.

Cell body almost constricted
into two.

Fig. 5.13 Drawing of transverse fission in *Paramecium*.

Fig. 5.14 Drawing of conjugation in *Paramecium*.

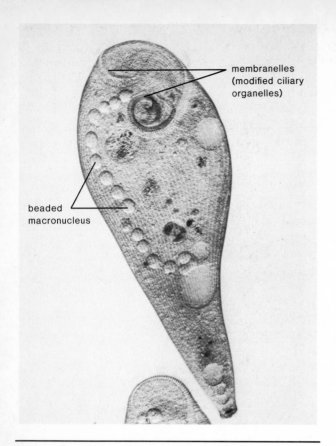

membranelles
(modified ciliary
organelles)

beaded
macronucleus

Fig. 5.15 *Stentor.* (Courtesy Carolina Biological Supply Company.)

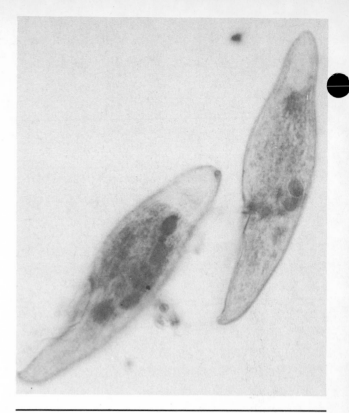

Fig. 5.17 *Blepharisma.* (Photograph by Barbara Grimes.)

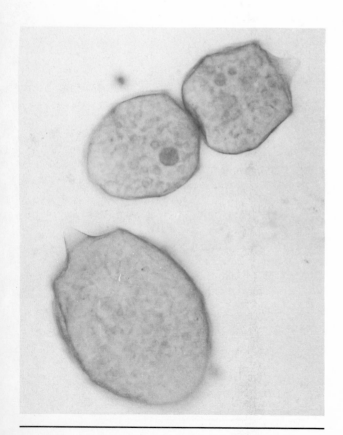

Fig. 5.16 *Didinium.* (Photograph by Barbara Grimes.)

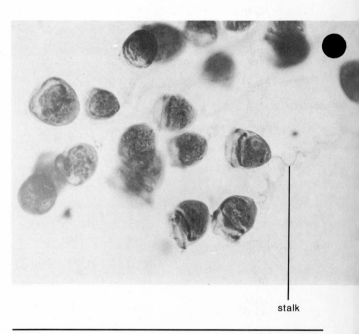

stalk

Fig. 5.18 *Vorticella.* (Photograph by Barbara Grimes.)

Chapter 5

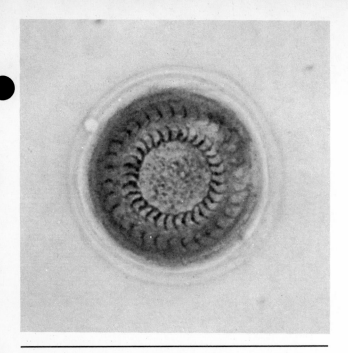

Fig. 5.19 *Trichodina*, showing aboral disc with denticles ("teeth"), which aid in attachment to the host. (Photograph by Barbara Grimes.)

Podophyra is a suctorian, a specialized group of sessile ciliates that have **suctorial tentacles** in their mature stages and are predators of other ciliates. Cilia are found only on juvenile stages of the suctorians.

Demonstrations

1. Models and charts of *Paramecium*.
2. Stained slide showing pellicle of *Paramecium*.
3. Stained slide to show discharged trichocysts.
4. Representative members of the Class Ciliata, such as *Stentor, Euplotes, Tetrahymena, Vorticella, Didinium, Blepharisma, Trichodina,* and *Podophyra.*

Phylum Apicomplexa

Members of this group were formerly included among the Sporozoa, but recent investigations have indicated that they should be considered a separate phylum. All members of this phylum are parasitic on other organisms.

Many species in this group parasitize invertebrate animals such as earthworms, crabs, and oysters, but the most important species are parasites of vertebrates, including humans. Among the most important sporozoans are *Plasmodium* and *Eimeria*. Several species of *Plasmodium* cause various forms of **malaria** in humans and other animals.

Eimeria is a genus of sporozoan parasites that causes **coccidiosis** in birds, rabbits, and other animals. This disease has great economic impact on the poultry industry.

Eimeria, like *Plasmodium*, has a complex life cycle that includes both sexual and asexual forms. Study the demonstrations illustrating the life history and importance of *Eimeria* and coccidiosis.

The Malaria Parasite: *Plasmodium*

More than 50 species of *Plasmodium* have been described. All are parasites of vertebrate animals, including amphibians, reptiles, birds, and mammals. Four species cause human malaria, *P. vivax, P. ovale, P. malariae,* and *P. falciparum.* The life cycles of these species are all similar.

Malaria, one of the most serious and debilitating of human diseases, has had an important role in history from the fall of the Roman Empire to the war in Vietnam. Although modern medicine has made some progress in eliminating malaria, the disease is yet to be conquered. It is most prevalent in tropical and semitropical areas, and costs millions of lives and trillions of dollars annually.

Life Cycle of *Plasmodium*

The life cycle of *Plasmodium* (see figure 5.20) is complex like those of other apicomplexans and includes several generations with both sexual and asexual reproduction. The life cycle can best be understood by starting with the zygote in the gut of a mosquito, one of the two hosts necessary for completion of the life cycle.

The zygote becomes motile and passes through the lining and wall of the stomach or midgut of the mosquito and is now called an ookinete. The ookinete then rounds up and encysts on the outside of the gut wall and is called an oocyst. The oocyst divides internally to form several hundred sporozoites. The sporozoites escape by rupturing the external wall of the oocyst and migrate through the hemocoel to the salivary glands of the mosquito.

When a mosquito bites a human, the sporozoites and the mosquito's salivary secretions are injected into this host. Sporozoites that find their way into the human bloodstream are eventually carried to the liver, where the sporozoites enter host cells. Inside the host liver cells the sporozoites transform into amoeboid multinuclear schizonts and feed upon the contents of the host liver cells. The schizonts reproduce asexually to form many merozoites (the next stage in the life cycle), escape from the liver cells, and, in some cases, invade other liver cells to repeat the process.

Merozoites escaping into the bloodstream penetrate erythrocytes to initiate the erythrocytic phase of the life cycle. Parasitic stages in the liver prior to entry into the erythrocytes are termed the exoerythrocytic phase.

The developing *Plasmodia* inside the erythrocytes exhibit a characteristic morphology, as seen in Giemsa-stained microscope slide preparations, and are recognized by their red nucleus and blue ring-shaped cytoplasm. This

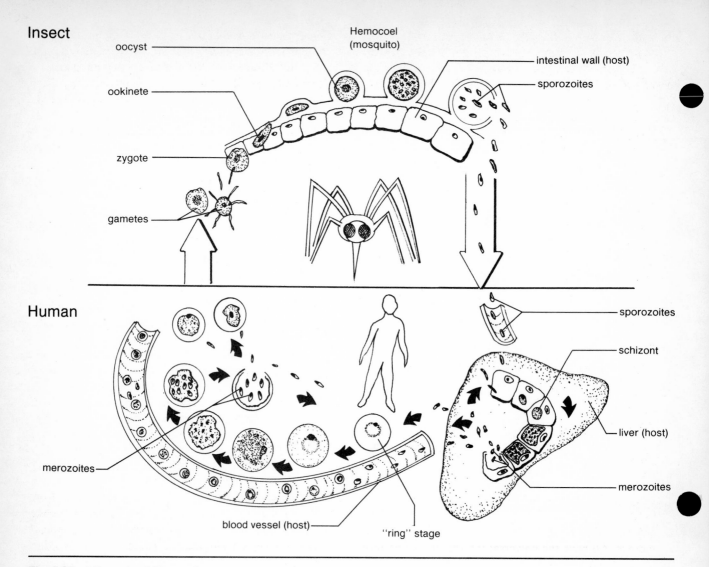

Insect

oocyst

ookinete

zygote

gametes

Hemocoel
(mosquito)

intestinal wall (host)

sporozoites

Human

sporozoites

schizont

liver (host)

merozoites

merozoites

blood vessel (host)

"ring" stage

Fig. 5.20 Life cycle of *Plasmodium*.

characteristic morphology is very helpful in the laboratory diagnosis of malaria. The merozoites in the erythrocytes undergo further asexual reproduction within the erythrocytes. Later, the merozoites rupture the wall of the erythrocytes, escape into the blood, and enter most erythrocytes. This multiplication process may be repeated several times, so that an enormous number of parasites are produced within the host. The rupture of the erythrocytes by the merozoites also releases accumulated toxic wastes from the parasites and causes the symptomatic chills and fever commonly associated with human malaria.

Some of the merozoites develop into sexual forms instead of repeating the asexual merogony. These sexual forms develop within the erythrocytes and become microgametocytes (male) and macrogametocytes (female). These stages are the progenitors of the male and female gametes and represent the start of the sexual portion of the life cycle.

If a mosquito bites an infected host and ingests infected erythrocytes, the gametocytes pass into the mosquito's stomach, where they mature into microgametes and macrogametes. The microgametocytes develop flagella-like outgrowths, which break free, become motile, and fertilize macrogametes. The fertilized macrogametes, or zygotes, then invade the gut wall of the mosquito to start the cycle.

Study the microscope slides with blood smears prepared with blood from humans infected with *Plasmodium vivax* or a similar species. Identify as many stages in the life cycle of *Plasmodium* as you can from your own slides and the demonstration slides. Observe the changes in morphology of the parasite during its development in the human erythrocytes. With the aid of figure 5.20, try to relate the portion of the *Plasmodium* life cycle in human erythrocytes to the other parts of the life cycle completed in the mosquito and in the exoerythrocytic stages in the

Chapter 5

human. Why do you think *Plasmodium* requires two hosts to complete its life cycle? What special adaptations for life as a parasite can you identify in *Plasmodium*? List several of these adaptations for parasitism in table 5.1.

Demonstrations

1. Microscope slides with selected stages in the life cycle of *Plasmodium*.
2. Chart illustrating the life cycle of *Plasmodium*.
3. Microscope slide of *Eimeria*.
4. Chart illustrating the life cycle of *Eimeria*.

Table 5.1 Adaptations for Parasitism Exhibited by *Plasmodium*

\
\
\
\
\
\
\

Key Terms

Cilia cylindrical cytoplasmic extensions from the surface of certain Protozoa and of some metazoan cells. Serve in locomotion and feeding of ciliated Protozoa. Basically similar to flagella, but shorter and more numerous. All have the universal internal 9 + 2 pattern of microtubules. Singular: cilium.

Coccidiosis disease of birds, rabbits, and other animals. Caused by members of the protozoan genus *Eimeria*.

Conjugation a specialized type of mating, nuclear exchange, and nuclear reorganization characteristic of ciliated Protozoa. A form of sexual reproduction.

Contractile vacuole an organelle found in many freshwater protozoans that serves in osmoregulation (water balance).

Cytostome the "cell mouth" found in many ciliated protozoans. Often surrounded by specialized ciliary feeding organelles.

Flagella cylindrical cytoplasmic extensions from the surface of certain Protozoa and some metazoan cells. Function in locomotion and feeding of mastigophorans. Similar to cilia but longer and usually fewer per cell. Singular: flagellum.

Gonidia specialized reproductive cells in *Volvox*. Singular: gonidium.

Macronucleus the large metabolic nucleus typical of ciliates. Often has a characteristic shape. Divides amitotically by pinching in two. Contains many duplicated sets of genes (polyploid).

Malaria disease of humans and other animals. Caused by members of the protozoan genus *Plasmodium*, which invade the blood and other tissues of the hosts.

Micronucleus the small reproductive nucleus in ciliates. Some ciliates have more than one micronucleus. Usually divides by ordinary mitosis.

Nuclear dimorphism having two distinct types of nuclei in the same cell. Characteristic of the ciliates; e.g., with macronucleus and micronucleus.

Pseudopodia the protoplasmic extensions, "false feet," of the Sarcodina used for locomotion and in feeding. Various types of sarcodines have pseudopodia specialized for certain purposes or modes of life. Singular: pseudopodium.

Stigma a light-sensitive spot found in certain flagellated protozoans, such as *Euglena* and *Volvox*.

6
Porifera

Objectives

After completing the laboratory work in this chapter, you should be able to perform the following tasks:

1. Briefly characterize the Phylum Porifera.
2. Describe the basic organization of the sponge body.
3. Explain the skeletal elements of sponges and their composition.
4. List and distinguish the four classes of sponges.
5. Differentiate between ascon, sycon, and leucon body types.
6. Explain the importance of choanocytes and describe where they are found in asconoid, syconoid, and leuconoid sponges.
7. Describe the pattern of water flow in a syconoid sponge and explain its importance.
8. Explain the role of gemmules in sponges.
9. Describe an experiment that demonstrates the ability of certain sponges to reorganize from dissociated sponge cells. Explain the significance of this experiment.

Introduction

The Phylum Porifera includes the sponges, a group of peculiar sedentary animals so different from other types of animals that they were long thought to be plants. Sponges are among the most primitive multicellular animals. They have a simple type of body organization, with a porous body permeated by a system of water canals through which water is pumped by the action of special flagellated cells (choanocytes). The bodies of sponges consist of a loose aggregation of cells embedded in a jellylike matrix; there are no distinct tissues, except for inner and outer epithelial (lining) layers made up of pinacocytes and choanocytes, respectively.

An inner layer of flagellated choanocytes (collar cells) generates water currents through the internal canal systems. These water currents are essential in the life of the sponge, for they carry food particles and oxygenated water into the sponge and waste products as well as gametes and/or larvae out of the sponge.

Sponges are sedentary filter feeders, able to capture tiny food particles measuring from about 0.1 to 50 μm from the seawater. The capture of food, which consists chiefly of fine, suspended organic particles and tiny planktonic organisms, is accomplished mainly by the choanocytes and internal amoebocytes. All digestion is intracellular (within individual cells).

Sponges lack digestive, nervous, muscular, respiratory, excretory, circulatory, and reproductive systems. Because of their lack of differentiated organs and also because of their various other morphological and developmental peculiarities, sponges are generally believed to represent an early offshoot from the main line of animal evolution and not to be closely related to any more advanced animal types.

Of the more than 5,000 described species of sponges, most are marine, except for approximately 150 species found in freshwater streams, ponds, and lakes. Formerly, certain types of marine sponges were of considerable economic importance, and sponge fishing industries flourished in several areas, such as the Gulf of Mexico, the Caribbean Sea, and the Mediterranean Sea, where the warm, shallow waters and rocky bottoms were favorable for the growth of bath sponges (figure 6.1). Overfishing and sponge diseases have taken their toll, however, and increasing competition from synthetic sponges has further diminished the commercial importance of natural sponges.

Skeleton

The skeleton of sponges is relatively complex in comparison to the general organization of the body of sponges.

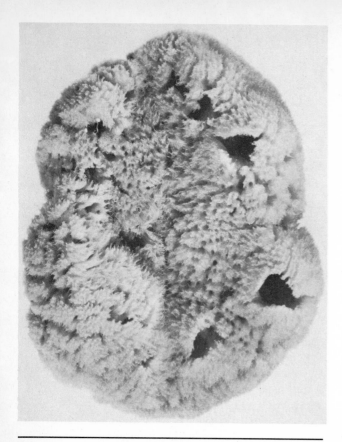

Fig. 6.1 Bath sponge, Class Demospongiae. (Courtesy Carolina Biological Supply Company.)

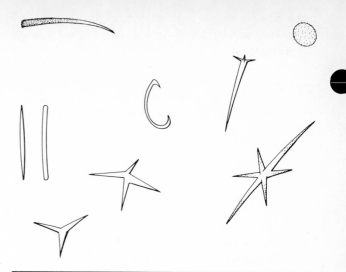

Fig. 6.2 Examples of spicule shapes.

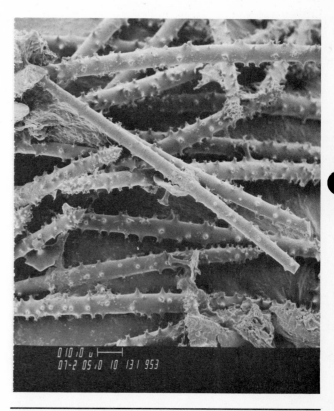

Fig. 6.3 Spicules from a freshwater sponge. (Scanning electron micrograph by F. W. Harrison.)

The skeleton is internal and consists of individual mineralized elements, called **spicules,** made up of calcium carbonate, silicon salts, and/or a network of organic fibers composed of fibrillar collagen and/or spongin fibers.

Spongin is a tough, fibrous protein chemically similar to collagen but unique to the sponges. Spicules occur in a variety of forms (figures 6.2 and 6.3) and are important in the classification of sponges.

Classification

The division of the phylum into classes is based largely on the nature of the skeleton. For many years three classes of living sponges were recognized. Recently, however, a fourth class, Class Sclerospongiae, was described from sponges discovered in the waters off Jamaica. This important discovery provided a link with some long extinct fossil types and gave us new insights into the possible evolution of the sponges.

Class Demospongiae

Horny sponges. Members of this class possess a skeleton made up of a network of spongin fibers (a structural protein secreted by certain sponge cells), siliceous spicules, both, or neither. Most members of this class are marine, but two families are found in freshwater streams, ponds, and lakes. All commercial sponges belong to this class. Examples: *Spongia, Haliciona, Microciona* (all marine), and *Spongilla* (freshwater).

Class Sclerospongiae

Corraline sponges. Marine sponges with a skeleton composed of siliceous spicules, spongin fibers, and a massive basal skeleton of calcium carbonate (aragonite or calcite).

Chapter 6

Only a few living species are known, all are marine, and all appear to be relics of groups common in the Mesozoic and Paleozoic eras.

Class Calcarea (Calcispongiae)

Calcareous sponges. Sponges with a skeleton consisting of many small spicules made of calcium carbonate embedded in a loose, jellylike matrix. All species are marine. Examples: *Scypha, Leucosolenia*.

Class Hexactinellida (Hyalospongiae)

Glass sponges. Sponges with a skeleton composed of siliceous spicules, usually with six rays as the class name implies. The spicules are often fused together into a continuous network. All glass sponges are marine, and most species are found in deep areas of the world oceans. Little is known about their biology. Examples: *Euplectella* (Venus's flower basket) and *Hyalonema*.

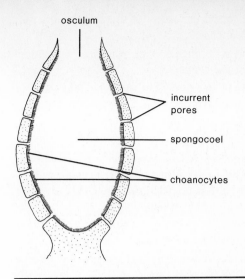

Fig. 6.4a Ascon body type.

Materials List

Preserved Specimens
  Leucosolenia
  Scypha
  Bath sponge
  Glass sponge (Demonstration)
  Assorted marine sponge types (Demonstration)
  Freshwater sponges (Demonstration)
Prepared Microscope Slides
  Leucosolenia, whole mount
  Scypha, cross section
  Scypha, eggs and embryos (Demonstration)
  Sponge spicules (Demonstration)
  Spongin fibers (Demonstration)
  Sponge gemmules (Demonstration)

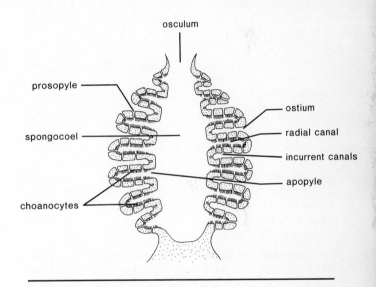

Fig. 6.4b Sycon body type.

Body Organization

Morphologically, the bodies of sponges exhibit three distinct types based on the organization of their internal canal systems. These three morphological types are designated as the **ascon, sycon,** and **leucon** types (figures 6.4 a, b, and c). It is important to recognize that these three body forms represent morphological types and are not directly related to the three classes of sponges. In fact, only a few species of sponges exhibit the ascon and sycon body types. The majority of sponges are of the leucon type.

Ascon-type Sponge

Leucosolenia is a small colonial sponge of the ascon type. Examine a preserved specimen or a microscopic whole mount with your stereoscopic microscope and observe the following features.

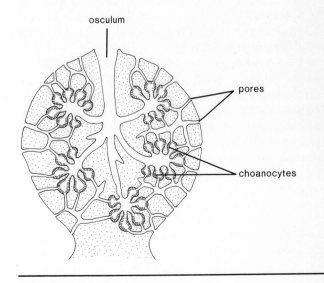

Fig. 6.4c Leucon body type.

1. A system of horizontal tubes that bear numerous upright branches.

2. The upright branches that represent individual sponges of the colony.

3. Buds formed on the sides of the individual sponges.

4. The terminal opening, or **osculum,** at the upper end of each sponge. Water passes out of the sponge through this opening.

5. The **spongocoel,** a large central cavity within the sponge. This cavity is lined by the specialized, flagellated collar cells (**choanocytes**) which create water currents within the sponge.

6. Water enters the sponge through many tiny pores that penetrate the body wall.

7. Numerous triradiate (3-rayed) **spicules** may be seen embedded in the wall.

In figure 6.5, make a two-inch drawing of *Leucosolenia.*

Sycon-type Sponge

Scypha (figures 6.6 and 6.7), formerly also called *Sycon* and *Grantia,* is a small, slender sponge of the sycon type. Obtain a preserved specimen and note the size and shape of the body. Examine specimens cut in half longitudinally and observe the spongocoel, the radial canals leading from the spongocoel into the body wall, and the short collar region which leads to the osculum surrounded by a funnel of giant spicules. On the surface observe the numerous cortical spicules which extend outward through the epidermal layer which covers the external surface of the sponge.

Study a stained cross section of *Scypha* and identify the **spongocoel,** the **radial canals,** and the **incurrent canals,** which open to the exterior of the sponge through small incurrent pores. The openings between the incurrent and excurrent canals are the **prosopyles;** the excurrent canals empty into the spongocoel through the **apopyles.** Much of the tissue within the sponge body consists of a loosely packed mesenchyme. Details of the cross section of *Scypha* are illustrated in figure 6.7.

The outer surface of the sponge is covered by a layer of thin, flat cells, called **pinacocytes.** A similar layer of pinacocytes lines the spongocoel. The **choanocytes** (figure 6.8) are found lining the radial canals, which empty into the spongocoel through the apopyle (figure 6.9). These choanocytes are small and difficult to identify in most microscopic preparations. You should be able to find some large undifferentiated amoeboid cells within the body wall. Eggs and developing embryos are also found within the body wall. There are no differentiated sex organs.

Fig. 6.5 Drawing of *Leucosolenia.*

Leucon-type Sponge

Leuconoid sponges are structurally the most complex and also the most common body type among the living sponges. All freshwater sponges and most marine sponges are leuconoid.

Study a portion of a preserved specimen of a bath sponge and note its rubbery texture and the complex system of branching canals. Attached to the radial canals are numerous small, spherical, flagellated chambers. Collar cells are found lining only these tiny flagellated chambers in leuconoid sponges. Study also a dried specimen of a bath sponge and note how its texture differs from that of the preserved specimen. Only the network of **spongin fibers** remains in the dried sponge. Observe the microscopic demonstration slide of spongin fibers.

Demonstrations

1. Eggs and developing embryos of *Scypha* (microscope slide).
2. Assorted spicules (microscope slide).
3. Spongin fibers (microscope slide).
4. Glass sponge and examples of other types of sponges.
5. Skeletons of glass sponges.
6. Preserved and dried samples of bath sponges and other types of sponges.

Fig. 6.6 *Scypha*, typical cluster of sponges. (Courtesy Carolina Biological Supply Company.)

Freshwater Sponges

Although most sponges are marine, a few species live in freshwater streams, ponds, and lakes. Freshwater sponges are much less prominent than their ubiquitous marine relatives. The freshwater sponges grow as tufts or small irregular masses encrusting sticks, stones, or submerged plants. Most species are yellow or brown in color, but a few species are green because of symbiotic algae that live within the sponge.

All freshwater sponges (and a few marine sponges) form internal asexual buds called **gemmules** (figure 6.10). Gemmules are resistant stages which serve to carry the sponge through the winter or aid in survival during a drought. Observe the microscope slide of gemmules on demonstration. What important features of the gemmule aid in the survival of the species?

Regeneration and Reconstitution (Optional Exercise)

Sponges are especially noted for their powers of regeneration. A small part of a sponge can regenerate a complete sponge. Many years ago, H. V. Wilson showed that pieces of a living sponge pressed through a fine cloth mesh to separate the sponge into individual cells and clusters of cells could reassemble and develop into a complete new sponge. More recent experiments with the separation and reassociation of sponge cells have provided important information about the nature and mechanism of cell recognition processes and the organization of tissues during development.

If living sponges are available in your laboratory, you may be able to repeat this famous experiment. *Microciona,* a common red sponge on the Atlantic coast of the United States, was used in the original experiments. You will need a small finger bowl, a watch glass, a clean microscope slide, a pipette, some silk mesh cloth, and some seawater for this experiment.

Procedure

1. Fill the finger bowl about two-thirds full of seawater of the proper ionic strength (depending on the source of the sponges) and place the watch glass on the bottom of the dish. Place the microscope slide on top of the watch glass.

2. In a separate bowl of seawater, prepare a suspension of cells and fragments of *Microciona* by pressing small pieces of sponge through a fine silk bolting cloth.

3. Pipette a small amount of this cell suspension onto the slide and allow the cells to settle.

4. Carefully lift out the slide and observe the cells. Make similar observations at intervals during the next 24–48 hours. Watch for the initiation of cellular aggregation and the thin protoplasmic extensions (filopodia) put out by the small clumps of aggregating cells. Sketch the cells and clumps of cells as a record of your observations.

Demonstrations

---

1. Gemmules (microscope slide).
2. Living or preserved freshwater sponges.

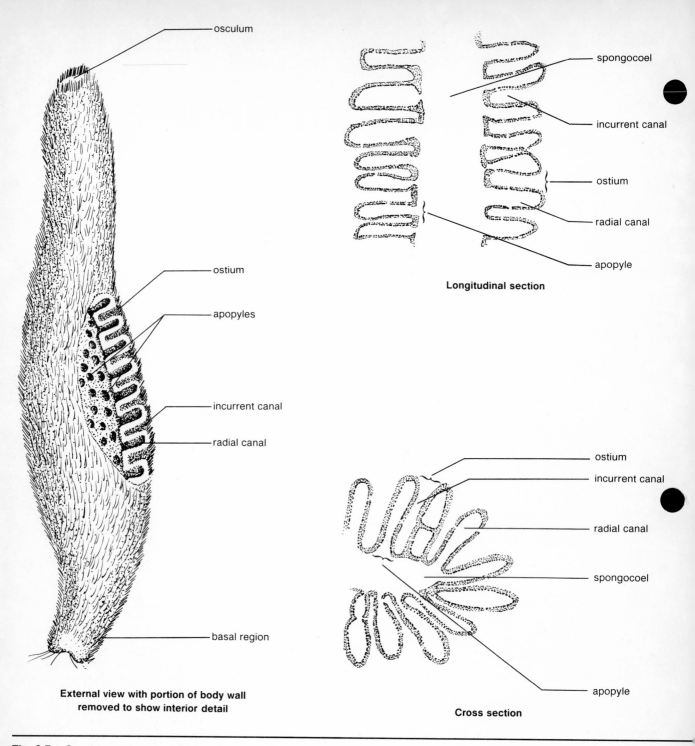

osculum

ostium

apopyles

incurrent canal

radial canal

basal region

**External view with portion of body wall
removed to show interior detail**

spongocoel

incurrent canal

ostium

radial canal

apopyle

Longitudinal section

ostium

incurrent canal

radial canal

spongocoel

apopyle

Cross section

Fig. 6.7 *Scypha.*

Chapter 6

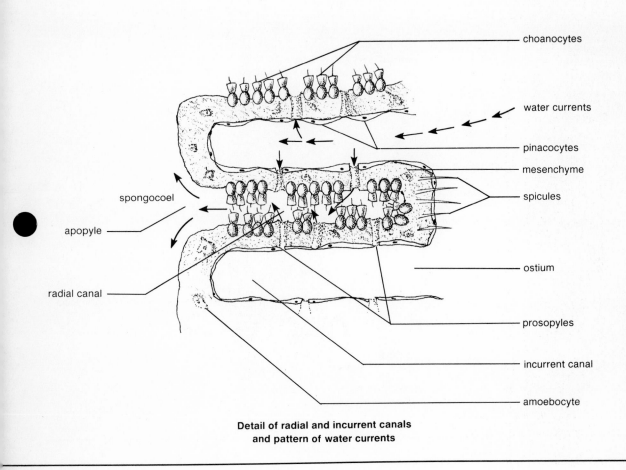

choanocytes

water currents

pinacocytes

mesenchyme

spicules

ostium

prosopyles

incurrent canal

amoebocyte

spongocoel

apopyle

radial canal

**Detail of radial and incurrent canals
and pattern of water currents**

flagellum collar

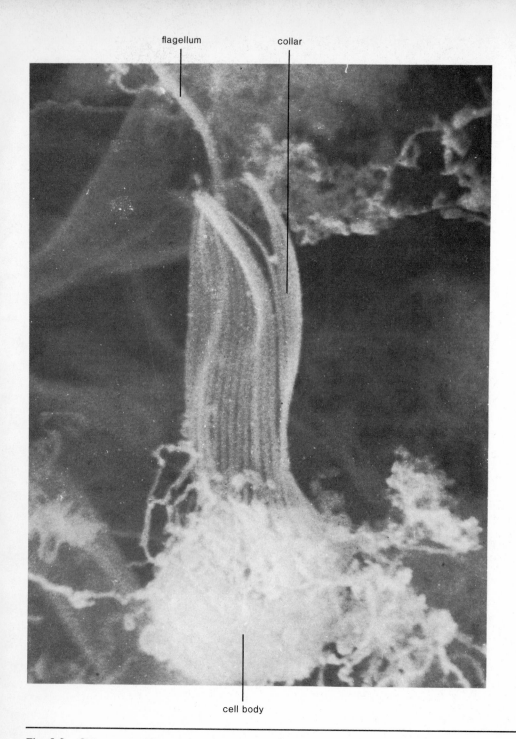

cell body

Fig. 6.8 Choanocyte. (Scanning electron micrograph by
Louis de Vos.)

Chapter 6

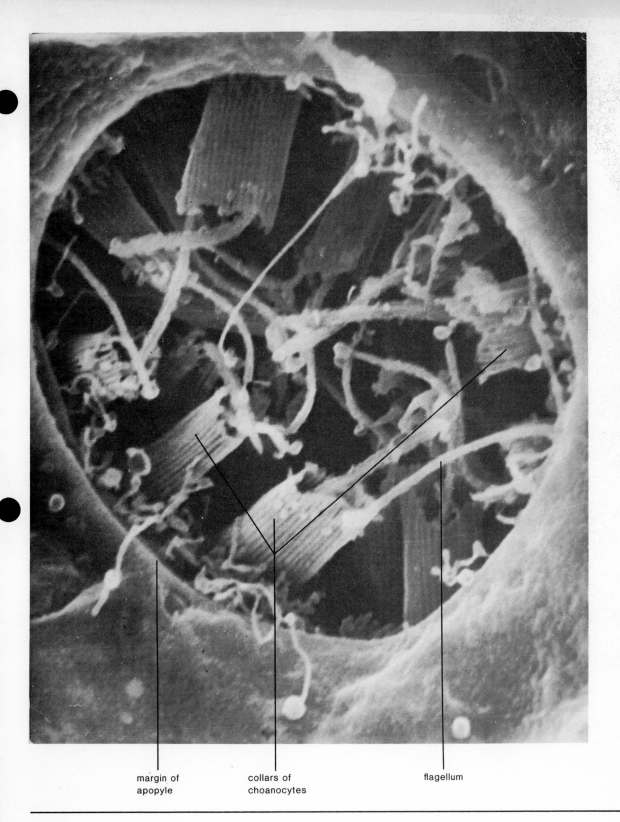

margin of   collars of   flagellum
apopyle   choanocytes

Fig. 6.9 Opening of apopyle showing arrangement of
choanocytes within. (Scanning electron micrograph by Louis
de Vos. 21,000×.)

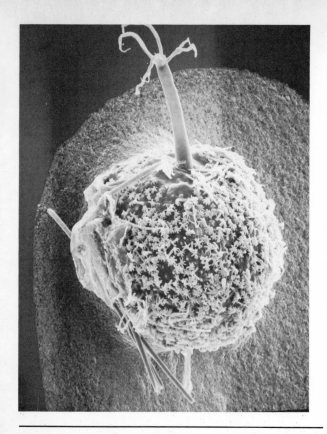

Fig. 6.10 Gemmule of freshwater sponge. Many specialized spicules are embedded in the resistant covering of the spicule. (Scanning electron micrograph by F. W. Harrison. 16,000×.)

Key Terms

Ascon the simplest type of sponge body form with a central spongocoel lined with choanocytes and with incurrent pores opening directly into the spongocoel.

Choanocyte a special type of flagellated collar cell characteristic of sponges.

Gemmule a dormant stage in the life cycle of freshwater sponges and a few species of marine sponges. Formed as internal asexual buds, they aid the sponges in surviving adverse environmental conditions.

Leucon a complex type of sponge body form with an intricate system of internal water canals. Choanocytes line certain small chambers within the system.

Pinacocyte a thin, flattened type of cell that lines the outer surface of sponges and most inner surfaces not lined by choanocytes.

Spicules individual skeletal elements that make up the skeleton of most sponges. Consist mainly of calcium carbonate or silicon salts and exhibit many different shapes.

Spongin fibers a loose network of protein fibers that form part or all of the skeleton in most horny sponges (Class Demospongia).

Sycon a type of sponge body that exhibits a central spongocoel into which numerous radial canals lined with choanocytes empty.

Chapter 6

7
Cnidaria (Coelenterata)

Objectives

After completing the laboratory work in this chapter, you should be able to perform the following tasks:

1. List the three classes of cnidarians and briefly characterize each class.
2. Describe the general morphology of *Hydra*.
3. Identify the main morphological features of *Hydra* on a microscope slide or illustration.
4. Discuss the structure and function of nematocysts and explain their importance in *Hydra* and other cnidarians.
5. List five cell types found in *Hydra* and explain their function.
6. Describe the feeding reaction of *Hydra* and its control.
7. Describe the reproductive processes of *Hydra* and be able to identify mature male and female individuals.
8. Identify the chief morphological features of the medusa of *Gonionemus* and explain the function of each part.
9. Describe the life cycle of *Gonionemus*.
10. Identify the main morphological features of a mature *Obelia* colony and explain the function of each.
11. Describe the life cycle of *Obelia* and explain how it illustrates the alternation of generations commonly found in cnidarians.
12. Describe the main morphological features of the scyphozoan medusa *Aurelia* and explain the function of each.
13. Describe the life cycle of *Aurelia*.
14. Describe the structure of the sea anemone *Metridium* and explain the function of its principal parts.
15. Compare the feeding mechanisms of *Hydra, Gonionemus,* and *Metridium*.
16. Describe the life cycle of *Metridium*.
17. Briefly discuss the structure of stony corals.

Introduction

Cnidarians (also called coelenterates) are the simplest animals with definite tissues; the cnidarian body consists fundamentally of two well-defined tissue layers with an intervening layer of gelatinous material, the **mesoglea.** Thus, the cnidarians are **diploblastic** or two-layered animals. The outer **epidermis** layer covers the external surface of the body, and the inner **gastrodermis** layer lines a single internal body cavity.

The cnidarians and the related Phylum Ctenophora are especially noted for their prominent **radial symmetry.** For this reason these two groups are commonly referred to as the radiate phyla.

The name Cnidaria is derived from the **cnidocytes,** special cells which produce the nematocysts or "stinging capsules" characteristic of this phylum. The alternate phylum name, Coelenterata, refers to the large **coelenteron,** or gastrovascular cavity, also characteristic of the group. The related Phylum Ctenophora derives its name, meaning "comb-bearer," from its characteristic eight rows of **comb plates,** or ctenes, which serve for locomotion.

Two basic body forms are exhibited by the Cnidaria: an attached **polyp** stage and a free-swimming **medusa** stage. Many species exhibit both a polyp stage and a medusa stage, and their life cycles involve an **alternation** of these two body forms or "generations." Two other important distinguishing characteristics of the phylum include **tentacles** around the mouth and a diffuse **nerve net,** which provides a modest degree of nervous coordination.

Classification

The phylum is divided into three classes:

Class Hydrozoa (Hydroids and Siphonophores)

Animals usually with both polyp and medusa in the life cycle, medusae with a velum, gonads on radial canals of medusae. Freshwater and marine species. Examples: *Obelia, Hydra, Gonionemus,* and *Physalia* (Portuguese man-of-war).

Class Scyphozoa (True Jellyfish)

Mainly large marine jellyfish with abundant mesoglea, polyp stage (scyphistoma) reduced or absent, no velum in medusae. All marine. Examples: *Aurelia, Chrysaora* (sea nettle), and *Cyanea*.

Class Anthozoa (Sea Anemones and Corals)

Solitary or colonial animals with polyp stage only, medusa absent, pharynx or gullet present, gastrovascular cavity partitioned by septa. All marine. Examples: *Metridium* (sea anemone), *Astrangia* (coral), *Gorgonia* (sea fan), and *Renilla* (sea pansy).

Materials List

Living Specimens
  Hydra
  Daphnia or *Artemia* larvae (food for *Hydra*)
  Sea anemones in aquarium (Demonstration)
Preserved Specimens
  Gonionemus, medusa
  Metridium
  Physalia
  Aurelia, medusae
  Ctenophores (Demonstration)
  Representative Hydrozoa, Anthozoa, and Scyphozoa (Demonstrations)
  Representative anemones and corals (Demonstration)
Prepared Microscope Slides
  Hydra, cross section, longitudinal section
  Obelia, whole mount of polyp, whole mount of medusa
  Nematocysts, discharged (Demonstration)
  Gonionemus, statocysts (Demonstration)
  Hydra, male with testes, whole mount (Demonstration)
  Hydra, female with ovary, whole mount (Demonstration)
  Green *Hydra* with symbiotic algae, whole mount (Demonstration)
  Planula larva, whole mount (Demonstration)
  Aurelia marginal sense organs, scyphistoma, ephyra larva, strobila (Demonstrations)
Chemicals
  1% Acetic acid
  0.01% Methylene blue solution
  Pond water

A Polyp: *Hydra*

Hydra (figure 7.1) typifies the polyp form of a cnidarian. *Hydra* lives in freshwater streams, lakes, and ponds, where it is usually attached to submerged sticks, stones, or vegetation and feeds on various small aquatic animals. About fourteen species of *Hydra* are known to occur in the United States.

Although *Hydra* serves as a good example of the polyp form of a cnidarian, it differs in several important ways from most members of the Class Hydrozoa: (1) *Hydra* is solitary, while most hydrozoan polyps are colonial. (2) *Hydra* has no separate medusa stage, which most hydrozoan species have. (3) The polyp of *Hydra* bears gonads, unlike most hydrozoan polyps. (4) *Hydra* is mobile, unlike most hydrozoan polyps, which are sessile. (5) *Hydra* lives in fresh water, while most hydrozoans are marine.

General Appearance and Morphology

Examine a living specimen of *Hydra* in a dish of pond water. Be sure to use pond water and not tap water, since most tap water contains trace amounts of copper and other substances toxic to *Hydra*.

Locate the following structures on your specimen: (1) the **basal disc** at the lower end which serves for attachment; (2) the cylindrical **body;** (3) a circle of **tentacles** at the free end (how many?); (4) the **hypostome,** an elevation between the bases of the radially arranged tentacles; (5) the **mouth** in the center of the hypostome; (6) **buds,** the products of asexual reproduction, may also be present; and (7) **ovaries** or **testes** are also present on the lower portion of the body in mature specimens.

Behavior

Does your specimen change in shape? Touch one of the tentacles with the tip of your dissecting needle. What is the reaction? What methods of locomotion are used by *Hydra?* Observe *Hydra* feeding in an aquarium, or add a few *Daphnia* or washed brine shrimp (*Artemia*) larvae to your dish near the specimen to observe feeding.

Cnidocytes and Nematocysts

After you have studied the basic form and behavior of your specimen, place it on a clean microscope slide in a drop of water and carefully add a cover glass. Observe the numerous **cnidocytes,** which appear as swellings on the tentacles. Each cnidocyte is a cell containing a **nematocyst,** or stinging capsule.

The cnidocytes of *Hydra* can be stained to aid in your observations by adding a drop of 0.01% methylene blue solution to the edge of the coverslip.

Hydra has four different kinds of **nematocysts,** each with a distinctive structure and function. Nematocysts are important in the capture of food, in locomotion, and in

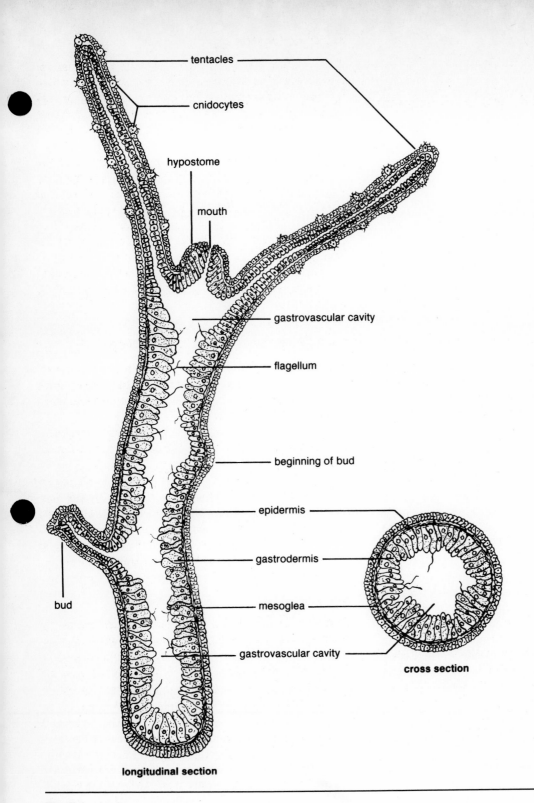

tentacles

cnidocytes

hypostome

mouth

gastrovascular cavity

flagellum

beginning of bud

epidermis

gastrodermis

mesoglea

gastrovascular cavity

bud

cross section

longitudinal section

Fig. 7.1 *Hydra.*

Fig. 7.2   Drawing of nematocysts.

attachment. Nematocysts are complex **secretory products** formed within developing cnidocytes. Each nematocyst consists of an outer protein **capsule,** and a long, coiled **tube** often armed with **spines** or **barbs.** Toxins, enzymes, and other chemicals are also contained within the capsule of certain kinds of nematocysts.

Tap on the coverslip of your wet mount of *Hydra* tentacles to induce discharge of the nematocysts. When properly stimulated, nematocysts empty their contents with a rapid discharge as the coiled tube is shot out. In addition to tapping the coverslip to induce nematocyst discharge, you may also wish to try adding a drop of dilute acetic acid (1% solution) to stimulate nematocyst discharge.

Study the discharged nematocysts under high power or with an oil immersion lens, if available. Observe the outer capsule, the long thread or tube, and the large spines or barbs at the base of the tube. How many kinds of nematocysts can you find on your slide?

Observe also the microscopic demonstration of discharged nematocysts and make drawings of at least two different types in the space provided in figure 7.2.

Histological Structure

An outstanding characteristic of the Phylum Cnidaria is the diploblastic (two-layered) structure of its members. This fundamental plan of structure is clearly illustrated in *Hydra* (see figure 7.1.) Examine the stained mounts of cross and longitudinal sections of *Hydra,* and note the following: (1) **epidermis,** the outer, thinner epithelial layer of cells; (2) **gastrodermis,** the inner layer of cells; (3) **mesoglea,** a very thin, noncellular layer between the epidermis and gastrodermis; (4) **gastrovascular cavity,** or enteron, the internal cavity lined by the gastrodermis.

Cellular Structure

Hydra shows distinct advances compared with a sponge, not only in its general structure but also in the degree of differentiation of cellular structure and functions. Several different types of cells may be distinguished in the stained preparations. In the epidermis, under high power, try to distinguish the following cell types: (1) The large and abundant **epitheliomuscular cells,** which possess contractile processes or fibers at their base, all running lengthwise. What specific functions do these fibers perform? (2) The small **interstitial cells** at the bases of the epitheliomuscular cells. (3) The **cnidocytes.** (4) The mucus-secreting **gland cells** abundant on the pedal disc. Also among the epidermal cells are many small nerve cells as described in the next section. They can be seen only in specially stained slides.

In the inner gastrodermal layer, note the following cell types: (1) The abundant, large, vacuolated **digestive cells.** These cells bear one or two flagella and ingest food particles for intracellular digestion. They are also epitheliomuscular in character and possess contractile fibers which run transversely along their bases, thus providing a circular musculature. What, therefore, is the function of the contractile fibers in the gastrodermal cells? Compare this with the function of the fibers in the epidermis. (2) The **gland cells,** which secrete either mucus or digestive enzymes. Mucus-secreting cells are abundant in the hypostome region. (3) The **interstitial cells** at the bases of the gastrodermal cells. Several other types of cells are also present but are difficult to observe except in specially stained preparations.

Nervous System

Specialized nerve cells, or **neurons,** are located among the cells of both the epidermal and gastrodermal layers of *Hydra*. These neurons may have two or more processes that connect via **synaptic junctions** with other neurons or with various types of receptor or effector cells.

Figure 7.3 shows an insolated multipolar neuron from the body wall of *Hydra*. Special techniques are required to see cnidarian neurons with either a light or an electron microscope.

The interconnecting network of neurons in *Hydra* form a **nerve net** that lacks concentrations of neurons into ganglia or a brain. This diffuse nerve net is characteristic of *Hydra* and other cnidarians. Nervous impulses tend to spread in a radiating pattern from the point of origin or stimulation because of the structure of synapses between adjacent neurons. Many of these synapses are symmetrical and are nonpolarized, thus allowing impulses to flow in both directions. Some polarized synapses also occur in cnidarians. Recent studies have suggested that remnants of the diffuse nerve net of cnidarians may be represented in the nervous system of certain higher animals, including the digestive systems of annelid worms and of humans. The rhythmic peristaltic movements of your stomach and intestine after you eat are coordinated by a similar nerve net.

Neurons also have processes connecting via synapses with epithelial muscular cells, gland cells, and nematocysts. Substantial evidence now suggests that nematocyst discharge is at least partly under nervous control.

Feeding Behavior

Hydra is a carnivore and feeds on living crustaceans, rotifers, insect larvae, and other small animals. When properly stimulated, *Hydra* exhibits a characteristic feeding response. You can observe the feeding behavior by placing a healthy, unfed *Hydra* in a clean watch glass containing about 10 ml of pond water (or *Hydra* culture solution). Add a few (6–12) washed brine shrimp (*Artemia*) larvae near the *Hydra* and record your observations. Note how the food organisms are captured, what happens to them after their capture, and the movements of various parts of the *Hydra*. Use the second hand on your watch (or your

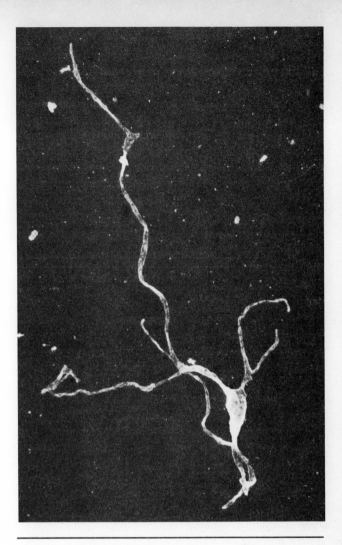

Fig. 7.3 Isolated multipolar neuron from body column of *Hydra*. (Scanning electron micrograph by J. A. Westfall and L. G. Epp from *Tissue and Cell* 17(2):165.)

neighbor's watch!) to time various parts of the complex behavioral reaction of the *Hydra*. Make a record of your observations on the pages provided for *Notes and Sketches* at the end of this chapter.

Scientists have shown that the feeding reaction of *Hydra* is normally caused by body fluids oozing from the body of its prey, which has been pierced by nematocysts. Further experiments have shown that a feeding reaction can also be elicited by a solution of a tripeptide, reduced glutathione.

Reproduction

Hydra reproduces asexually by budding and sexually by the production of eggs and sperm. In addition to the testes and/or ovaries that you may have seen on your living specimen (refer to figure 7.4), study also the demonstration materials. Most species of *Hydra* are **dioecious** (sexes separate) although a few species are **monoecious** (both sexes in one individual).

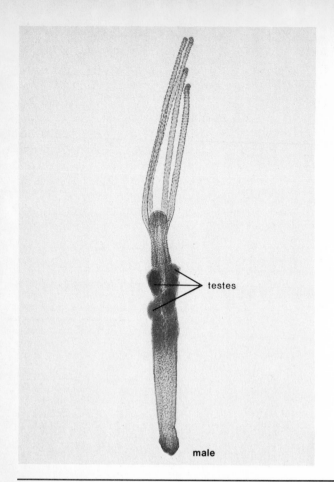

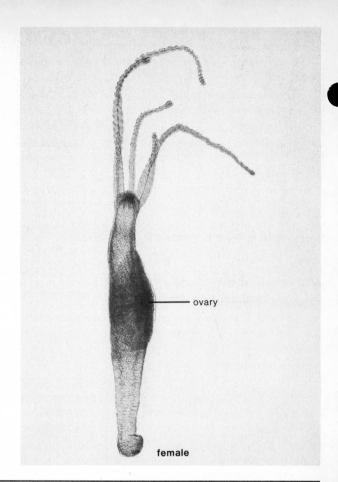

Fig. 7.4. Male and female *Hydra*. (Courtesy Carolina Biological Supply Company.)

For some reason not yet understood, males are more common than females in most laboratory cultures of *Hydra*. On female specimens, the ovary may bear an unfertilized ovum, or a later stage of development, since early embryonic development may occur while the embryo is still attached to the parent.

Regeneration (Optional Exercise)

Hydra and other cnidarians have great powers of regeneration. Place a *Hydra* in a watch glass and cut across the middle of the body with a sharp scalpel or razor blade. Separate the two pieces into different watch glasses and observe their development during the next several days. Observe the formation of a new hypostome, mouth, and tentacles by the basal half of your *Hydra*. What happens to the other half? Keep a record of your observations in the pages provided for *Notes and Sketches* at the end of this chapter.

Demonstrations

1. *Hydra* with testes (microscope slide).
2. *Hydra* with ovary (microscope slide).
3. Symbiotic algae in green hydra, *Chlorohydra* (living or microscope slide).

A Medusa: *Gonionemus*

Gonionemus is a small marine medusa, or jellyfish, common in many parts of the world. The medusa of *Gonionemus* (figure 7.5) is the adult or sexually mature stage and serves to illustrate the typical structure of a cnidarian medusa (figure 7.6).

Examine a preserved specimen in a watch glass partly filled with water and note its umbrellalike form. The specimens are delicate and must be handled with care. Note the jellylike consistency of the medusa, due to the thick layer of **mesoglea** within the body. Most of the bulk of the medusoid body consists of mesoglea.

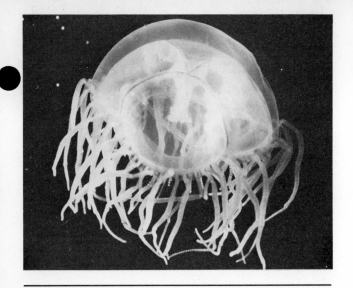

Fig. 7.5 *Gonionemus* medusa. (Courtesy Carolina Biological Supply Company.)

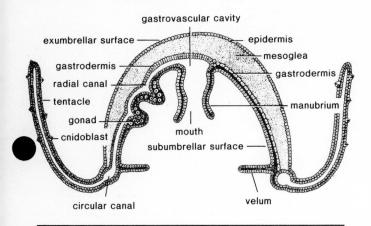

Fig. 7.6 *Gonionemus* medusa, vertical section.

The medusa usually swims with its convex **exumbrellar surface** upward and the concave **subumbrellar surface** downward. Observe the numerous **tentacles** around the margin of the bell. Extending downward from the center of the subumbrellar surface is the **manubrium,** with the **mouth** at its tip surrounded by four **oral lobes.** Around the inner margin of the bell, extending inward, is a thin circular flap of tissue, the **velum,** which is believed to aid in swimming. At the base of the manubrium is an expanded portion of the manubrium, the "stomach." Find the four **radial canals** extending from the stomach to the **circular canal** at the margin of the bell. The hollow tentacles connect with the circular canal, and their cavities thus represent a continuation of the gastrovascular cavity. Observe the numerous **nematocyst batteries** and the **adhesive pads** on the tentacles.

At the bases of the tentacles are round, pigmented structures believed to be light sensitive photoreceptors. Between the tentacle bases are the **statocysts,** which serve as balancing organs.

Observe the folded gonads attached to the subumbrellar surface of the radial canals. The gonads on the specimen you are studying are either ovaries or testes, since *Gonionemus,* like most cnidarians, is dioecious. Gametes produced by the gonads are released into the sea, and the fertilized eggs develop into a **ciliated planula larva.** Later, the planula larva settles and attaches to some submerged object in the sea and transforms into a microscopic **polyp stage.** The polyp stage may reproduce asexually by budding and, under certain conditions, may form tiny medusa buds that detach and grow to develop into mature medusae, thus completing the life cycle.

Demonstrations

1. Statocysts at margin of *Gonionemus* (wet mount or microscope slide).
2. Planula larva (microscope slide).

A Colonial Polyp: *Obelia*

Obelia is a colonial marine cnidarian which illustrates the complex life cycle with alternating polyp and medusa stages found in many cnidarians (figure 7.7). Examine a stained whole mount of the polyp or asexual stage of *Obelia* and study its organization. Like many colonial animals, *Obelia* exhibits **polymorphism** or morphological specialization of its members. Therefore you can distinguish two different kinds of individuals in an *Obelia* colony, feeding polyps or **hydranths** and reproductive polyps or **gonangia.** The feeding polyps bear tentacles armed with nematocysts, a mouth, a hypostome, and a delicate outer covering, the **hydrotheca** (an extension of the perisarc). Reproductive polyps consist of a central **blastostyle** on which **medusa buds** develop, and a thin outer covering, the **gonotheca.** At the distal end of the gonotheca is an opening, the **gonopore,** through which the newly liberated medusae escape. Note that the gonangia have no mouth or tentacles. How do they receive their nutrition?

The hydranths and gonangia are attached to a main stem, or **hydrocaulus,** which consists of a cylindrical tube of living tissue, the **coenosarc,** and an outer secreted covering, the **perisarc.** In cross section, the coenosarc resembles a cross section of *Hydra* with an outer epidermis, an inner gastrodermis, and a thin intervening layer of mesoglea.

Alternation of Generations

The life cycle of *Obelia* is fundamentally similar to that of *Gonionemus* and illustrates the alternation of generations characteristic of the Phylum Cnidaria (figure 7.7).

The alternation of the asexual polyp generation and the sexual medusa generation (also called metagenesis) is basically different from the alternation of generations in

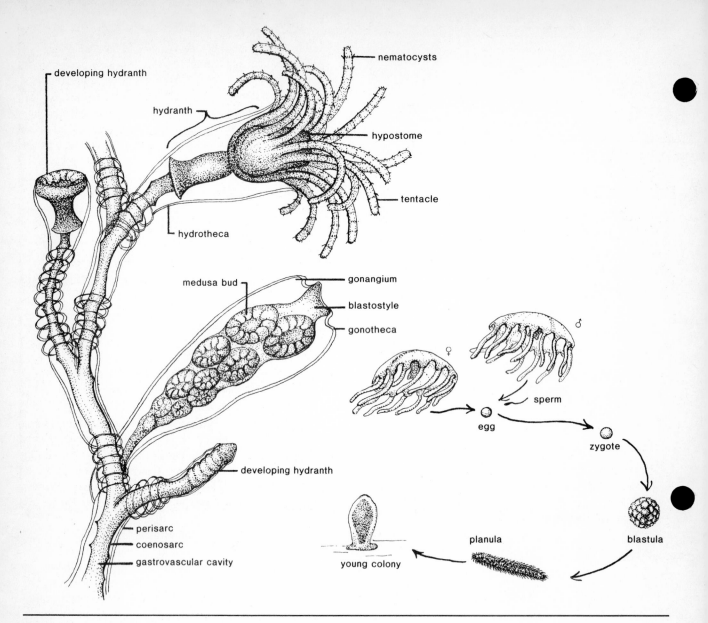

Fig. 7.7 *Obelia*, life cycle.

plants. Both the polyp and medusa generations of cnidarians are **diploid** (2N), while in plants the alternation is between a **haploid** (1N) gametophyte generation and a **diploid** (2N) sporophyte generation.

In the life cycle of *Obelia*, the polyp generation produces medusa buds within its gonangia. The tiny, short-lived medusae escape into the plankton and produce either eggs or sperm. Fertilized eggs develop into ciliated **planula larvae,** which swim about in the sea for a time and settle to transform into a new polyp. Buds transformed by asexual reproduction of the polyp do not detach, thus forming a colony.

The sexual stage of *Obelia* is a free-swimming medusa somewhat similar to *Gonionemus,* but much smaller in size. Observe a stained whole mount of an *Obelia* medusa and observe its structure. Draw a picture of an *Obelia* medusa in figure 7.8.

A Portuguese Man-of-war: *Physalia*

The Portuguese man-of-war (figure 7.9), *Physalia,* is a complex colonial hydrozoan (Class Hydrozoa, Order Siphonophora) exhibiting a high degree of **polymorphism.** A single colony may consist of as many as 1,000 individuals and several types of polypoid and medusoid forms. The familiar iridescent colonies of *Physalia* are commonly found along the beaches of Florida, the South Atlantic, the Gulf of Mexico, and, sometimes, as far north as Cape Cod.

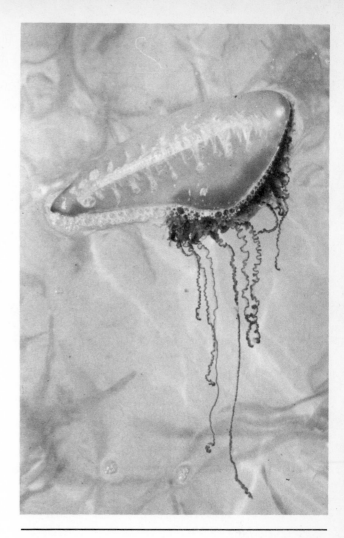

Fig. 7.9   *Physalia*, Portuguese man-of-war. (Photograph by C. F. Lytle.)

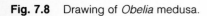

Fig. 7.8   Drawing of *Obelia* medusa.

The most prominent feature of *Physalia* is a gas-filled **float** above which is a sail-like **crest.** *Physalia* is transported by winds and oceanic currents, and its normal habitat is the open sea rather than the sandy beach where it is most often seen (and sometimes felt!) by bathers. The sting from the nematocysts on the tentacles of *Physalia* can be painful when touched but is rarely dangerous, except to highly sensitive individuals.

Below the float are suspended numerous **tentacles** and other structures made up of several kinds of modified polyps and medusae. Thus, *Physalia* is an unusual cnidarian since a single colony contains both polypoid and medusoid individuals of several types closely joined together, in contrast to separate polypoid and medusoid generations.

Observe a preserved or a plastic-embedded specimen of *Physalia* and identify the float, the crest, and the tentacles. See how many types of individuals you can identify among the structures attached to the float.

Velella and *Porpita* are two related colonial hydrozoans often seen on our Pacific coast and less often in the Gulf of Mexico and South Atlantic. Both exhibit polymorphism and habits similar to *Physalia*.

A Scyphozoan Jellyfish: *Aurelia*

Aurelia (figure 7.10) is a common, widely distributed marine jellyfish. Large specimens may reach twelve inches (30 cm) in diameter. The polyp form, called a **scyphistoma,** is small, sessile, and lives attached to rocks and other submerged objects in shallow coastal waters.

Study a preserved specimen of *Aurelia* to learn about the organization of a scyphozoan medusa. You will not need to dissect the specimen, since the transparent body readily shows most important features. Handle the specimen with care and return it for study by another student.

Observe the four-part **radial symmetry** and locate the four long **oral arms** arising from the corners of the square mouth. Along the arms find the many short **oral tentacles** that help to capture food (small planktonic animals), which are then moved toward the mouth along the **ciliated groove** on the oral side of each arm. After passing

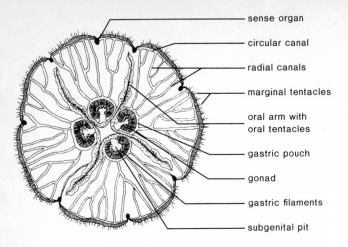

sense organ

circular canal

radial canals

marginal tentacles

oral arm with
oral tentacles

gastric pouch

gonad

gastric filaments

subgenital pit

Fig. 7.10 *Aurelia* medusa, oral view.

through the mouth, the food enters the gastrovascular cavity. Internally the gastrovascular cavity is divided into four **gastric pouches.** A ring of **gastric filaments** within each gastric pouch immobilizes or kills any food organisms still active. The gastric filaments bear many nematocysts.

Four horseshoe-shaped **gonads** surround the ring of gastric filaments within the four gastric pouches. Depressions on the subumbrellar surface of the bell beneath the gonads are called **subgenital pits.** Their function is unknown.

Observe the complex branching system of **radial canals** that distribute food materials from the gastric pockets to other parts of the bell, and an outer **circular canal** around the margin of the bell. Also, around the margin of the bell, locate the eight marginal **sense organs.** These marginal organs are sensitive organs of touch and balance.

Reproduction and Life Cycle

Mature *Aurelia* medusae release gametes from the gonads into the gastrovascular cavity. How does this differ from the hydrozoan medusae? The gametes exit from the mouth and fertilization is external.

The fertilized eggs or zygotes develop into ciliated **planula larvae,** which may be retained for a time on the oral arms of the medusa and later settle to the sea bottom. There the larvae develop into a small, trumpet-shaped polyp, called a **scyphistoma.**

Under appropriate environmental conditions the scyphistoma transforms into a **strobila** (figure 7.11). The strobila develops and releases by transverse fission a series of saucer-shaped **ephyra larvae,** which bear marginal sense organs and other medusoid features. The ephyra larvae gradually transform into adult jellyfish to complete the life cycle.

Study the demonstration materials and draw a scyphistoma in figure 7.12.

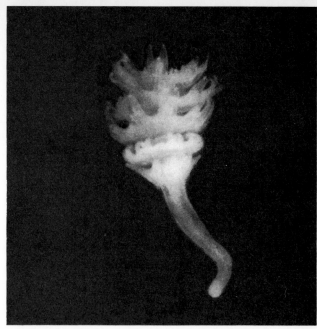

Fig. 7.11 *Aurelia* strobila. (Photograph by C. F. Lytle.)

Fig. 7.12 Drawing of an *Aurelia* scyphistoma.

Demonstrations

1. Marginal sense organ of *Aurelia* medusa (microscope slide).
2. Scyphistoma of *Aurelia* or other scyphozoan (preserved or microscope slide).
3. Planula and ephyra larvae of *Aurelia* (microscope slide).
4. Strobila stage of *Aurelia* (microscope slide).

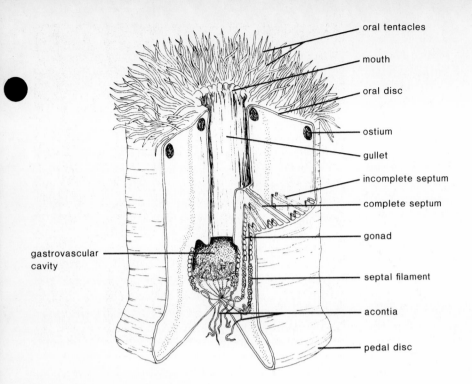

oral tentacles

mouth

oral disc

ostium

gullet

incomplete septum

complete septum

gonad

septal filament

acontia

pedal disc

gastrovascular cavity

Fig. 7.13 *Metridium,* partly dissected.

A Sea Anemone: *Metridium*

Sea anemones are typically sessile cnidarians that attach to rocks, shells, pilings, and other hard substrates in the sea. Some species, however, burrow in soft bottoms or are even free-swimming. All nonetheless represent the polyp form of the cnidarians; the medusa generation is totally lacking in this class. The anemones and other Anthozoa represent the highest degree of specialization of the cnidarian polyp. The basic features of the anthozoan polyp are well illustrated by the common North Atlantic anemone, *Metridium* (figure 7.13). The same or a similar form also occurs in the Pacific Ocean.

Select a preserved specimen of *Metridium* and identify the **mouth** in the center of the **oral disc** surrounded by many short **oral tentacles,** and the **basal disc,** which attaches to rocks or other hard substrates.

Most of the internal anatomy can be observed in specimens that have been bisected longitudinally or horizontally. It is not necessary, therefore, to dissect the specimens further. Study precut preserved specimens and locate the following structures. Find the tubular **gullet** leading internally from the mouth to the large **gastrovascular cavity.** One or more **ciliated grooves** (siphonoglyphs) should be found along the edge of the gullet. Usually the cilia within these grooves beat inward to provide a respiratory current of water into the gastrovascular cavity. Cilia on the remainder of the gullet wall beat outward to remove wastes and foreign particles from the gastrovascular cavity. During feeding, however, the beat of the cilia along the gullet wall is reversed and aids in moving food particles into the gastrovascular cavity. Observe the feeding of living anemones in an aquarium (if available) to better understand the interaction of the tentacles and ciliary currents on the oral disc and in the gullet during the feeding process.

The gastrovascular cavity is partially divided into sections by thin vertical walls of tissue called **septa.** Some septa attach to the central gullet; others (partial or incomplete septa) bear thickened **septal filaments** on their free inner margins. The septal filaments extend below the partial septa as thin, twisted, threadlike **acontia.** The acontia bear numerous nematocysts and act to subdue living prey taken into the gastrovascular cavity. Distal to the septal filaments on the partial septa, find the **gonads,** which appear as thickened ridges parallel to the septal filaments. On the septa near the oral disc, locate the **ostia,** round openings that allow circulation of fluid between adjacent sections of the gastrovascular cavity.

Reproduction and Life Cycle

Reproduction in *Metridium* and most other anemones is both asexual and sexual. Some anemones reproduce asexually by splitting longitudinally (**longitudinal fission**), but the main asexual means of reproduction in *Metridium* is **pedal laceration.** Bits of tissue from the pedal disc are split from the anemone as the animal moves along the substrate. These tissue pieces later regenerate an entire small anemone, literally in the footsteps of its parent.

Sexual reproduction occurs seasonally when gametes are released from the gonads on the partial septa into the gastrovascular cavity. The gametes are released and are fertilized in the sea. The fertilized eggs develop into free-swimming **planula larvae.** After a period as planktonic larvae, the planulae settle on some hard substrate and metamorphose into an anemone.

Corals

Most anthozoans are corals, the largest and best known of which are the **stony** or **scleractinian corals** (Order Zoantharia) and the **octocorals** (Subclass Alcyonaria), which includes the soft and horny corals. Corals, like all anthozoans, have only a polyp form. There is no medusa stage in the life cycle.

The polyp of a stony coral resembles a sea anemone, although the individual polyps are generally smaller than those of anemones. The coral polyp sits in a **cup** on the surface of a calcareous exoskeleton secreted by the lower portion of the column and the basal disc. Extending inward from the wall of the cylindrical cup are several **calcareous septa,** which extend from the sides and base of the cup into folds of the basal tissue of the polyp. These tissue-covered septa partially subdivide the gastrovascular cavity inside the polyp into several chambers.

Most stony corals are colonial, and adjacent polyps are connected by lateral extensions of the body wall which cover the intervening stony skeleton of the colony. These sheets of lateral tissue also contribute to the formation of the skeleton by their secretions and serve to connect the gastrovascular cavities of adjacent polyps.

Many growth forms occur among the stony corals, and most species exhibit characteristic skeletons. A few species are solitary and occur as large individual polyps, like *Fungia* from the Pacific Ocean.

Some stony corals contain **symbiotic algae** (zooanthellae), and many interesting studies have been conducted on the physiological and biochemical interactions of these symbionts. Among the stony corals are several **reef-building species** which are largely responsible for the formation of many coral reefs in warmer parts of the oceans, including those in the Bahama Islands, off the Florida Keys, and the Great Barrier Reef of Australia.

Some examples of stony corals are *Fungia,* a solitary coral; *Astrangia danae,* the Atlantic star coral; *Oculina,* the eyed coral; and *Diploria* and *Meandrina,* brain corals.

The octocorals exhibit strong **eight-part** (octamerous) **radial symmetry** and have an **endoskeleton** consisting of separate microscopic pieces (spicules). A tough, horny organic material is also present in some species. This group is especially prominent in tropical waters and includes the sea fans, sea whips, the sea pens, the sea pansies, the organ pipe coral, and the precious red coral *(Corallium)* used for jewelry.

Study the demonstration materials illustrating several types of coral and draw several different types in figure 7.14. Identify each type of coral that you draw.

Demonstrations

1. Living sea anemones and/or corals in a marine aquarium.
2. Representative preserved anemones.
3. Assortment of preserved corals and dried coral skeletons.

Key Terms

Alternation of generations   the alternation of the sessile polyp and free-swimming medusa generations typical of the life cycle of the cnidarians.

Cnidocytes   specialized cells of cnidarians that produce and contain the nematocysts.

Coenosarc   the living portion of the tubular connecting portions of colonial cnidarians like *Obelia*. Consists of a simple cylinder of an outer epidermal tissue layer, an inner gastrodermal tissue layer, and an intermediate mesoglea surrounding a central gastrovascular cavity.

Dioecious   condition of an animal with sex organs borne in different individuals.

Diploblastic construction   the two-layered construction typical of the cnidarians. Consists of an outer epidermal tissue layer and an inner gastrodermal layer with an intervening noncellular mesoglea layer.

Epidermis   outer tissue layer protecting the surface of an animal from its environment.

Gastrodermis   inner tissue layer of animals bordering the digestive cavity.

Gastrovascular cavity   a central cavity of an animal that serves both for digestion and circulation, with a single mouth opening that serves both for entrance and exit. A type of incomplete digestive system.

Gonangium   a type of reproductive individual in colonial Hydrozoa, such as *Obelia;* produces free-swimming medusae.

Hydranth   feeding individual in a colonial hydrozoan, as in *Obelia.*

Medusa stage   free-swimming stage in the life cycle of many cnidarians. Usually bears gonads and produces gametes.

Mesoglea   gelatinous layer intermediate between the epidermal and gastrodermal layers of cnidarians.

Monoecious   condition of bearing both sex organs in one individual.

Nematocysts   the stinging capsules produced by the cnidocytes of cnidarians. Characteristic of the phylum.

Nerve net   the diffuse, interconnected network of nerve cells in the cnidarians. Lacks ganglia or other nervous centers. A very primitive type of nervous system.

Perisarc   the nonliving outer covering secreted by the coenosarc of colonial hydrozoans. Surrounds the interconnecting coenosarc.

Planula larva   a simple, ciliated, sausage-shaped larval form produced by many cnidarians. Develops from the zygote or fertilized egg.

Polyp stage   the sessile (and usually asexual) stage in the life cycle of many cnidarians.

Radial symmetry   a body plan in which all body parts are arranged symmetrically around a central axis.

Scyphistoma   the inconspicuous polyp stage in the life cycle of certain scyphozoan medusae.

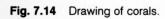

Fig. 7.14 Drawing of corals.

8
Platyhelminthes

Objectives

After completing the laboratory work in this chapter, you should be able to perform the following tasks:

1. Briefly outline the characteristics of the Phylum Platyhelminthes and identify some major advances in organization over the Phylum Cnidaria.

2. List and briefly characterize each of the three classes of the Phylum Platyhelminthes.

3. Describe the behavior of a free-living flatworm (planarian) such as *Dugesia* or *Planaria* and relate this behavior to the function of its sense organs.

4. Describe the feeding and nutrition of a planarian such as *Dugesia* and relate these processes to the organization of its digestive system.

5. Discuss the structure of the epidermis of a planarian and explain the location and function of cilia, rhabdites, and gland cells.

6. Identify the major structures that can be seen in microscopic cross sections at various levels of the body of a planarian; for example, anterior, pharyngeal region, and posterior.

7. Explain the structure and function of the reproductive system of a planarian and identify its principal reproductive organs.

8. Discuss the general morphology of the trematode *Clonorchis* and compare it with a free-living turbellarian such as *Planaria* or *Dugesia*.

9. Describe the life cycle of *Clonorchis* and identify in microscopic preparations the principal stages.

10. Describe the general morphology of the liver fluke *Fasciola* and identify its principal organs.

11. Identify in microscope slides the scolex, proglottids, and strobila of a tapeworm and explain the basic organization of a tapeworm.

12. Describe the life cycle of a tapeworm such as *Taenia* or *Dipylidium* and identify its principal stages in microscopic preparations.

Introduction

The Platyhelminthes, or flatworms, are soft, wormlike animals with flattened, elongated bodies. They exhibit several important structural advances over the cnidarians, including **three distinct tissue layers** (triploblastic construction), **bilateral symmetry,** and several well-developed **organ systems.**

The body parts of bilaterally symmetrical animals are arranged symmetrically along a central anterio-dorsal plane, so that the left and right sides are approximately mirror images of each other (see figure 9.1 in the next chapter). This type of symmetry is characteristic of most higher metazoans, except for adult echinoderms. Triploblastic body construction with **endoderm, mesoderm,** and **ectoderm** tissues is also characteristic of higher metazoans from Platyhelminthes to Chordata.

Flatworms lack the large central body cavity found in most higher animals. Instead, the interior of flatworms is typically filled with loosely packed **parenchyma** tissue with irregular spaces between the cells and clumps of cells. Since flatworms have no circulatory system or heart, body fluids percolate through these irregular interior spaces to bring nutrients and oxygen to the cells and to remove wastes from them. The body fluids are moved in part by muscular contractions. This type of organization without a central body cavity is called **acoelomate construction.**

Free-living flatworms (figure 8.1) have well-developed digestive, excretory, reproductive, nervous, and muscular systems, but, as noted previously, they have no circulatory system nor central body cavity. Some flatworms exhibit spiral cleavage and determinate development that suggest some affinities with the molluscs, annelids, and arthropods, animals that make up the group known as the protostomes.

Most flatworms, however, have become parasites and show many adaptations for parasitism, which include reduction and/or modification of some of their organ systems, and thus obscure clues to the possible evolutionary relationships of the phylum.

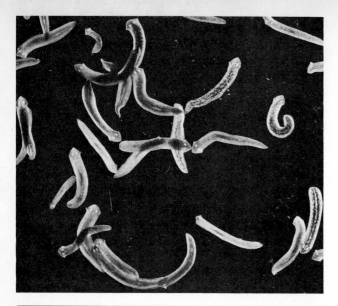

Fig. 8.1 Free-living flatworms. (Courtesy Carolina Biological Supply Company.)

Classification

The phylum includes three distinct classes:

Class Turbellaria (Free-living Flatworms)

Mainly free-living flatworms with bodies flattened dorsoventrally and a ciliated epidermis. Mouth usually ventral, leading into gastrovascular cavity; anus lacking. Freshwater and marine forms. Examples: *Dugesia, Planaria, Stenostomum, Leptoplana* (marine polyclad flatworm).

Class Trematoda (Flukes)

Parasitic animals with external tegumen (cuticle) secreted by underlying cells. Ovoid body with one or two suckers for attachment to host; incomplete digestive system, usually with two main branches. With complex life cycle, involving larval stages and alternate hosts. Examples: *Clonorchis* (human liver fluke), *Fasciola* (sheep liver fluke), *Schistosoma*.

Class Cestoda (Tapeworms)

Elongate body with specialized scolex with hooks and/or suckers for attachment to host; body divided transversely into a series of proglottids; thick external tegument (cuticle); mouth and digestive tract absent. Usually with complex life cycle involving alternate hosts. Examples: *Dipylidium caninum* (dog tapeworm), *Taenia pisiformis* (dog and cat tapeworm), *Dibothriocephalus* (fish tapeworm).

Materials List

Living Specimens
  Dugesia or *Planaria*
  Stenostomum
Preserved Specimens
  Whole tapeworms
Prepared Microscope Slides
  Planaria, whole mount, representative cross sections
  Stenostomum, whole mount
  Clonorchis, whole mount
  Fasciola, whole mount
  Dipylidium, whole mount
  Taenia pisiformis, scolex and representative sections
  Taenia solium, scolex and representative sections (Demonstration)
  Clonorchis, miracidium, sporocyst, redia, cercaria, metacercaria (Demonstrations)
  Dibothriocephalus latus, scolex and representative sections (Demonstration)
  Onchosphere (six-hooked) larva (Demonstration)
  Cysticercus larva (Demonstration)
Plastic Mounts
  Dipylidium, whole mount (Demonstration)
  Taenia pisiformis, representative sections (Demonstration)
Miscellaneous
  Methyl cellulose solution
  Beef liver

Free-living Flatworms: Class Turbellaria

A Planarian: *Dugesia*

Freshwater flatworms are often found on the underside of rocks, leaves, and sticks submerged in lakes, ponds, and streams. Although they are usually referred to as "planarians," the most common American freshwater flatworms are members of the genus *Dugesia* rather than the genera *Planaria* or *Euplanaria*.

Behavior and External Anatomy

Observe a living specimen and note its general size and shape. Is the worm uniformly pigmented, or does it have a distinctive pattern of pigmentation? Locate the **head,** the **eyes,** and the **auricles** (lateral projections of the head). The eyes are light receptors but are not capable of forming real images. The auricles are well equipped with touch and chemical receptors. The head region also contains a concentration of nerve ganglia, which have some function in the processing of sensory information and serve as a primitive "brain."

Observe the smooth, gliding locomotion of the worm. This form of locomotion is due to the action of cilia on the ventral surface of the body, coordinated with rhythmic muscular contractions of the body. Note the behavior of the head and the auricles during locomotion. How is thi

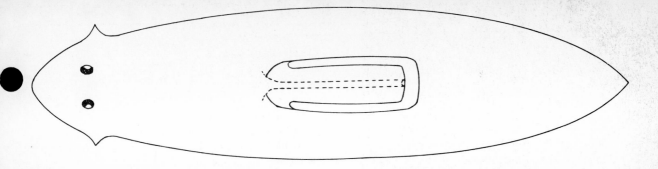

Fig. 8.2 Drawing of *Dugesia*, digestive system.

behavior related to the function of sensory and nervous structures of the head? Touch the head of the worm gently with a clean dissecting needle or a toothpick. How does the worm react? Touch other body regions in a similar way and compare the reaction. Turn the worm over on its back. How does it react? Can you relate your observations to the structure of the head and the concentration of sense organs and nervous elements there? Can you make any conclusions about the possible advantages to an animal of having an anterior head with a concentration of sense organs and nervous tissue?

Place a small piece of fresh beef or pork liver in a dish containing one or more planarians. Observe their feeding behavior. Note the ventral **mouth** and the extension of the protrusible **pharynx** through the mouth when food is located. How does the worm locate the food? What sensory structures may be involved?

Internal Anatomy

Obtain a microscope slide with a stained whole mount of a planarian to study the anatomy further. Review the structures previously noted in the living specimen and also observe the **three-branched gastrovascular cavity.** Note the one anterior and two posterior branches of the cavity and the many smaller lateral branches, or diverticula. Draw the digestive system of the flatworm in the outline of figure 8.2.

Study also a microscope slide with cross sections through the anterior, middle, and posterior portions of the body. Figure 8.3 illustrates such a section through the anterior portion. Locate (1) the **epidermis,** the external layer of cells surrounding the body; (2) the large, vacuolated cells of the **gastrodermis** lining the digestive tract; (3) the layers of **longitudinal** and **circular muscles** lying just inside the epidermis; (4) the large, irregularly shaped cells, the **parenchyma** tissue, which fills most of the interior space of the body; and (5) the two large **ventral nerve cords.**

The external covering of the body is composed of a single layer of epidermal cells, many of which contain densely staining **rhabdites.** Rhabdites are small, rodlike bodies; their function is not fully understood, but they are believed to play a role in the secretion of mucus that aids in the smooth gliding movements of the worm. Also in the epidermis are numerous mucus **gland cells.** Electron micrographs reveal that the epidermal cells also bear cilia and have numerous microvilli extending from their surface.

Study the cross sections of the worm carefully. In figures 8.4 and 8.5 draw the structures that you observe in cross sections through the pharyngeal and posterior regions of the worm. Label all important structures.

Reproduction and Excretion

The reproductive system of *Dugesia* and other freshwater flatworms is small and difficult to observe, except in special microscopic preparations. A few of the reproductive structures are indicated in figure 8.3 (yolk glands, testes, oviducts), but they may be difficult to identify in your slide. Each worm has both male and female sex organs and is, therefore, **monoecious.** Animal species in which the sexes are separate are termed **dioecious.**

Although planarians are monoecious, they are not normally self-fertilizing. In sexual reproduction, sperm is transferred from the male system of one worm by the male copulatory organ, the **penis,** to the **seminal receptacle** of the partner. The sperm subsequently moves to the oviduct of the female reproductive system where fertilization occurs. The fertilized eggs are later deposited outside the body where they develop directly into young worms.

In many species of planarians, however, the most common form of reproduction is **asexual.** A worm separates into two parts, and each part regenerates the missing structures. Planarians have great powers of regeneration, and even relatively small parts of a worm can develop into a complete animal. If time and materials permit, your instructor may be able to help you set up an experiment with regeneration in planarians.

The excretory/osmoregulatory system of *Dugesia* and other planarians is primitive. It consists of a system of **flame bulbs** (a type of protonephridia) interconnected by a system of collecting ducts leading to a posterior excretory pore. The excretory structures are difficult to observe, except in special microscopic preparations. The flame bulbs appear to function mainly in **osmoregulation.** The body fluids and cellular contents are hypertonic to the environment (contain more dissolved salts, etc.); thus, a

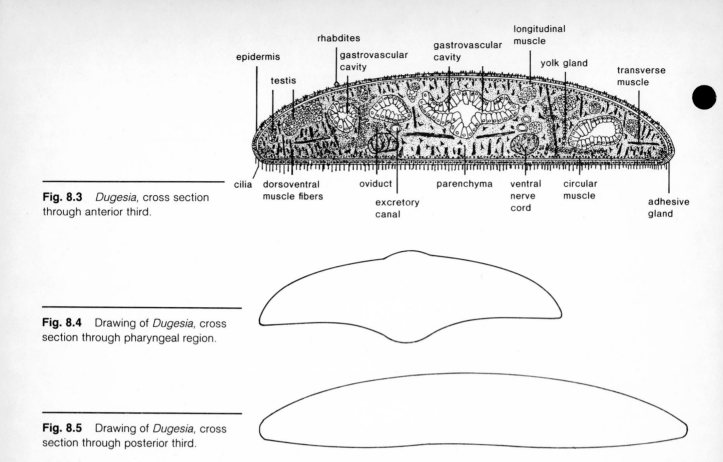

Fig. 8.3 *Dugesia*, cross section through anterior third.

Labels: epidermis, testis, rhabdites, gastrovascular cavity, gastrovascular cavity, longitudinal muscle, yolk gland, transverse muscle, cilia, dorsoventral muscle fibers, oviduct, excretory canal, parenchyma, ventral nerve cord, circular muscle, adhesive gland

Fig. 8.4 Drawing of *Dugesia*, cross section through pharyngeal region.

Fig. 8.5 Drawing of *Dugesia*, cross section through posterior third.

planarian must constantly eliminate excess water from the body. Nitrogenous wastes, resulting from the breakdown of proteins and other nitrogenous matter in the food, are excreted, mainly as ammonia (NH_3), directly from the body cells.

Feeding and Digestion

Planarians, such as *Dugesia,* are chiefly carnivores and typically feed on protozoans and small animals, such as rotifers and small crustaceans. Food is ingested by the protrusible pharynx, which is extended through the midventral mouth. Do not confuse the opening of the muscular pharynx, which is extended through the mouth in feeding, with the actual midventral mouth opening. Proteolytic enzymes, secreted from glands near the tip of the pharynx, aid in penetration of a prey organism, such as a crustacean. The contents of the prey (a *Daphnia*, for example) can then be sucked into the muscular pharynx and passed into the gastrovascular cavity. The digestive system of a planarian is a gastrovascular cavity with a single opening that serves as both the entrance for food and the exit for waste materials. How does this basic organization of the digestive system compare with that of a cnidarian or a higher animal, such as an earthworm or a frog? Which type of system would be more efficient? Why?

Digestion in a planarian is both **extracellular** (outside of the digestive cells) and **intracellular** (inside of the digestive cells). Digestive enzymes are secreted by **gland cells** in the gastrodermis that lines the gastrovascular cavity to assist in the breakdown of food materials. Later, small bits of food are engulfed by **phagocytic cells** in the gastrovascular lining. How many cell types can you identify in the gastrodermis in your cross section?

Muscular System

Other structures you should identify in the cross sections include the **longitudinal** and **circular muscle layers** just beneath the epidermis. Which of these layers lies closer to the epidermis? How can you relate these muscle layers to the locomotion that you observed in the living worms? Contraction of which layer would increase body length? How would this be helpful in locomotion?

Locate also in the cross sections the **dorsiventral muscle bands.** What is their function? Find the loosely packed parenchyma cells that fill most of the interior spaces. How are interior cells nourished? How are wastes removed from them? Near the ventral epidermis find the **two ventral nerve cords.**

The Flukes: Class Trematoda

The Human Liver Fluke: *Clonorchis (Opisthorchis) sinensis*

Members of the Class Trematoda are all parasitic and have well-developed **suckers** for attachment—one located in the region of the mouth, and one located on the ventral surface. The outer covering of the trematode body is highly modified and lacks cilia. The outer layer, the **tegument** (formerly called the cuticle), is a syncytial extension of underlying cells embedded in the body wall. Electron microscope studies have revealed that the tegument has a complex structure. Tapeworms (Class Cestoda) also have a tegument with a similar structure.

The tegument is a nonciliated syncytial tissue that serves an active role both in protecting trematodes from the digestive enzymes of the hosts and in the uptake of nutrients from the host gut. The tegument is an excellent example of morphological and physiological adaptation of a parasite for its very special mode of life.

Clonorchis sinensis (figure 8.7), sometimes also called *Opisthorchis sinensis,* is a common and important human parasite in certain parts of the world, particularly in the Orient. Like many other trematodes, this species has a complex life cycle involving several hosts and a series of larval stages (see figure 8.8).

Obtain a prepared microscope slide with a stained whole mount of an adult fluke and observe its size, shape, and general morphology under your stereoscopic microscope. Observe the **oral sucker** surrounding the mouth at the anterior end. Behind the mouth is a muscular **pharynx**, a short **esophagus**, and two **intestinal caeca**. Note that the digestive tract has a single opening, the **mouth**, and thus is an **incomplete digestive system**. A bilobed **cerebral ganglion** ("brain") lies on the dorsal side of the pharynx (small and difficult to see in most slides). Near the branching of the intestinal caeca, note the **posterior sucker**. Immediately behind the posterior sucker, along the midline of the body, is the long, coiled **uterus** containing many eggs.

Posterior to the uterus lie the many-branched **testes** where the sperm are produced. Along the lateral margin of the body in its midregion, observe the many small **yolk glands**. The yolk glands connect with the ovary by means of two delicate **yolk ducts**. The **ovary** is a single, small structure located near the center of the body. It is connected with the **seminal receptacle**, which serves to store sperm received during copulation.

Study figure 8.8 and the demonstration materials provided to illustrate the life cycle of *Clonorchis*. Note that the life cycle includes parasitic stages in three different hosts: *man, snail,* and *fish.* In order to survive, a parasite with such a complex life cycle including several hosts must have some effective means of transfer from one host to the next. Unless the proper host is available at the appropriate time, the life cycle cannot continue, and the

---

Fig. 8.6 Drawing of *Stenostomum.*

A Simple Microscopic Flatworm: *Stenostomum*

Many microscopic flatworms also commonly live in freshwater ponds, lakes, and streams. These smaller flatworms (Order Rhabdocoela) are more simple in structure than the larger planarians (Order Tricladida) and superficially resemble ciliated protozoans.

Stenostomum is one of the common microscopic flatworms of this type, and members of this genus can be found in almost any pond among leaves, algae, or decaying plant materials. Study some living specimens in a wet mount, or if living material is not available, study a stained whole mount.

Observe the general **size** and **form** of the body, the ciliated **epidermis**, and the anterior **mouth** leading into an oval **pharynx**. Behind the pharynx is a simple tubular intestine; note that the rhabdocoel flatworms do not have the three-branched intestine characteristic of the planarians (triclads). *Stenostomum* reproduces rapidly by transverse fission (asexual reproduction), and you may find chains of two or three animals still linked together.

Make a four-inch drawing of *Stenostomum* in the space provided in figure 8.6 and label all of the principal structures.

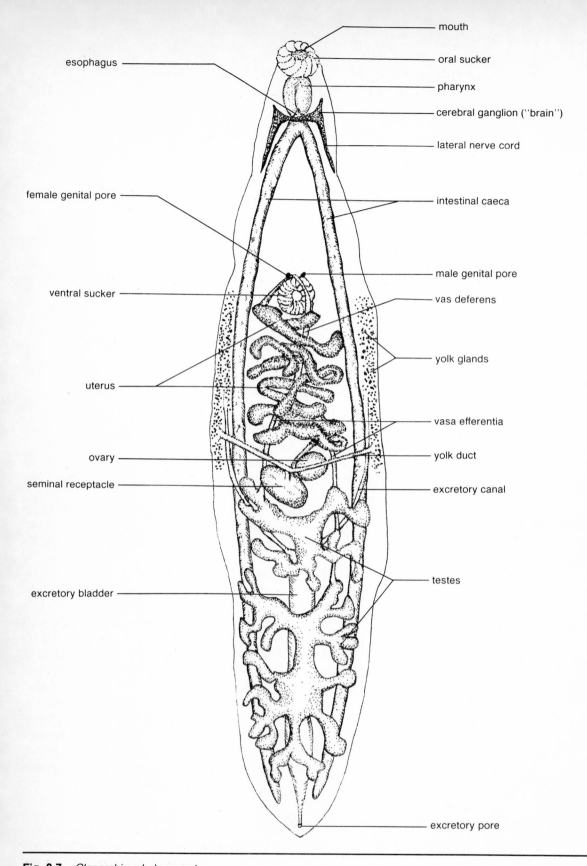

mouth

oral sucker

pharynx

esophagus

cerebral ganglion ("brain")

lateral nerve cord

female genital pore

intestinal caeca

male genital pore

ventral sucker

vas deferens

yolk glands

uterus

vasa efferentia

ovary

yolk duct

seminal receptacle

excretory canal

testes

excretory bladder

excretory pore

Fig. 8.7 *Clonorchis*, whole mount.

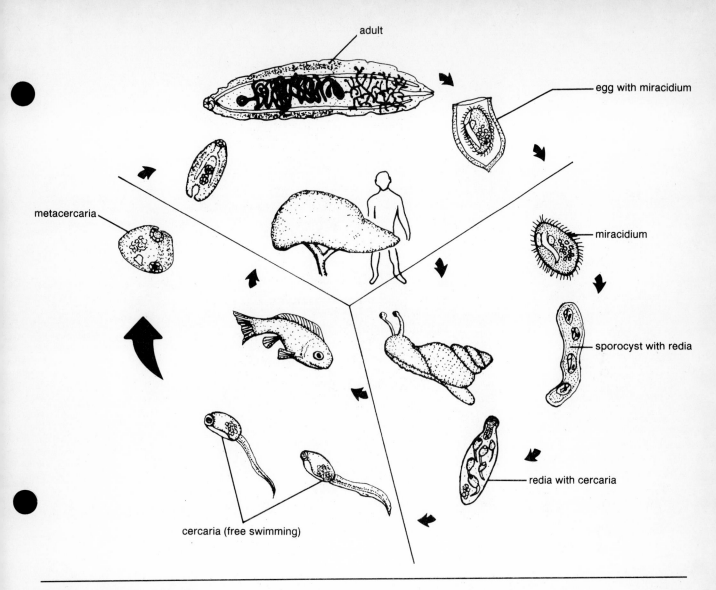

Fig. 8.8 *Clonorchis,* life cycle.

parasite will die. This principle is used as the basis for the control of many parasitic diseases of man and animals, such as malaria and schistosomiasis.

The adult liver fluke lives in the bile duct of man and of certain other carnivorous animals. The host in which the adult stage of a parasite resides is designated as the **definitive host**. All other hosts in the life cycle are termed **intermediate hosts**.

Human infections of *Clonorchis* occur from eating raw fish. Adult worms live in the bile ducts of the liver, and fertilized eggs are released into the bile duct. The eggs pass into the small intestine and are later voided in the feces of the host. If the feces get into water, the eggs may be eaten by certain species of snails (**first intermediate host**). Inside the digestive system of the snail, the egg hatches into a larval form called a **miracidium**. The miracidium lives in the tissues of the snail, passing through several other larval stages (**sporocyst, redia,** and **cercaria**) and reproducing asexually to produce thousands of new

larvae. The last larval stage, the **cercaria,** escapes from the snail and swims in the water until it contacts the **second intermediate host**, certain species of fish. When the fish is contacted, the cercaria burrow through the skin, shed their tails, and encyst to form still another stage, the **metacercaria**. If raw or improperly cooked fish containing metacercaria is eaten by man or another appropriate **definitive host**, the cyst walls are digested, and the metacercaria are released. Subsequently, they migrate into the bile ducts of the liver where they develop into adult flukes to complete the life cycle.

The Sheep Liver Fluke: *Fasciola hepatica*

This large trematode is generally similar in structure to *Clonorchis*, although it is considerably larger in size, and its reproductive system is slightly more complex (figure 8.9). *Fasciola hepatica* is a common parasite of sheep and cattle, but it also occasionally parasitizes other mammals,

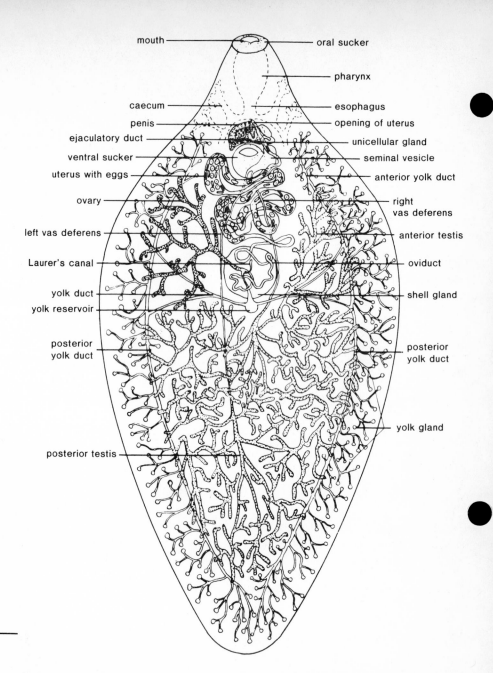

Labels (clockwise from top):
mouth — oral sucker
pharynx
caecum — esophagus
penis — opening of uterus
ejaculatory duct — unicellular gland
ventral sucker — seminal vesicle
uterus with eggs — anterior yolk duct
ovary — right vas deferens
left vas deferens — anterior testis
Laurer's canal — oviduct
yolk duct — shell gland
yolk reservoir
posterior yolk duct — posterior yolk duct
— yolk gland
posterior testis

Fig. 8.9 *Fasciola,* whole mount.

including man. The adult flukes live mainly in the bile ducts and cause the breakdown of the adjacent liver tissue, producing the disease called "liver rot."

Obtain a slide with a stained whole mount of an adult *Fasciola* and note its general shape. How does its shape differ from that of *Clonorchis*? Locate the **anterior sucker** around the mouth and the nearby **ventral sucker**. Between the two suckers, find the **genital pore**.

The digestive system includes the anterior **mouth**, a muscular **pharynx**, a short **esophagus**, and two branches of the **intestine**, each of which has many smaller lateral branches, the **intestinal caeca** (singular: caecum). What would you expect to be the main purpose of the intestinal caeca? Both male and female reproductive systems are present as in *Clonorchis*. With the aid of figure 8.9, locate the principal organs of each system.

The female organs (mainly found in the anterior half of the body) include: **ovary, yolk glands, yolk ducts, yolk reservoir, shell gland, oviduct, uterus** with eggs, and the **opening of the uterus** just inside the genital pore.

The male organs include: **testes** (one anterior and one posterior), **vas deferens** (how many?), **seminal vesicle** (how many?), **ejaculatory duct**, and a muscular **penis**.

___

Demonstrations

1. Slides with adult stages of some other representative trematodes.
2. Slides representing the stages in the life cycle of *Clonorchis* and/or *Fasciola*.
3. Living cercaria.

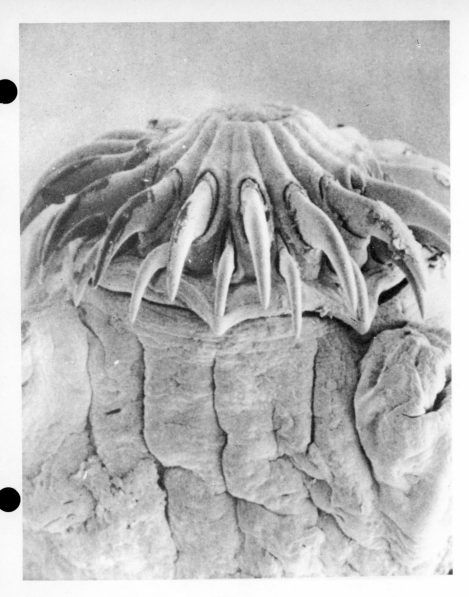

Fig. 8.10 *Taenia taeniaeformis*, from cat. Magnification 300×. (Scanning electron micrograph by Fred. H. Whitaker.)

The Tapeworms: Class Cestoda

The Dog Tapeworm: *Dipylidium caninum*

Tapeworms are highly specialized internal parasites and show several important adaptations for their parasitic mode of life. The adult worms inhabit the intestines of various species of vertebrate animals, and the larvae live in the tissues of some alternate host. In general, the life cycles of tapeworms, or cestodes, are less complicated than those of the trematodes.

The flat, ribbonlike body of a tapeworm is typically divided into many sections called **proglottids**, and the body is divided into three major regions: an anterior **scolex**, a specialized holdfast organ with suckers and/or hooks (figure 8.10); followed by a narrow **neck**, which contains the budding zone where, at the posterior end, new proglottids are produced asexually; and the **strobila**, the rest of the long body, usually consisting of many proglottids.

Although the tapeworms superficially appear to be segmented, the proglottids are not generally believed to represent true body segments because of the way in which they are formed and because each proglottid is a complete reproductive unit within itself. Many zoologists therefore view the body of a tapeworm as comparable to a colony, or a chain of individuals, rather than as being actually segmented.

Obtain a prepared microscope slide or a plastic mount with the scolex and some representative proglottids of the dog tapeworm, *Dipylidium caninum* (figure 8.11). Locate and study the following structures: the **scolex**, the **neck,** and the **strobila**. On the scolex, find the **rostellum**, the raised tip of the scolex, which bears several rows of **hooks**, and four lateral **suckers**. The hooks and suckers aid in attachment to the intestinal wall of the host intestine. Note the absence of a **mouth, pharynx,** and **digestive system.**

Where would you find the youngest proglottids in an intact tapeworm? The oldest? Study several proglottids

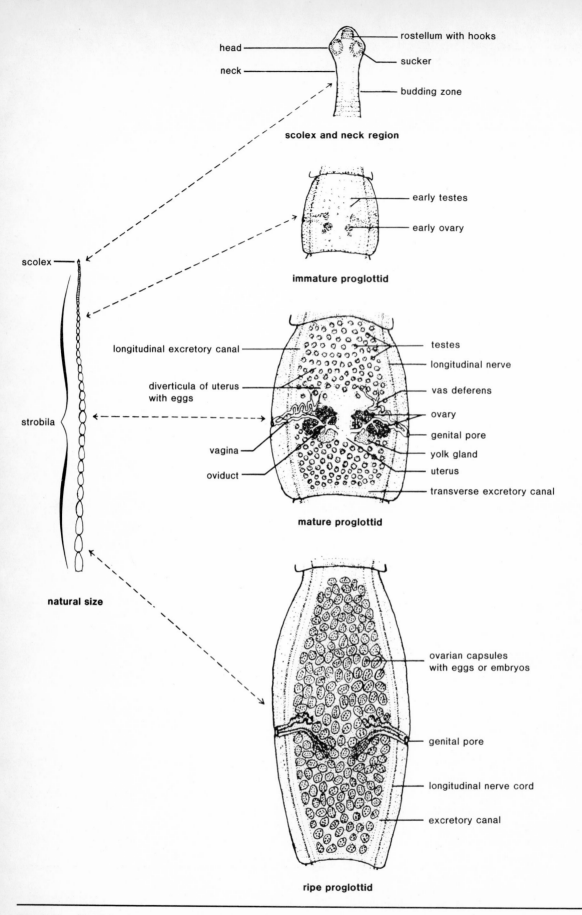

Fig. 8.11 *Dipylidium,* representative sections of body.

Chapter 8

in different stages of development and note the different degrees of elaboration of the reproductive systems. Identify **immature, mature,** and **ripe proglottids** (see figure 8.11).

Select a mature proglottid for more detailed study and identify the **genital pore, vagina, oviduct, yolk glands, uterus, testes, vasa deferentia** (singular: vas deferens), **excretory canals,** and **longitudinal nerve cords.** Note the complete absence of any digestive structures in the tapeworm. How do you suppose tapeworms obtain their nourishment?

Study also some of the ripe proglottids and observe the many ovarian capsules containing several eggs or embryos. Is the number of eggs per capsule always the same? Locate some of the transitional segments between the mature and the ripe proglottids to observe the progressive atrophy of certain of the reproductive organs.

The life cycle of *Dipylidium* includes two larval stages, a six-hooked **onchosphere larva** and a **cysticercoid larva** ("bladder worm"). Ripe proglottids containing eggs and embryos pass out of the host intestine in the feces. The eggs are ingested by flea larvae and hatch into onchospheres inside the intestinal wall. Later they develop into cystercoid larvae as the fleas become adults. A dog or cat may be infected by nipping a flea. Children sometimes become infected with *Dipylidium,* presumably after being licked by a dog or by ingesting eggs deposited in the soil.

A Dog and Cat Tapeworm: *Taenia pisiformis*

Taenia pisiformis (figures 8.12 and 8.13) is another tapeworm commonly studied in zoology laboratories. Adult tapeworms of this species occur in the small intestine of dogs and cats. The larval stages are found in the liver and mesenteries of rabbits.

Obtain a microscope slide and/or plastic mount with a scolex and representative proglottids of *Taenia pisiformis.* Identify the **scolex, neck, hooks,** and **lateral suckers.** Observe that *Taenia pisiformis,* like other tapeworms, has no mouth or digestive system. How does this tapeworm obtain its nourishment?

Locate and identify **immature, mature,** and **ripe proglottids.** Select a mature proglottid on your slide and study its internal structure. Find the **uterus, yolk glands, oviduct, vagina, seminal receptacle,** and **genital pore** comprising the female reproductive system. Among the male reproductive structures, locate the **testes, vas efferens, vas deferens,** and the copulatory organ, the **cirrus.** Also find the lateral **excretory canals.** Two **longitudinal nerve cords** should be visible lateral to the excretory canals.

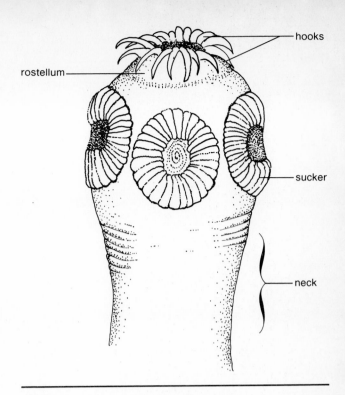

Fig. 8.12 *Taenia pisiformis,* scolex.

Adaptations for Parasitism

Tapeworms provide a good example of the adaptations of animals for a special mode of life. Tapeworms exhibit several structural and physiological features that increase their chances of survival as internal parasites of vertebrates. List six different adaptations related to their parasitic mode of life that you have been able to observe or to identify from your reading.

1. _____

2. _____

3. _____

4. _____

5. _____

6. _____

Demonstrations

1. Microscopic preparations of other tapeworms such as *Taenia solium* (pork tapeworm), *Taeniarhynchus saginatus* (beef tapeworm), and *Dibothriocephalus latus* (broad or fish tapeworm).
2. Onchosphere (six-hooked) larva.
3. Cysticercus larva.
4. Preserved whole tapeworms.

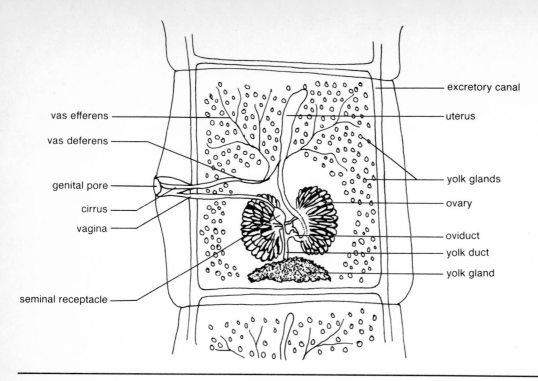

Fig. 8.13 *Taenia pisiformis*, mature proglottid.

Key Terms

Bilateral symmetry body form with parts arranged symmetrically along a central plane.

Cercaria motile, free-swimming, tadpolelike larval stage in the life cycle of certain trematodes. Body resembles miniature fluke. Plural: cercariae.

Definitive host final host in the life cycle of a parasite with more than one; host in which sexually mature parasite develops.

Intermediate host host in the life cycle of a parasite which bears a larval or immature stage of that parasite.

Miracidium ciliated larval stage in the life cycle of many trematodes. Develops from the fertilized egg and gives rise to the sporocyst stage.

Proglottid reproductive body division of a tapeworm. Since each proglottid contains a complete set of male and female organs, these units are often considered analogous to a complete individual, and thus the tapeworm as a colony.

Redia larval stage in the life cycle of certain trematodes. Formed by the sporocyst and produces many cercaria.

Scolex specialized holdfast organ of the cestodes (tapeworms); with hooks and/or suckers for attachment. Plural: scoleces or scolices.

Sporocyst saclike larval stage in life cycle of some trematodes. Develops from miracidium and produces several redia.

Strobila the major portion of the body of a tapeworm excluding the scolex. Includes all of the proglottids.

Tegument living syncytial protective outer covering of the cestode and trematode body; formerly called the cuticle; secreted by large underlying cells with numerous tubular cytoplasmic connections with the tegument.

Triploblastic construction body type with three distinct tissue layers: endoderm, mesoderm, and ectoderm. Characteristic of flatworms and all higher metazoan phyla.

9
Hints for Dissection

Objectives

After completing the laboratory work in this chapter, you should be able to perform the following tasks:

1. Explain the objectives of dissection and the importance of good dissection techniques.
2. List the basic instruments needed for dissection, explain the use of each, and demonstrate their proper use and care.
3. Distinguish between the anatomical terms anterior and posterior; dorsal and ventral; transverse plane, frontal plane, and sagittal plane.

Introduction

Many of the animals to be studied in the following exercises will require dissection for the study of internal structures. To gain the most from these studies it is important that you follow directions carefully and that you develop good dissection skills. Both attributes are essential to your training as a scientist. You must take special care in each dissection to avoid damaging important structures before you have had an opportunity to complete your study.

Some important anatomical terms are illustrated in figure 9.1 and are defined among the **Key Terms** at the end of this chapter. Some understanding of basic anatomy

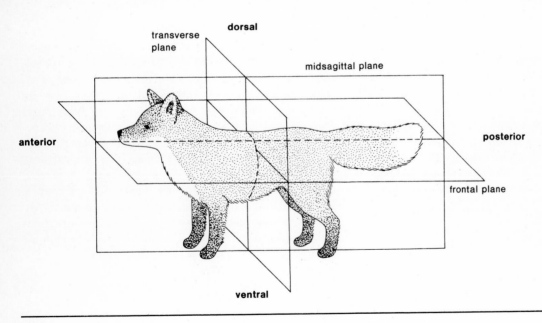

Fig. 9.1 Basic anatomical directions and planes.

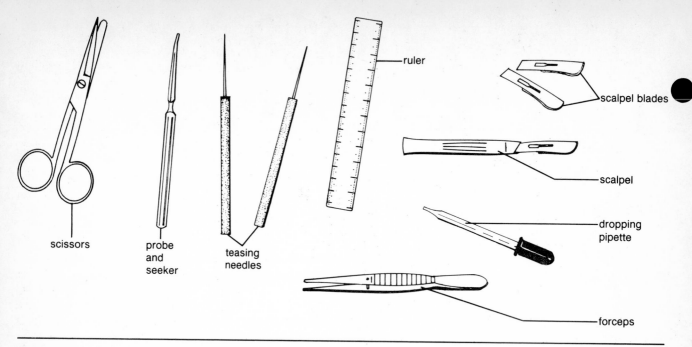

Fig. 9.2 Dissecting instruments.

is essential for success in zoology, regardless of your future specialization. You should review figure 9.1 and the list of **Key Terms** frequently, to reinforce your understanding of anatomy and proper terminology, as you proceed through the remaining exercises in this course.

The chief objective in the dissection of a specimen is to expose body parts for the study of their structure and their relationships to other parts. Therefore, it is important to proceed carefully with each dissection and to avoid cutting or removing anything, unless so instructed in the printed directions or by your laboratory instructor. Read through the instructions completely before you begin each dissection. This will save you time later and will often prevent disappointing results.

Body parts should be parted carefully along natural lines of separation and attachment wherever possible. This can often be done best with forceps, probe and seeker, the handle of a scalpel, or with a finger, without cutting.

Good quality dissection instruments are essential to good laboratory work. Your dissection kit should include the following items:

Scissors, fine points (or one fine and one blunt point), about 4–6 inches long
Forceps, fine points, about 4–5 inches long
Scalpel with replaceable blade
Extra scalpel blades
Two teasing needles
Probe and seeker, about 6 inches long
Plastic centimeter ruler
Dropping pipette

Scissors should be first-quality chrome or stainless steel and have the blades joined by a screw rather than a rivet. Some common dissecting instruments are illustrated in figure 9.2.

Dissecting instruments should be kept clean and sharp at all times. A small oilstone and a piece of emery cloth or fine sandpaper should be used to keep cutting edges and needle points sharp. Two types of scalpel are in common use today. One type has a permanent blade that can be resharpened, and the second type has a separate handle and replaceable blades that are not usually sharpened. Blades on scalpels with replaceable blades should be replaced often to ensure a good edge. Sharp scalpels and scissors allow clean cuts and minimize damage to adjacent tissues. Always wash and fully dry your dissecting instruments with a paper towel after each use. With proper care, a good set of instruments will last for many years.

The scalpel is used for making incisions in the body wall and occasionally for sectioning interior structures. Hold the scalpel upright with the handle nearly perpendicular to the surface to be cut (figure 9.3) and make a clean forward incision while supporting surrounding tissues with your fingers. Be conscious of internal structures and avoid cutting too deeply. If you have to saw with the scalpel to cut the tissue, your blade is too dull.

Forceps and dissecting needles should be used for loosening, lifting, and moving various structures to facilitate study and to expose underlying parts. It is important to gain some idea of the thickness of the covering or surface layer to be cut and its relationship to other important structures nearby before starting your dissection. Consult

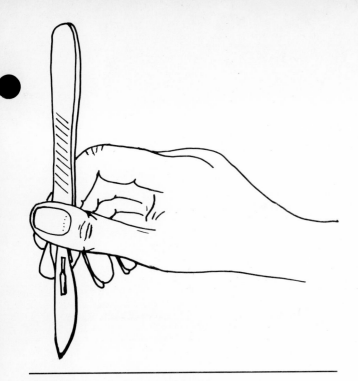

Fig. 9.3 Proper use of the scalpel.

any available charts, photographs, models, and demonstration dissections of the animal to be studied. A few minutes spent in such preliminary orientation often will prevent later disappointment. Also remember that your laboratory work will be evaluated by your instructor partly on the basis of your care and skill in dissecting.

Read the dissection instructions in the laboratory manual and follow them carefully. Always remember that **direct study** of internal anatomy gained from your dissections of the animals is the primary objective. The illustrations in this manual are intended to **help** you in your dissections and study but are **not intended to substitute** for the study of real specimens. Refer to the illustrations frequently, but focus your study primarily on the animal, not on the pictures.

In some of the exercises, you are also asked to make drawings of the animals you study and dissect. Take care to be accurate and neat in your drawings so that they will be helpful to you for later study and review. Remember that your laboratory notes and drawings serve a purpose similar to your lecture notes. They provide a record to help you to recall actual structures during your later study and reviews. They help you to retain and to recall mental pictures of the things actually seen through the microscope and during your dissections.

You should also use the pages provided for *Notes and Sketches* at the end of each chapter for any additional things that you observe or need to remember.

Key Terms

Aboral   away from or opposite to the mouth.

Anterior   toward the front or head end; the opposite of posterior.

Asymmetry   an irregular arrangement of body parts; without a central point, axis, or plane of symmetry.

Bilateral symmetry   an arrangement of body parts on opposite sides of a central plane (midsagittal plane), which divides the body into two symmetrical halves (mirror images).

Caudal   toward the tail or tail end; the opposite of cephalic.

Cephalic   of or pertaining to the head; the opposite of caudal.

Cranial   relating to the skull or cranium.

Cross section   sections of the body cut on any transverse plane; such sections are perpendicular to the sagittal and frontal planes.

Deep   pertaining to structures away from the surface of the body; the opposite of superficial.

Distal   away from the center or point of attachment; the opposite of medial.

Dorsal   relating to the back or upper surface; the opposite of ventral.

Frontal plane   plane parallel to the dorsal and ventral surfaces of the body, which bisects a bilaterally symmetrical animal into upper and lower halves.

Lateral   toward the side; the opposite of medial.

Longitudinal   lengthwise; parallel to the long axis.

Medial   toward the sagittal plane or center of the body; the opposite of lateral.

Median   located in or near the sagittal plane.

Oral   relating to the mouth.

Peripheral   toward the outer surface.

Posterior   the hind part (rear) of the body; the opposite of anterior.

Proximal   toward the center or point of attachment; the opposite of distal.

Radial symmetry   arrangement of body parts symmetrically around a central axis; any plane through the central axis divides the body into symmetrical halves (mirror images).

Sagittal plane   any longitudinal plane passing from the head to tail. The midsagittal plane bisects a bilateral animal into two symmetrical halves (mirror images).

Superficial   located near the surface of the body; the opposite of deep.

Transverse plane   any plane perpendicular to the sagittal and frontal planes. Sections of the body cut on a transverse plane are called cross sections.

Ventral   relating to the belly or underside; the opposite of dorsal.

10
Pseudocoelomate Animals

Objectives

After completing the laboratory work in this chapter, you should be able to perform the following tasks:

1. List the seven animal phyla commonly grouped together as pseudocoelomate animals and list five of the characteristics they have in common.
2. Describe the external morphology of the roundworm *Ascaris* and locate the chief external features in preserved specimens. Distinguish between mature male and female specimens.
3. Describe the male and female reproductive systems of *Ascaris* and identify the principal reproductive organs in preserved specimens.
4. Describe the anatomy of the rotifer *Philodina* and locate its principal organs in a specimen.
5. List and briefly explain the importance of five other pseudocoelomate animals.

Introduction

The roundworms, Phylum Nematoda (Nemathelminthes), are sometimes studied together with several other animal groups in a large heterogeneous Phylum Aschelminthes, but recent investigations have led most zoologists to consider these animal groups as separate phyla because of their many differences. The most prominent similarity among these groups is the pseudocoelom, a central body cavity lying between the endoderm and mesoderm layers, derived from the embryonic blastocoel. Other common features include a cylindrical body with an external cuticle, microscopic size, a complete digestive tract with a mouth and an anus, a specialized pharynx, a protonephridial excretory system, and a fixed number of cells (or nuclei) in the adult. The latter condition results from the cessation of mitoses in the adult, thus limiting the number of cells (or nuclei) present in each mature individual.

Numerous exceptions to these "common" features have been found and have led to the present uncertainty about the true relationships among these animal groups. For instance, recent electron microscopic investigations have shown that some of these groups have a much reduced body cavity, and in some cases this cavity may not actually be a pseudocoelom. Much more study is needed on these enigmatic animals.

Seven groups (or phyla) of animals are commonly recognized in this assemblage, including the Nematoda, Rotifera, Gastrotricha, Kinorhyncha, Nematomorpha, Acanthocephala, and the Entoprocta (see figure 10.1). An eighth phylum, the Gnathostomulida, is sometimes included, but many zoologists believe that the Gnathostomulids are actually more closely related to the flatworms (Phylum Platyhelminthes) than to the pseudocoelomates.

Classification

The seven phyla most commonly grouped among the pseudocoelomates are briefly described below.

Phylum Nematoda (Nemathelminthes or Roundworms)

Free-living and parasitic worms with an **elongate cylindrical body tapered at both ends**, a **triradiate pharynx**, and a **specialized excretory system.**

Phylum Rotifera (Rotifers)

Microscopic animals, found primarily in freshwater habitats, with an anterior ciliated locomotory and feeding organ, the **corona**, and a specialized internal grinding organ, the **mastax.**

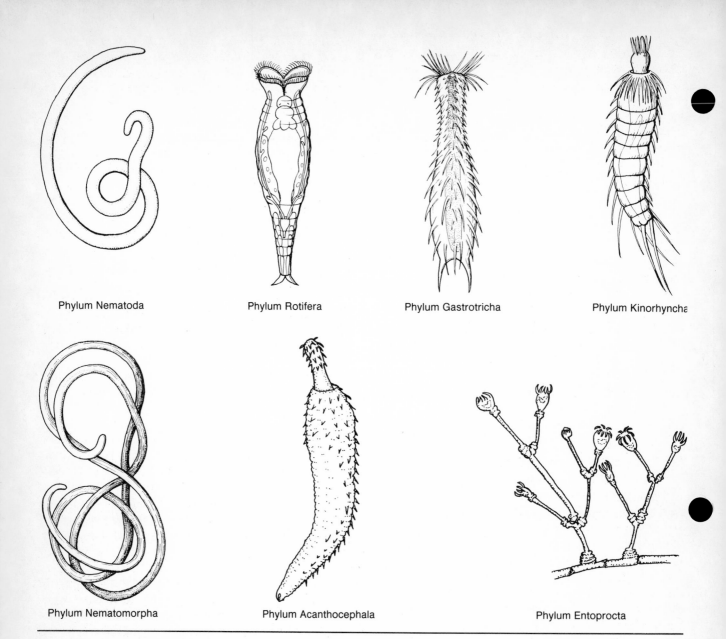

Phylum Nematoda

Phylum Rotifera

Phylum Gastrotricha

Phylum Kinorhyncha

Phylum Nematomorpha

Phylum Acanthocephala

Phylum Entoprocta

Fig. 10.1 Chief phyla of pseudocoelomate animals.

Phylum Gastrotricha (Gastrotrichs)

Microscopic aquatic animals with an **external cuticle** covered by scales or spines, and **patches or tracts of cilia** on the epidermis of the ventral surface.

Phylum Kinorhyncha (Kinorhynchs)

Microscopic marine animals with strong **superficial segmentation** (segmented cuticle), a **retractile head with spines,** and **lateral spines** along the body.

Phylum Nematomorpha (Gordiacea or Horsehair Worms)

Slender, elongate worms with a **cylindrical body rounded at the ends**, a **reduced digestive tract**, and **larvae parasitic in insects** or other arthropods.

Phylum Acanthocephala (Spiny-headed Worms)

Parasitic worms with adults living in the intestines of vertebrates and larvae in arthropod hosts; **anterior proboscis with hooks** for attachment to gut wall of host; **no digestive tract.**

Phylum Entoprocta (Entoprocts)

Tiny, sessile animals with a **cup-shaped body** attached to the substrate by a **stalk,** solitary or colonial, with a **U-shaped digestive tract** and a **crown of ciliated tentacles** surrounding the mouth and the anus. Most species are marine.

Chapter 10

Living Specimens
  Anguilla aceti
  Philodina
Preserved Specimens
  Ascaris
Microscope Slides
  Ascaris, cross sections of male and female
  Ancylostoma, whole mount
  Chiloplacus, free-living nematode (Demonstration)
  Trichinella, whole mount (Demonstration)
  Enterobius, pinworm, whole mount (Demonstration)
  Representative Rotifera (Demonstration)
Plastic Mounts
  Ascaris, male and female worms (Demonstration)
  Macracanthorhynchus, acanthocephalan, parasite of swine
  (Demonstration)
  Ancylostoma, hookworm (Demonstration)

Phylum Nematoda

The Nematoda and Rotifera are the largest, best known, and most important among these phyla. Some of the common representatives of these two phyla are included in this laboratory exercise to illustrate some of the important features of pseudocoelomate animals.

A Parasitic Roundworm: *Ascaris lumbricoides*

External Anatomy

Ascaris lumbricoides (figure 10.2) is a large roundworm parasite of humans and swine. The adult worms range in length from about six to sixteen inches. Living specimens are yellow or pink in color, but preserved specimens usually range from pink to brown. In this exercise, you will study a preserved specimen to determine some of the characteristic features of roundworms, members of the Phylum Nematoda.

Observe the elongate cylindrical body that tapers at both ends. The anterior end is more pointed than the posterior, which tends to be slightly blunt. The triangular **mouth**, surrounded by three distinct **lips**, is located at the anterior end of the body (see figure 10.2). Under your dissecting microscope observe the three lips and the mouth. Also locate the **anus**, which appears as a transverse slit on the ventral side of the body near the posterior end. The ventral location of the anus is a convenient means of distinguishing the dorsal and ventral surfaces of the worm. You should be able to find four longitudinal lines extending lengthwise along the body wall: one dorsal, one ventral, and two lateral in position. These longitudinal lines are more apparent in living specimens than in preserved specimens, and their significance will be better understood after your study of the microscopic anatomy of the worm later in this exercise.

A further examination of the external features of your specimen should reveal its sex. Most species of nematodes are **dioecious** (sexes separate), and the males and females usually exhibit certain external morphological differences. Male specimens of *Ascaris lumbricoides* typically exhibit a curved posterior end and a pair of **copulatory spicules** extending from the anus (see figure 10.2). Female worms have a ventral **genital pore** about one-third the distance from the anterior end. From your observation of the location of the copulatory spicules in the male, what can you conclude about the location of the male genital opening? Examine several specimens of *Ascaris lumbricoides* to determine whether you can find any other morphological differences between the sexes.

Internal Anatomy

After you have completed your study of the external features of *Ascaris*, carefully dissect the worm by making a longitudinal slit down its dorsal side. Take care with your incision to avoid damaging the internal organs. Pin down the body wall on each side of the worm in a wax-bottom dissecting pan to expose fully the internal structures, and cover your specimen with a little water to keep the internal organs moist and flexible.

Caution: Although the *Ascaris* specimens you will receive in the laboratory have been preserved in formalin, the eggs found in the female specimens are enclosed in a very resistant "shell." They may not all be killed, even after many weeks in formalin. Be careful to avoid putting your hands in your mouth during your dissection and wash your hands thoroughly after you complete your work.

Observe the long, ribbonlike **intestine** extending most of the length of the body and the numerous tangled threadlike coils surrounding the middle part of the intestine (see figure 10.2). The female reproductive system is Y-shaped, and each branch of the Y consists of a long, threadlike **ovary** connected with a larger **oviduct** and a still larger **uterus**. The distinction among these three organs is not apparent externally but is easy to observe in microscopic cross sections. The two uteri join to form a single midventral muscular tube, the **vagina**, which opens to the exterior via the female genital pore, or **vulva**, observed previously.

The male reproductive system is not branched, as in the female reproductive system, and it consists of a single, continuous tubular structure, including a threadlike **testis** connected with a **vas deferens**, or sperm duct, larger in diameter, a still-larger-in-diameter **seminal vesicle**, and a short, muscular **ejaculatory duct**, which empties into the **cloaca** at the posterior end of the digestive tract.

The digestive system consists of the ribbonlike **intestine**, which is attached anteriorly to a short muscular **pharynx** into which the mouth opens, and posteriorly to the **cloaca**.

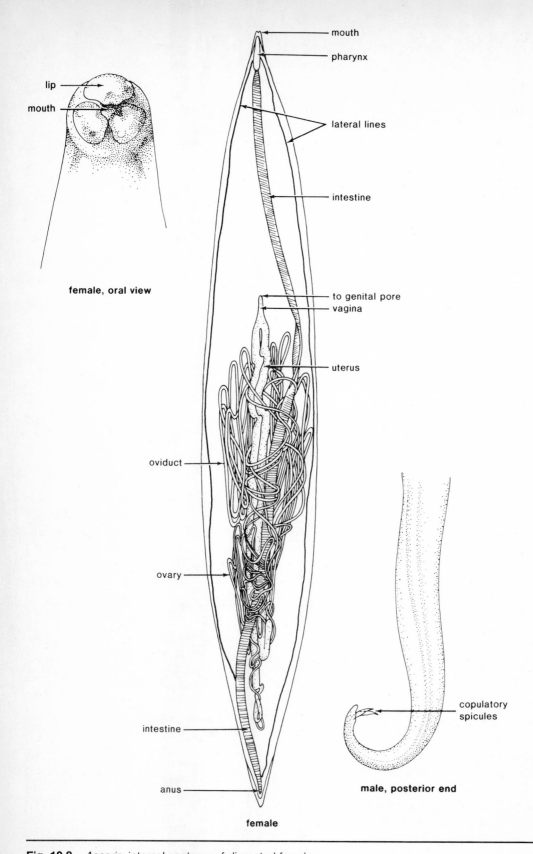

female, oral view

female

male, posterior end

Fig. 10.2 *Ascaris,* internal anatomy of dissected female plus selected external features of male and female.

Chapter 10

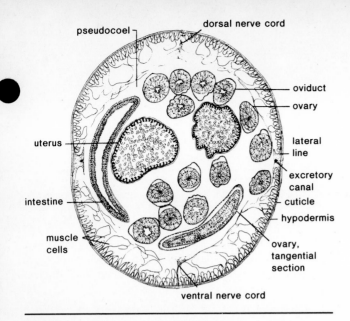

Fig. 10.3 *Ascaris*, cross section, female.

Microscopic Study

Prepared microscope slides with cross sections through male and female worms illustrate further aspects of nematode structure (figure 10.3). Examine a cross section of a female worm and identify the thick **cuticle** covering the body and the underlying layer of **hypodermis**, which secretes the cuticle. Observe the many **longitudinal muscle bands**; there are no circular muscles present in roundworms. This anatomical feature produces the peculiar whiplike movements often exhibited by living roundworms. Free-living nematodes are sometimes called "whipworms." Also associated with the body wall are the **dorsal** and **ventral nerve cords** and the two **lateral lines**. Locate within each of the lateral lines the **excretory canal**, part of the peculiar specialized excretory system of the nematodes. Observe the thin-walled intestine and note that the intestinal wall consists of a single layer of large **gastrodermal cells**. The body cavity of *Ascaris* is a **pseudocoelom**, since it lacks a mesodermal layer around the gut.

In the cross section of the worm, the female reproductive system should be represented by numerous round or oval structures of various sizes. The **ovaries** are small in diameter and bear some resemblance to tiny wagon wheels. A layer of columnar ovarian cells surrounds a central core, or **rachis**. The **oviducts** are slightly larger in diameter than the ovaries and have a central opening, or lumen, for the passage of eggs. If the section was cut through the region of the vagina, you could observe a single, large median structure, usually filled with eggs. Sections posterior to the vagina region usually show two large uteri, also filled with eggs.

Cross sections of male specimens (figure 10.4) exhibit features similar to female specimens except for the

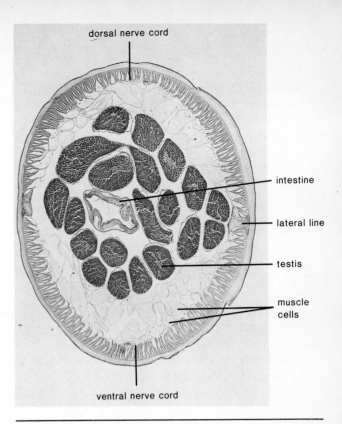

Fig. 10.4 *Ascaris*, cross section, male. (Courtesy Carolina Biological Supply Company.)

reproductive structures. The reproductive system is usually represented by several sections through the **testes**, which are small round structures containing many spermatogonia. Several sections through the larger **vas deferens** should also be present in the section. You should find numerous spermatocytes in the vas deferens of mature male specimens. A large **seminal vesicle** containing mature spermatozoa should also be present.

Life Cycle

Ascaris lumbricoides has a direct life cycle with no intermediate hosts. Mature female worms in the intestine of the host produce eggs that are expelled in the feces. The fertilized eggs begin to develop in the soil and may be ingested by a new host animal. Hatching occurs in the small intestine and the larvae emerge, burrow through the intestinal wall, and are carried to the liver of the host via the hepatic portal vein. From the liver the larvae migrate to the lungs. Later they move up through the bronchi and trachea and reach the oral cavity. From there the worms are swallowed and are carried to the small intestine. Sexual maturity is reached within about two months, when fertilized eggs begin to be released. Fertilization occurs after copulation between male and female worms in the host's intestine.

A Parasitic Roundworm: *Trichinella spiralis*

Another common parasitic nematode is *Trichinella spiralis*, the trichina worm. *Trichinella* is the causative agent of the disease **trichinosis** and occurs commonly in humans, swine, rats, and other mammals. The disease is usually contracted by humans through eating raw or inadequately cooked pork.

Observe a microscope slide showing encysted worms in the skeletal muscles of a pig or a rat. Encysted worms in infected pork may be released through digestion of the cyst walls by enzymes in the human digestive tract. The juvenile worms mature sexually in about two to three days and then mate. The females burrow into the wall of the small intestine and shortly begin to produce living offspring. A single female may produce up to 1,500 young in her three-month life span. These juvenile worms enter the host's lymphatic system and are distributed through the body via the circulatory system. The young worms burrow mainly in the active skeletal muscles, where they grow to about 1mm in length and become enclosed in cysts. Although such encysted worms cannot continue to develop unless the infected tissue is eaten by another suitable host, the encysted worms may remain alive within the cysts for several years.

Hookworms: *Ancylostoma duodenale* and *Necator americanus*

Ancylostoma duodenale and *Necator americanus* are the two most common and most serious human hookworms. *A. duodenale* is predominantly a northern species common in Europe, North Africa, northern China, and Japan, while *N. americanus* is primarily a tropical species occurring in many warmer areas of the world. Until recent years *N. americanus* was a very serious public health menace in the southeastern United States. Improved sanitation and control measures in the past few decades, however, have greatly reduced the severity of the hookworm problem.

Both species are principally human parasites, but are sometimes also found in other animal species. Both species also exhibit specialized **teeth** and/or **cutting plates** around the mouth (see figure 10.5), which aid in the attachment of the adult worms to the wall of the small intestine. The body of *N. americanus* is smaller and more slender than that of *A. duodenale*.

Life Cycle

The life cycles of the two species are basically similar. The adults live in the small intestine, where they attach to the intestinal mucosa and feed on blood and tissue fluids. Adult females release large numbers of eggs each day (5,000–20,000). The eggs are released in the host feces, and the embryonated eggs develop and hatch in the dung. The young, free-living larvae feed on bacteria and other debris, grow, molt several times, and reach the infective stage in about five days. During this time they live in the

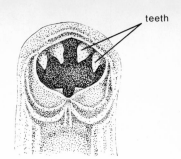

Fig. 10.5 *Ancylostoma*, mouth and buccal cavity showing teeth.

upper layer of soil and frequently migrate to the surface. They are attracted to heat and are stimulated by physical contact. These attributes facilitate contact and penetration of exposed human skin—like a bare foot.

The larvae burrow through the bare skin until a blood or lymph vessel is reached. In the bloodstream they are carried to the heart and then to the lungs. There they burrow into the alveoli. From the lungs they are passed up the bronchi and trachea to the throat. From the throat they may be swallowed and pass through the digestive tract to the small intestine, where they undergo a final molt and attach to the mucosa. Sexual maturity is reached in about six weeks.

A Free-living Roundworm: *Anguillula aceti*

The vinegar eel (figure 10.6), *Anguillula aceti (Turbatrix aceti)*, is a tiny, free-living nematode sometimes occurring abundantly in vinegar. It was more common in the past than it is today, since most commercial vinegar is now pasteurized and also has preservatives added to prevent the growth of vinegar eels and other undesirable flora and fauna. The worms are most abundant in the bottom sediments of unpasteurized vinegar and other fermented fruit juices. Vinegar eels thrive in such acid conditions and feed on the yeast and bacteria growing in the sediment. This combination of yeast and bacteria is often called the "mother" of the vinegar.

Examine a sample from the culture of the worms and observe their active, whiplike swimming movements. How can you relate their mode of locomotion to what you have learned about the anatomy of the nematodes? What kind of muscles are involved?

Select a few large worms for further study and place them in a small drop of vinegar on a clean microscope slide. If the worms are too active for study, they may be killed by briefly warming the slide over an incandescent lamp or a small flame. The specimens may then be observed unstained, or they may be lightly stained with hematoxylin or Nile blue sulphate in 70% alcohol.

Observe the blunt anterior end with its terminal mouth and the pointed posterior end. What is the location of the anus? Identify the straight digestive tract consisting of the **mouth, pharynx, pharyngeal bulb** containing

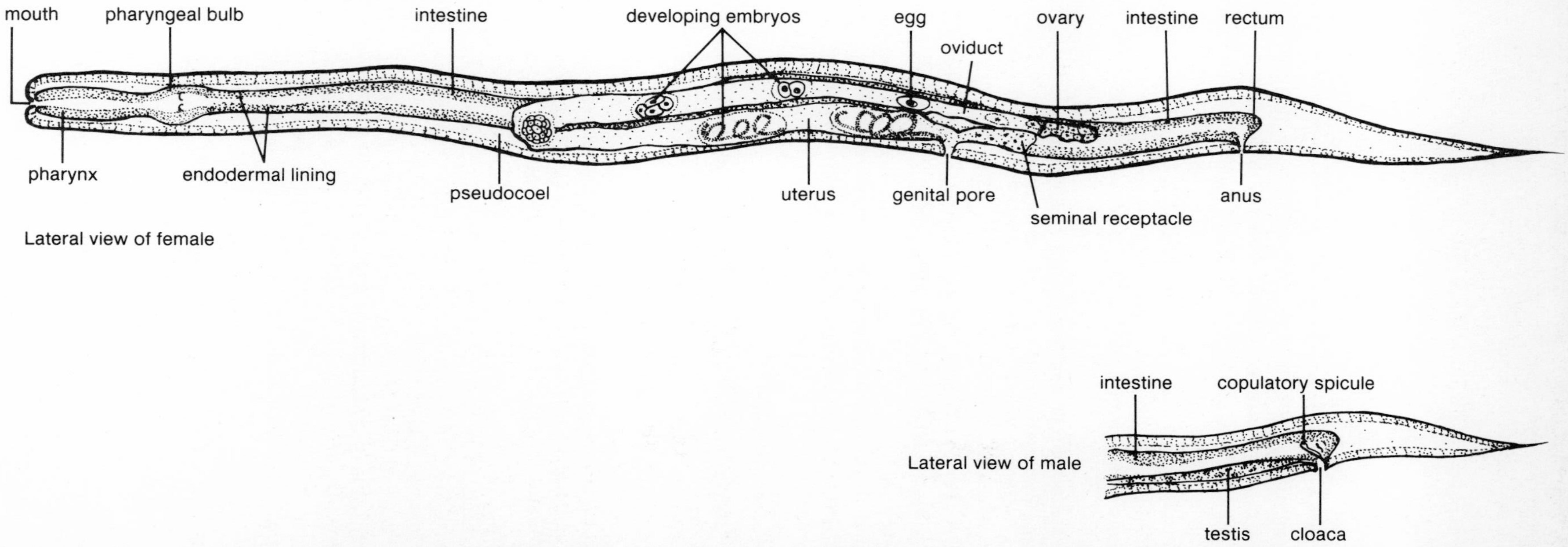
mouth
pharyngeal bulb
intestine
developing embryos
egg
oviduct
ovary
intestine
rectum
pharynx
endodermal lining
pseudocoel
uterus
genital pore
seminal receptacle
anus
Lateral view of female
intestine
copulatory spicule
Lateral view of male
testis
cloaca

Fig. 10.6 *Anguillula aceti,* the vinegar eel.

a three-part pharyngeal valve, an **intestine**, a thin-walled **rectum**, and a subterminal **anus**. In male specimens, the rectum opens into a **cloaca**, which in turn opens through the anus.

Select a large female worm for the study of the reproductive system. Identify the **female genital pore** on the ventral side of the body. This genital opening serves both for receiving sperm during copulation and for the release of the offspring, which are born live in this species. Extending posteriorly from the genital opening is a blind sac, the **seminal receptacle**, believed to serve for the temporary storage of sperm after copulation. Anterior to the genital pore is a single large **uterus**, which may contain several developing young. At its anterior end (about one-third the distance from the anterior end of the body), the uterus is attached to a smaller **oviduct**, which bends posteriorly and connects with the threadlike **ovary**, where the eggs are produced.

Males may be recognized by their **copulatory spicules** and by the different structure of their reproductive system. Observe the opening of the **genital pore** into the **cloaca**, the small spherical spermatozoa in the **vas deferens**, and the filamentous **testis**, which may extend almost to the middle third of the body, where it terminates in a slight enlargement. Observe a demonstration of a male worm that has been treated with strong formalin that causes the copulatory spicules to project from the anus.

Other Free-living Roundworms

Many species of free-living roundworms live in soil, decaying vegetation, and in many aquatic habitats, such as lakes, ponds, and streams. An easy way to find aquatic roundworms is to collect small bits of decaying plants from the edge of a pond or stream. Place the plant material in a small finger bowl and gently tease apart the material with dissecting needles and/or forceps. You should be able to identify the nematodes under a stereoscopic microscope by observing their shape and characteristic whiplike movements. A stereoscopic microscope equipped with a substage lamp to provide transmitted light through the sample works best.

You may also wish to examine some small samples of soil rich in organic material, "black dirt," in a similar manner for soil nematodes. Pieces of a raw or boiled potato may be used for bait. Place the potato pieces in a moist place in a garden, under a board or rock, or beneath a log in the woods for a few days. You can then tease apart the potato bait to search for nematodes immediately or incubate the pieces for a few days in a sterile petri dish or test tube in a warm place. The latter procedure will enable any eggs and/or larvae to develop.

Phylum Rotifera

A Rotifer: *Philodina*

Rotifers are common and abundant freshwater animals. They are important constituents of the plankton of lakes, ponds, and streams, and are significant as a food source for many species of fish and other filter-feeding animals.

Philodina (figure 10.7) is a common North American rotifer that illustrates many of the typical features of the Phylum Rotifera.

Obtain a living specimen of *Philodina* and prepare a wet mount on a microscope slide to study its behavior and appearance. Living specimens are much superior to fixed and stained specimens for the study of general anatomy of rotifers because of the serious contraction and distortion almost always encountered during their fixation and staining.

Observe the elongate cylindrical body, which can be divided into three general regions—the **head**, the **trunk**, and a posterior **foot**. Superficially, the body is divided into several segments, usually about sixteen. The rotifers have no true segmentation, but the cuticle covering the body is often divided into a number of superficial segments. Observe the telescoping of the segments when the animal retracts its head. Note also the large ciliated corona at the anterior end. Most rotifers have a conspicuous **corona**, which serves both for locomotion and for feeding. In *Philodina*, the corona consists of two large ciliated **trochal discs** and a posterior band of cilia, the **cingulum**. Special **retractor muscles** serve to retract the corona. Observe also, on the head, the dorsal fingerlike process, the **rostrum**. The rostrum is believed to be a sensory structure. Observe the long, tapering **foot** with two **spurs** and four retractable **toes** at the posterior end of the body.

The body wall is covered externally by a thin **cuticle** secreted by a syncytial **hypodermis**. Cell membranes are generally lacking in the organs and tissues of adult rotifers, and specific organs typically have a fixed number of nuclei. Although there are no definite muscle layers associated with the body wall of rotifers, several distinct **muscle bands** connect various parts of the body. As in members of related phyla, the central body cavity of rotifers is a **pseudocoelom**.

The **mouth** is a funnel-shaped opening located at the base of the corona. It receives food particles swept down by ciliary currents created by the corona. Connecting with the mouth is a muscular **pharynx**, which encloses a conspicuous grinding apparatus, the **mastax**. The mastax is a distinctive feature of the rotifers and can be easily identified in a living rotifer because of its active, almost constant movement. In *Philodina*, the mastax is specialized for grinding plankton, periphyton, and detritus, but other

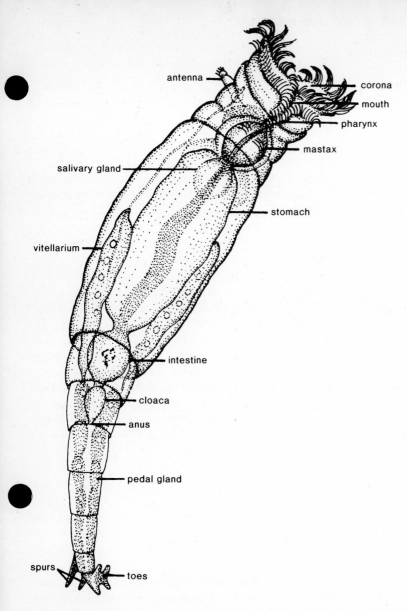

Labels on figure:
antenna
corona
mouth
pharynx
mastax
salivary gland
stomach
vitellarium
intestine
cloaca
anus
pedal gland
spurs
toes

Fig. 10.7 *Philodina*, a rotifer.

types of mastax are found in different species of rotifers, and their structure appears to be closely related to the food habits of the various species.

Posterior to the pharynx is a short **esophagus** surrounded by the large **salivary glands**. The esophagus opens into a large, thick-walled **stomach** where digested food is absorbed by the gastrodermal lining. Behind the stomach is a short **intestine** leading into the **cloaca**. The cloaca empties through the **anus**.

Most of the reproductive system of *Philodina* can be seen only in special preparations, but a pair of large, yolk-forming glands, the **vitellaria**, are readily visible in living specimens. Observe the two large vitellaria in the posterior part of the trunk region of your specimen. How many transparent nuclei can you identify? The vitellaria are syncytial like most other rotifer organs and have a fixed number of nuclei. Other parts of the reproductive system

include two small anterior **ovaries** (difficult to see), which connect via small ducts to the large vitellaria, and two tiny oviducts that lead from the vitellaria to the cloaca.

Males are unknown in the group of rotifers to which *Philodina* belongs (Order Bdelloidea), and the eggs produced develop exclusively by parthenogenesis. Certain other types of rotifers, however, do exhibit normal sexual reproduction.

Other Pseudocoelomates

Phylum Nematoda

Enterobius vermicularis, pinworm. A microscopic nematode especially common as a parasite of young children.

Rhabditis maupasi. A small nematode found in the soil and often as a parasite in earthworms.

Wuchereria bancrofti, filaria worm. Important human parasite in many warm parts of the world. Causative agent of the disease elephantiasis, which is characterized by a massive accumulation of fluids after the blockage of lymphatic circulation, causing great enlargement of the limbs or other parts of the body.

Heterodera schactii, the sugar beet nematode. A specialized plant-eating nematode that feeds on the root of the sugar beet.

Phylum Rotifera

Epiphanes senta. A common, relatively large rotifer often cited in textbooks as a representative rotifer. Exhibits a single median ovary and vitellarium; males are small and simplified in structure.

Asplanchna. This genus includes several species of planktonic rotifers with a saclike body, an incomplete digestive tract, and a well-developed corona.

Phylum Nematomorpha

Paragordius. Widely occurring genus of horsehair worms. Adult worms found in marshes, ponds, streams, and lakes, often in tangled masses of several worms. Larvae occur as internal parasites of beetles, grasshoppers, and crickets.

Demonstration

Specimens of representative pseudocoelomates.

Key Terms

Cloaca   a chamber at the posterior end of the digestive tract in certain types of animals that serves to receive wastes from the digestive tract, gametes from the reproductive tract, and/or wastes from the excretory system.

Corona   a specialized organ at the anterior end of rotifers, usually ciliated, which often serves for locomotion and in food gathering. The circular motion of the coronal cilia in many species of rotifers provides the basis for the common name, "wheel animals," often given to rotifers.

Cuticle   a thick, noncellular protective covering secreted by an underlying layer of epithelial cells, the hypodermis.

Hypodermis   a type of tissue made up of epidermal cells that secrete an overlying cuticle.

Mastax   a specialized grinding organ found in the pharyngeal region of rotifers.

Pseudocoelom   a type of internal body cavity that lacks a mesodermal lining around the digestive tract. It arises embryonically as a remnant of the blastocoel.

Trichinosis   a parasitic disease of humans and certain other carnivorous mammals caused by the nematode parasite *Trichinella spiralis*. Usually acquired by eating rare or uncooked pork.

11
Mollusca

Objectives

After completing the laboratory work in this chapter, you should be able to perform the following tasks:

1. List and briefly characterize the seven classes of living molluscs. Give an example of each.
2. Describe the morphology of the shell of a freshwater mussel and locate the principal structures on a specimen.
3. Identify the principal internal organs of a freshwater mussel on a specimen and briefly explain the function of each.
4. Explain the pattern of water flow through a freshwater mussel and its importance.
5. Explain the reproduction and life cycle of a freshwater mussel.
6. Describe the morphology and behavior of *Physa* or a similar freshwater gastropod.
7. Identify the chief external features of the squid and show four morphological features illustrating adaptation for its specific mode of life.
8. Explain the concept of adaptive radiation and how the molluscs demonstrate this concept.

Introduction

Molluscs, a large and diverse group of animals, include the chitons, clams, oysters, snails, slugs, squids, nautili, and octopi. Most of the 100,000 plus species of molluscs are marine, but there are also many freshwater species, as well as several species that have adopted a terrestrial mode of life. Molluscs are widely distributed and can be found at great depths in the ocean, in virtually all types of freshwater and estuarine habitats, and in many terrestrial environments.

Members of this phylum have soft, unsegmented bodies, which usually are enclosed, wholly or in part, by a thin fleshy layer, the **mantle.** The mantle usually secretes a hard **shell.** In some of the more specialized molluscs, the shell has been lost, reduced, or become embedded in the soft tissue.

Molluscs are triploblastic coelomate animals, as are the annelids and all other higher metazoans, including the chordates. The **coelom** is the principal internal body cavity in most of these groups and is completely lined by mesodermal tissue. The coelom arises during embryonic development as a cavity within the developing mesoderm. In most higher metazoans, the coelom expands greatly to become the principal internal cavity and serves many important functions. Among its typical functions are to provide space for the development of internal organs, to serve in the temporary storage of metabolic wastes, to provide space for the temporary storage of gametes, and to provide a hydrostatic (fluid) skeleton to facilitate the movements and burrowing of soft-bodied animals.

In modern molluscs, however, the coelom is reduced largely to the cavities surrounding the heart, gonads, and excretory organs (nephridia). Abundant evidence indicates that molluscan ancestors had a more prominent and spacious coelom. Modern annelids have a large and well-developed coelom and illustrate well the importance of the coelom in most higher metazoans.

The molluscan body typically consists of three major parts: an anterior **head,** a ventral **foot,** and a dorsal **visceral mass.** These basic parts are variously modified in different molluscs, and these morphological variations clearly illustrate the remarkable diversity that can be achieved by alterations on a relatively simple body plan.

Classification

This great diversity of form among the mollusca, an excellent example of **adaptive radiation,** is clearly represented in the seven classes of the phylum that present a diverse appearance. Figure 11.1 illustrates six of the major classes of living molluscs.

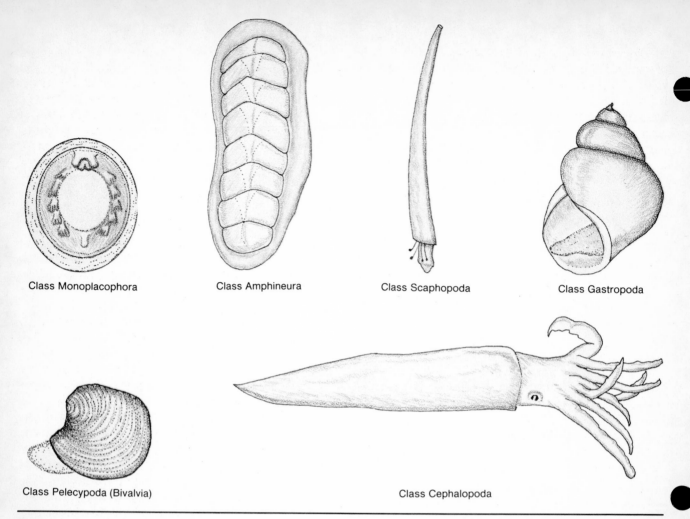

Class Monoplacophora   Class Amphineura   Class Scaphopoda   Class Gastropoda

Class Pelecypoda (Bivalvia)   Class Cephalopoda

Fig. 11.1 Six major classes of living molluscs.

Class Monoplacophora

Primitive molluscs with a conical, one-piece shell and showing evidence of segmentation in several organ systems. Known only from fossil forms and from a few recent specimens dredged from great depths in the sea at a few locations during the past few years. All marine. Example: *Neopilina.*

Class Amphineura

The chitons, or mail shells. Body oval shaped, with a large, flat foot and having eight dorsal calcareous plates. Algal feeders, mainly in the marine intertidal zone. Examples: *Chaetopleura, Chiton, Cryptochiton.*

Class Aplacophora

The solenogasters. A small group of peculiar wormlike marine molluscs with no head or shell and with a reduced foot; with an anterior mouth and a posterior anus. Some recent authorities suggest splitting the Aplacophora into two classes, the Solenogastres and the Caudofoveata. Examples: *Neomenia, Crystallophrisson (Chaetoderma).*

Class Scaphopoda

The toothshells. Body elongate dorsoventrally and encased in a tapered, tubular, one-piece shell open at both ends. Burrowing marine molluscs found in sandy and muddy sea bottoms. Scaphopod shells were once used by certain Native Americans on the American Pacific coast as money. Example: *Dentalium.*

Class Gastropoda

The snails, slugs, whelks, and limpets. Animals with a long, flat foot; a distinct head with eyes and tentacles; and a dorsal visceral mass usually housed in a spiral shell. The asymetrical growth of the visceral mass and the overlying mantle is responsible for the spiral shape of the shell. Gastropods comprise the largest and most successful class of molluscs. Ecologically, they are the most versatile molluscs with freshwater, marine, and terrestrial species. Some gastropods are carnivores (feed on animal tissues), some are herbivores (feed on plant material), and still others are parasites. Examples: *Helix* (European garden snail), *Littorina* (periwinkle), *Busycon* (whelk), *Physa,* and *Lymnaea* (freshwater snails).

Class Pelecypoda

The clams, mussels, oysters, and scallops. Body laterally compressed, small foot, no head, body contained in a bivalve (two-piece) shell hinged on the dorsal side. Most pelecypods are adapted for a sedentary life in marine or freshwater habitats. Typically they are filter-feeders and have specialized gills that serve to trap suspended food particles. Examples: *Anadonta* and *Unio* (freshwater mussels), *Mercenaria* (hardshell clam), *Crassostrea* (an oyster), *Pecten,* and *Aquipecten* (scallops).

Class Cephalopoda

The squids, cuttlefish, nautili, and octopi. Cephalopods are the most advanced molluscs and possess a large head with conspicuous eyes; mouth surrounded by eight or ten or more fleshy arms or tentacles; elongate body; shell often internal, reduced, or absent. They are typically active marine animals, preying on various fish, molluscs, arthropods, and worms. Examples: *Loligo* (squid), *Octopus,* *Nautilus* (chambered nautilus).

Because of the great diversity of form in the Phylum Mollusca, there is no "typical" mollusc. It is especially important, therefore, after you complete your study of the principal representative of this phylum (the freshwater mussel), that you make a careful study of the demonstration material to gain a better appreciation of the other kinds of molluscs.

Materials List

Living Specimens
    Physa or similar freshwater snail
    Representative molluscs (Demonstration)
Preserved Specimens
    Freshwater mussel
    Loligo
    Dextral and sinestral gastropod shells (Demonstration)
    Dried pen of squid (Demonstration)
    Nautilus shell (Demonstration)
    Representative molluscs (Demonstration)
Prepared Microscope Slides
    Mussel shell, cross section (Demonstration)
    Glochidium larva (Demonstration)
    Radula of gastropod (Demonstration)
Chemicals
    Carmine or carbon powder

Demonstration

Living and preserved representatives of various classes of molluscs.

The Freshwater Mussel: Class Pelecypoda

Freshwater mussels are sedentary animals that live in or on the bottoms of lakes, ponds, and streams. The particular species available for laboratory study vary from time to time and from place to place, and may include any of several large species in the genera *Anodonta, Lampsilis, Elliptio,* and *Quadrula.* The basic anatomy of the hardshell marine clam *Mercenaria* (formerly called *Venus*) is also very similar to the following description and can be used instead of the freshwater mussel if desired.

Obtain a mussel in a dissecting pan and note the hard **bivalve shell.** The two valves are joined along the dorsal surfaces by an elastic **hinge ligament.** The exterior surface of the valves is covered by a dark, horny material, the **periostracum.** Observe the concentric **lines of growth** on the exterior surface of the shell, which are formed as the mantle secretes new material at the edge of the shell.

The shell consists of three layers, the exterior **periostracum,** a middle **prismatic layer,** and an inner **nacreous layer.** The exterior periostracum is made up of a structural protein, conchiolin, which retards dissolution of the shell by the slightly acidic waters in which freshwater mussels are typically found. The middle prismatic layer consists of crystalline calcium carbonate ($CaCO_3$) and provides strength. The inner nacreous layer ("mother-of-pearl") consists of numerous layers of $CaCO_3$ and is iridescent. Near the anterior end of each valve is a raised portion, the umbo, which represents the oldest part of the valve. At the edge of the shell, between the valves, you should be able to observe the **incurrent siphon** (lower) and the **excurrent siphon** (upper), two openings between the edges of the mantle (see figure 11.2).

In figure 11.2, locate the two large muscles that hold the valves together, the **anterior** and **posterior adductor muscles.** They work antagonistically to the hinge ligament, a strong chitinous structure on the dorsal edge of the valves that tends to keep them open. You must cut through the anterior and posterior adductor muscles (carefully) to gain access to the interior organs.

Check to see whether your mussel has a wooden peg inserted to keep the valves spread apart. If not, it will be necessary for you to pry open the valves by inserting the handle of your forceps or the handle of your scalpel between the valves along the ventral edge. (Be careful of the blade!) Twist the handle, and when the valves are sufficiently separated, place a wedge between them so that they are separated about one-fourth to one-half inch. Carefully insert your scalpel, blade first, into the space between the left valve and the closely applied mantle in the region just below the anterior adductor muscle (see figure 11.2). Keep the blade close against the shell, loosen the mantle from the valve, and cut the large anterior adductor muscle. Repeat the procedure at the posterior end and cut the posterior adductor muscle. Now carefully lift the loosened left

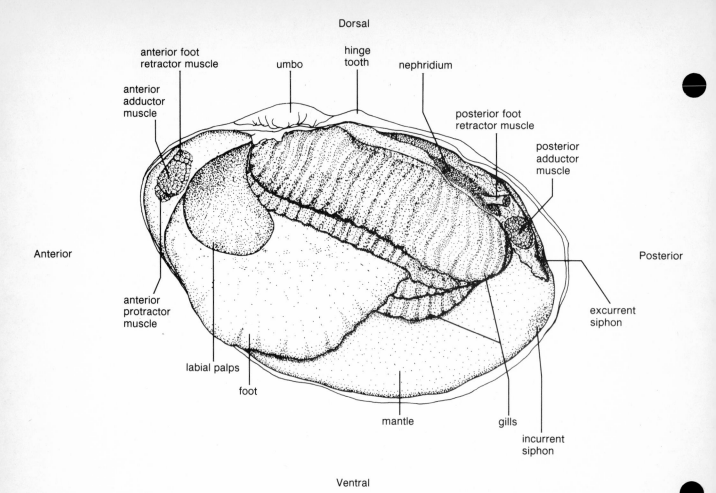

Dorsal

anterior foot
retractor muscle

anterior
adductor
muscle

umbo

hinge
tooth

nephridium

posterior foot
retractor muscle

posterior
adductor
muscle

Anterior

Posterior

anterior
protractor
muscle

excurrent
siphon

labial palps

foot

mantle

gills

incurrent
siphon

Ventral

Fig. 11.2 Freshwater mussel, partly dissected, lateral view.

valve, separating the mantle from it as you lift. **Warning:** The heart is located in the pericardial cavity near the dorsal side of the shell; take care to avoid damage to this region.

Feeding, Digestion, and Respiration

The thin layer of tissue that lines each valve is the **mantle.** It attaches to the inside of each valve along the **pallial line,** which is located about half of the distance from the edge of the valve to the center (see figure 11.3). Enclosed within the mantle is a space, the **mantle cavity,** which contains the other organs. Along the edge of the mantle identify the ventral **incurrent siphon,** with **sensory papillae** lining its borders, and the dorsal **excurrent siphon.** Locate the muscular **foot,** the **gills,** the **labial palps,** and the **mouth** located at the base of the palps.

The large **gills** play an important role in both respiration and feeding. Each gill consists of a double fold of tissue suspended in the mantle cavity (see figure 11.4). Each gill fold is a **lamella.** The lamellae of each gill (left and right) join ventrally to form a **food groove** and dorsally to form a **suprabranchial chamber,** which carries

water posteriorly to the excurrent siphon. The gills contain blood vessels and obtain some oxygen from the incoming water currents. The mantle is also vascularized and serves as a respiratory organ.

Cilia on the surface of the mantle and gills create water currents in the mantle cavity and draw in water through the incurrent siphon. Food particles contained in the incoming water are filtered from the water as it passes through the gills and are trapped in mucus secreted by glands in the gill tissue. The entrapped food particles are collected in the food tubes of the gills and transported anteriorly by highly coordinated ciliary movements to the labial palps and into the mouth. Along this route, nonfood particles are sorted out and eliminated.

If live mussels or clams are available in the laboratory, you can observe the coordinated ciliary movements on the surface of the gills by placing a few particles of carbon powder (or carmine powder) on the moistened surface of a gill. Make certain that the surface of the gills is moist (not wet) before you apply the carbon and take care to apply only a few particles of carbon as a marker. When working with live mussels or clams, be sure to add water to your specimen frequently to keep the soft internal tissues moist and flexible. The soft tissues dry out rapidly when they are exposed to the air.

Chapter 11

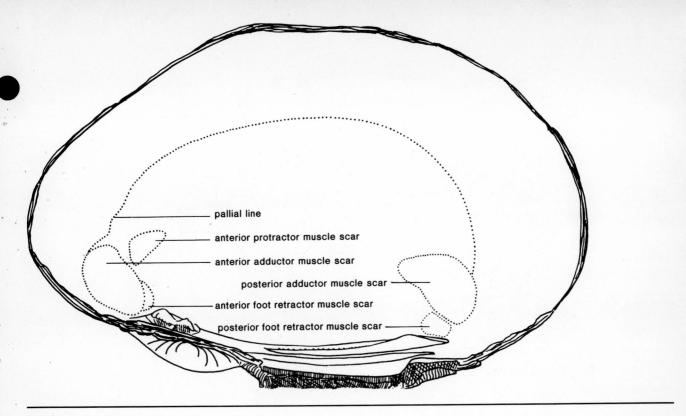

Fig. 11.3 Freshwater mussel, interior of shell.

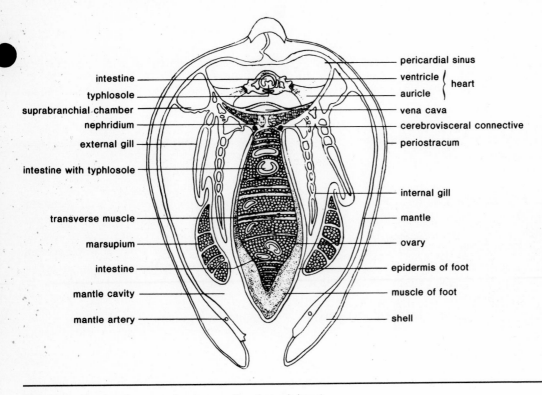

Fig. 11.4 Freshwater mussel, cross section through heart region.

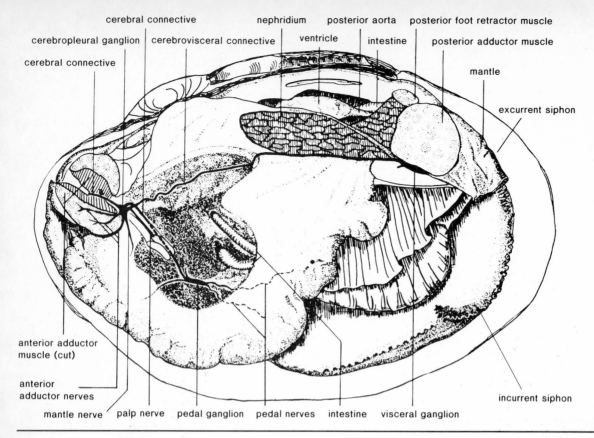

cerebropleural ganglion | cerebral connective | cerebrovisceral connective | nephridium | ventricle | posterior aorta | intestine | posterior foot retractor muscle | posterior adductor muscle

cerebral connective | mantle | excurrent siphon

anterior adductor muscle (cut)

anterior adductor nerves

mantle nerve | palp nerve | pedal ganglion | pedal nerves | intestine | visceral ganglion

incurrent siphon

Fig. 11.5 Freshwater mussel, partly dissected to show organs inside the foot, lateral view.

If the gills of your clam appear fat or swollen, they may contain eggs or larvae, since the gills of female freshwater clams also serve as a brood pouch or **marsupium** (see figure 11.4). Reproduction and the life history of the freshwater clam will be considered later in this exercise.

Most parts of the digestive system, including the **esophagus, stomach, digestive gland,** and **intestine,** are located within the foot and visceral mass (figure 11.5). To study the various structures enclosed within the **visceral mass,** you will need to cut along the ventral surface of the foot and bisect it into right and left halves. After you have bisected the foot, you should be able to locate the digestive structures mentioned above and the yellowish **gonad** tissue that surrounds a portion of the intestine.

Find the mouth, which is behind the anterior adductor muscle. Food particles, collected on the gills and transported to the labial palps, pass through the mouth and esophagus and into the stomach, where the process of digestion begins. The stomach floor is folded into numerous ciliary grooves that aid in sorting food particles from sediment and other nondigestible particles. Such sorting begins when particulate matter is trapped on the gills, and it continues enroute to the stomach. Particles rejected in the stomach are passed into the intestine for elimination.

Digestive enzymes are released into the stomach by the **crystalline style,** a gelatinous rod that extends into the stomach and releases enzymes as it is rotated by action of certain stomach muscles. Other enzymes are secreted by the digestive gland. Digestion is both extracellular and intracellular. Small food particles are engulfed by phagocytic cells in the digestive gland, and digestion is completed intracellularly. Undigested material passes through the intestine to the **rectum,** exits through the **anus,** and is flushed from the mantle cavity via the excurrent siphon.

Review the location and function of each part of the digestive system and observe how water currents and the gills play a central role in both digestion and respiration. To verify your understanding of the processes of feeding and digestion, trace the path of a group of food and nonfood particles from its entrance through the incurrent siphon to the ejection of the undigested material from the excurrent siphon. Be sure that you know the role of each structure along this path.

Muscles

You previously located two large and important muscles. Three other muscles facilitate movements of the foot. The **anterior protractor** aids in extending the foot, and the **anterior foot retractor** draws the foot into the shell. These two muscles are located near the mouth and the anterior adductor muscle. Often, they are partially fused with the anterior adductor and may be difficult to distinguish from it.

Chapter 11

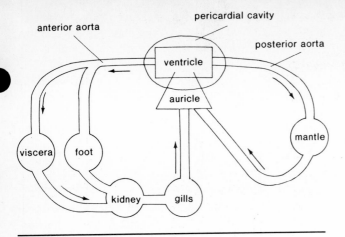

Fig. 11.6 Pattern of circulation in a mussel.

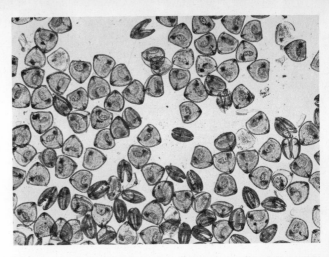

Fig. 11.7 Glochidia larvae of freshwater mussel. (Courtesy Carolina Biological Supply Company.)

The **posterior foot retractor** is found near the posterior end of the shell, slightly dorsal and anterior to this larger muscle. The posterior foot retractor draws the foot toward the posterior of the shell.

Circulation

The heart is located in the **pericardial cavity** found near the ventral surface (see figure 11.5). The pericardial sinus is enclosed by a thin membrane, the **pericardium**. The pericardial cavity represents the reduced coelom of the mussel. Locate within the pericardial cavity the muscular **ventricle** surrounding the intestine. Attached to it are two thin-walled **auricles**. Two major blood vessels carry blood away from the heart; the **anterior aorta** supplies the foot and most of the viscera, and the **posterior aorta** supplies the rectum and the mantle.

The circulatory system of the mussel is an **open system**. This means that the blood is not confined to a definite system of closed vessels, but that in the body tissues blood passes into large spaces or sinuses (see figure 11.6). From the visceral organs, blood passes to the nephridia for the removal of metabolic wastes and then to the gills for gas exchange before returning to the heart via the veins. The mantle also serves as a respiratory organ, and oxygen-rich blood from the mantle returns directly to the heart. What color is the blood of the mussel? Does it have hemoglobin?

Excretion, Osmoregulation, and Reproduction

Ventral to the heart and embedded in the mantle tissue is a mass of brownish or greenish tissue, the **nephridia**. The nephridia are the excretory organs, or "kidney," of the mussel. They remove nitrogenous and other wastes from the blood. In freshwater mussels the nephridia also excrete large amounts of water to maintain proper water balance in the body tissues.

The sexes are separate in freshwater mussels, but the male and female gonads are generally quite similar in appearance. The gonads consist of a yellowish mass of tissue and are located within the foot, surrounding a portion of the intestine (see figure 11.4).

Mature females of some species can be identified easily by the eggs and young larvae contained in the **marsupium,** a specialized portion of the gills that serves as a brood pouch (see figure 11.4). Most clams and mussels shed their gametes into the water where fertilization and embryonic development take place and result in the production of a free-swimming larva. Freshwater mussels (Family Unionidae), however, have a peculiar and highly specialized mode of reproduction.

The ripe eggs are released, and they pass into the suprabranchial chamber, where they are fertilized by sperm brought in by water currents. The fertilized eggs attach to the gills, which enlarge to form the **marsupia** (see figure 11.4). The eggs develop into tiny bivalve larvae, the **glochidia** (figure 11.7), which are released from the mussel to spend a portion of their lives as external parasites on certain species of fish. Later, the parasitic larvae detach from the host fish and move to the bottom, where they become free-living and develop into mature mussels (see demonstration).

If you have a mature female with a marsupium, cut away a portion of the gill and remove a few of the eggs or larvae. Examine the contents under a stereoscopic microscope. Can you identify the stages of development represented among the eggs and/or larvae?

Nervous Coordination

Although the mussels lack a differentiated head, the basic plan of their nervous system is similar to that of other molluscs and consists of three pairs of ganglia and their connecting nerves (see figure 11.5). A pair of **cerebropleural ganglia** (right and left) is located just posterior to the anterior adductor muscle; the two **visceral ganglia** lie just ventral to the posterior adductor muscle; and the **pedal ganglia** are embedded in the visceral tissue within the muscular foot.

A Freshwater Snail: *Physa*

Class Gastropoda

Physa is a small freshwater snail commonly found in North American ponds and streams.

Obtain a living snail in a watch glass partly filled with pond water and observe its spiral shell. Orient the shell with its **aperture** up and the pointed end or **spire** pointing away from you. The shell is said to be **dextral** if the opening is on your right, and **sinestral** if the opening is on your left when held in this position. In contrast to *Physa,* most shelled gastropods have dextral shells.

Allow the snail to attach to the dish and begin to creep along the substrate. Using your stereoscopic microscope observe the large, flat, muscular **foot** and the anterior **head.** On the head are two **tentacles** and two **eyes.** The shell encloses a dorsal **visceral mass.**

Note the gliding movement of the foot. This may be better observed if you place the snail on a wet glass plate or in half of a petri dish, allow the snail to attach, and invert the glass for observation under a dissecting microscope. The mechanism of this movement is not well understood, but it appears to involve a combination of rhythmic muscular contractions of the ventral surface of the foot and ciliary action. Also, the foot secretes a thin layer of mucus, which presumably aids in the gliding movement. You might also try to manipulate the snail carefully so its foot will attach to the surface film of the water. This will enable you to observe the waves of muscular contraction passing along the ventral surface of the foot.

While the snail is in this position, observe the **mouth** on the ventral side of the head, just anterior to the foot. Inside the mouth cavity is located the **radula,** a rasplike organ supported by the buccal cartilages. Also inside the mouth cavity are small sclerotized **jaws,** which are used to cut up food. The structure of the radula varies greatly in gastropods related to differences in feeding habits. This is an important diagnostic feature in taxonomy. Observe a slide of a gastropod radula on demonstration.

On the left side of the body, lateral to the foot, you can observe the opening to the pulmonary cavity, or "lung." *Physa* is a **pulmonate**; that is, it does not have gills like many marine snails, but obtains oxygen through a special, highly vascularized portion of the mantle. The opening, called the **pneumostome,** is located where the mantle and shell meet the foot.

In nature, *Physa* usually comes to the surface of the water from time to time and takes a fresh quantity of air into its pulmonary cavity. The respiratory behavior of freshwater snails varies greatly, however, depending upon specific environmental conditions.

On the right side of the foot (as seen while the ventral side is still directed upwards), observe the rapidly pulsating **heart** through the transparent shell. The heart consists of a single **atrium** and a **ventricle.** The ventricle pumps blood through an **aortic trunk** to the various organs. The blood then passes through various smaller **arteries, capillaries,** and collects in a number of **sinuses,** which collectively make up the **haemocoel.** How is this related to the coelom? Veins collect the blood from the haemocoel and return it to the atrium.

The respiratory pigment in gastropod blood is **hemocyanin,** a copper-protein complex. In gastropods, hemocyanin is dissolved in the blood rather than contained in blood cells and imparts a bluish hue to the blood. Hemocyanin also occurs in the blood of cephalopods, chitons, numerous arthropods, and in certain other smaller groups of animals.

Inside the aperture, observe the thin **mantle** closely applied to the shell. The space between the foot and the mantle is the **mantle cavity.** The **anus** opens into this cavity on the right side of the animal (when observed from the ventral side) near the edge of the mantle and shell.

Make a three-inch drawing of *Physa* in figure 11.8, showing the head and foot extended as during movement.

A Squid: *Loligo*

Squid are large, swift-moving, and highly specialized molluscs (figure 11.9). They exhibit a well-developed nervous system, complex sense organs, and sophisticated patterns of behavior. They are good representatives of the Class Cephalopoda, the most highly evolved class of molluscs. The name cephalopoda means "head-foot" and is quite appropriate because the foot, one of the basic anatomical elements of molluscs, is highly modified in this class and actually serves as a head. The "head-foot" contains several large nerve ganglia, which coordinate nerve impulses and several sense organs that provide important sensory information.

Fig. 11.8 Drawing of *Physa*.

Squid are often abundant in the surface waters of the oceans, where they may form large schools. Dozens or hundreds of squid may be captured together in trawl nets cast overboard in Atlantic and Pacific coastal waters. Other species of squid inhabit the ocean depths.

Several species of the genus *Loligo* are found in the coastal waters of the Atlantic and Pacific Oceans and in the Gulf of Mexico. *Loligo pealeii* is a common small Atlantic squid; *Loligo opalescens* is an small squid abundant in Pacific coastal waters. *Architeuthis*, the giant squid found in deeper waters of the Atlantic Ocean, is the largest known invertebrate animal, with some individuals reaching a length of 15–20 meters.

Examine a preserved squid to identify its major external features. Living squid in an aquarium are interesting to watch and to observe for swimming, feeding, and other types of behavior, but are difficult to obtain and keep in inland locations.

On the preserved specimen observe the **head, tentacles,** and two large **eyes.** The eyes are equipped with a pupil, iris, cornea, lens, and retina and are capable of forming clear images; they are remarkably similar to vertebrate eyes in many respects. This is an interesting and important example of **convergent evolution,** in which two organisms with quite different evolutionary histories (squid and humans) have evolved quite similar structures. Note that the body of the squid is slender and tapered, which

facilitates its swift movement through the water. The orientation of the squid is rather unusual, since morphologically the head and tentacles represent the **ventral surface** of the body and the pointed end opposite the head represents the **dorsal surface** (see figure 11.9). The funnel or siphon is located on the **posterior surface.** While swimming, a squid actually travels with its posterior end forward and its anterior surface up!

Squid swim by a type of jet propulsion in which water is squirted out of the funnel. Since the funnel is under muscular control, the direction of water ejection and thus the direction of movement can be swiftly changed. Movements of the arms and the two lateral fins also aid in steering the animal.

Locate the four pairs of **arms** on the head of your specimen and two pairs of elongated **tentacles.** The tentacles are retractile and play an important role in feeding. Study one arm and observe that it bears two rows of **suckers** (figure 11.10). Are the suckers all the same size? Which suckers are largest? Which are the smallest?

Each sucker is made up of a rounded cup attached to the arm by a stalk or **pedicle.** Remove a sucker from one of the arms and observe it under a stereoscopic microscope. Identify the **chitinous ring** with teeth, which supports the edge of the cup, and the small **piston** at the base of the cup. Note that the suckers on the two tentacles are found only near the distal ends.

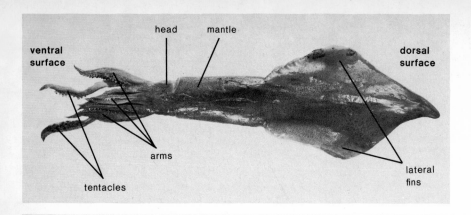

Fig. 11.9 Squid, *Loligo,* anterior view. (Courtesy Carolina Biological Supply Company.)

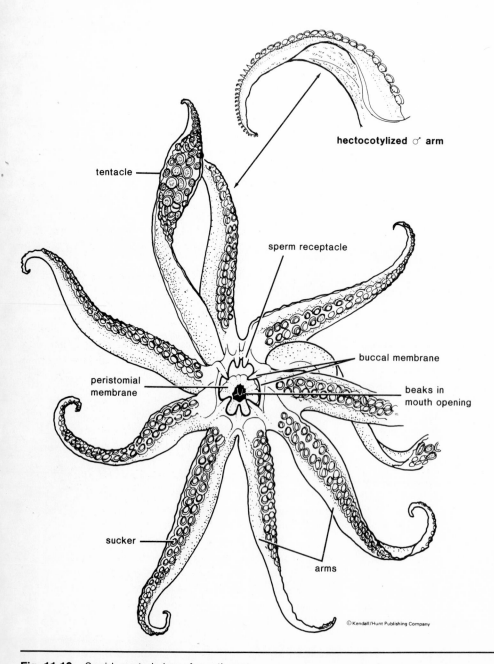

© Kendall/Hunt Publishing Company

Fig. 11.10 Squid, ventral view of mouth area.

A special modification of the suckers on one of the arms of mature males results in a reduction in the size of the suckers and an increase in the length of the pedicle supporting the terminal suckers (figure 11.10). This modified arm is called a **hectocotyl** and serves in the transfer of sperm bundles (spermatophores) to the female in mating. Such a modification of an anatomical structure by one sex (males in this case) and not the other is an example of **sexual dimorphism.**

Locate the muscular membrane at the base of the arms and tentacles. In the center of the membrane is the **mouth** opening (see figure 11.10). Inside the mouth find the two chitinous **beaks,** which are used to seize and tear the prey in feeding. Prey organisms are captured by swift movements of tentacles, which seize the prey, hold it fast with the suckers, and draw it toward the mouth with the aid of the other arms. Poison glands provide secretions, which aid in subduing active prey.

The fleshy mantle surrounds the remainder of the body and encloses the internal organs. The ventral edge of the mantle adjacent to the eyes is called the **collar.** Locate the **funnel** extending from the posterior side of the collar. On the anterior side of the collar you should find a projection of the mantle that represents the ventral tip of the **pen.** In the squid the skeleton is reduced to a translucent, internal, chitinous structure called the pen plus several cartilages, which protect the **cephalic ganglia** or "brain" within the head-foot, and other cartilages embedded in the collar region of the mantle and funnel.

Locate also the two large **lateral fins** extending from the mantle. As noted above, the fins are movable and serve as steering aids during swimming. On the surface of the mantle are numerous **chromatophores,** specialized pigment-containing cells that allow the squid to change the color of its integument. A chromatophore consists of a small sac of pigment surrounded by an elastic membrane attached to many small muscles. Contraction of these tiny muscles causes the pigment to spread out and add to the general coloration of the squid. A chromatophore whose pigment is not spread out contributes little to the apparent color of the squid.

Since squid have several types of chromatophores, each with a different colored pigment, many shades of color can be achieved. This aids greatly in the camouflage and protective coloration of the squid. Live squid show almost constant changes of color because of the continuous activity of the chromatophores, which are under nervous (and muscular) control.

Demonstrations

1. Dried pen of squid to illustrate internal skeleton.
2. Preserved or plastic mount of octopus.
3. Shell of chambered nautilus.

Key Terms

Adaptive radiation   evolutionary diversification of a group of animals with common ancestry to fill numerous ecological niches.

Chromatophore   specialized pigment-containing cells found in the integument of cephalopod molluscs and many crustaceans (Phylum Arthropoda).

Coelom   central body cavity lined by mesodermal tissues; develops from spaces arising within the mesoderm during embryonic development. Characteristic of annelids, molluscs, and all higher metazoans, although modern molluscs have a secondarily reduced coelom.

Convergent evolution   the evolution of two apparently similar structures by two or more groups of animals with different ancestral lines.

Foot   in molluscs the foot is a large ventral mass of muscular tissue; usually functions in locomotion; sometimes modified for other functions.

Glochidium larva   specialized larval form in the life cycle of certain freshwater pelecypod molluscs; enclosed in a transparent bivalve shell, these larvae spend part of their life cycle as parasites on the gills of freshwater fish. Later they detach and develop into free-living adult mussels. Plural: Glochidia.

Hemocyanin   a blue copper-containing respiratory pigment in the blood of molluscs and certain other invertebrate animals. Carries oxygen to the body cells.

Mantle   a thin outer layer of tissue largely enveloping the bodies of most molluscs. Usually secretes a shell.

Marsupium   a modified portion of the gills in females of certain freshwater clams and mussels; serves as a brood pouch for eggs and young larvae.

Notes and Sketches

12
Annelida

Objectives

After completing the laboratory work in this chapter, you should be able to perform the following tasks:

1. Briefly outline the characteristics of the Phylum Annelida and give five significant advances over the Phylum Platyhelminthes.

2. List and briefly characterize the three classes of Annelids. Cite an example of each.

3. Identify the principal external features of the polychaete worm *Nereis* and explain their functions.

4. List two other genera of polychaete worms; briefly describe and give the habits of each.

5. Describe the main features of the external anatomy of the earthworm and tell how the earthworm exhibits morphological adaptations for life in the soil.

6. Identify the main structures seen in a microscopic cross section of an earthworm and explain the function of each part.

7. Identify the parts of the digestive system in a dissected earthworm and give the function of each part.

8. Describe the basic pattern of circulation and identify the chief parts of the circulatory system of an earthworm.

9. Identify the main reproductive organs and discuss reproduction in earthworms.

10. Describe the external anatomy of a leech and tell how it differs from that of an earthworm.

11. Compare the internal anatomy of a leech with that of an earthworm.

Introduction

The Phylum Annelida includes more than 9,000 species of segmented worms, animals whose cylindrical, elongate bodies are comprised of a series of longitudinal segments (also called somites or metameres). In addition to their conspicuous segmentation, the annelids also exhibit several other important structural advances over the flatworms and roundworms that have been studied previously. Some important advances include: a spacious coelom, a closed circulatory system, an efficient excretory system, a highly developed muscular system, a well-developed central nervous system, and an increased concentration of nerve centers (ganglia) and sensory organs at the anterior end of the animal (cephalization).

Annelids are the first animals we have studied that exhibit a well-developed coelom. Although molluscs are also coelomate animals, the coelom of molluscs is reduced to the cavities in which the heart (pericardial cavity), nephridia, and gonads are located. The annelids, in contrast, have a large coelomic cavity in which most of the internal organs are suspended. Annelid worms illustrate well the basic body organization of a coelomate animal and the relationship of the coelomic cavity, its peritoneal lining of mesodermal origin, and the internal organs.

Classification

The phylum is divided into three classes, chiefly on the basis of **segmentation,** distribution of setae, and the presence or absence of a clitellum. Setae are thin, chitinous bristles or rods that are secreted by certain cells in the body wall. Setae are used in locomotion and, in many annelids, also aid in feeding. The number, morphology, and distribution of setae is often important in the classification of annelids (figure 12.1).

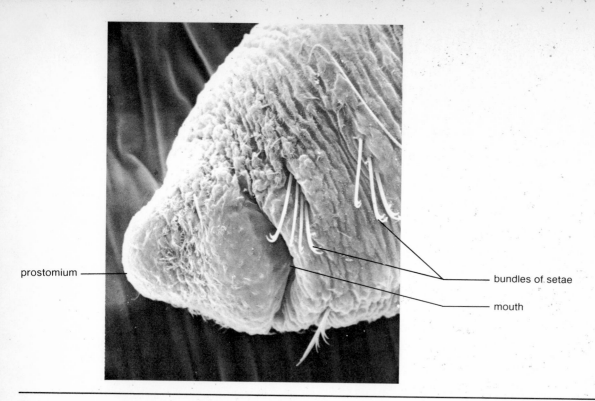

prostomium

bundles of setae

mouth

Fig. 12.1 Mouth region of an aquatic oligochaete showing characteristic arrangement and morphology of setae. Magnification 450X. (Scanning electron micrograph by Thomas Bouillon.)

The clitellum is a glandular swelling formed seasonally by certain segments. It secretes a cocoon into which eggs are deposited for fertilization and later incubation.

Class Polychaeta (Polychaete Worms)

Annelid worms with many segments, lateral appendages (parapodia) with many setae, a well-developed head region, no clitellum, lacking permanent gonads, usually dioecious, and typically having a trochophore larva. Most species are marine. This class also includes the archiannelids, a group of small marine worms formerly considered by many biologists to be a separate class. Examples: *Nereis* (formerly *Neanthes,* the clamworm); *Arenicola* (lug worm); *Chaetopterus; Hydroides* (feather duster or fan worm); *Diopatra* (plume worm); *Polygordius* (an archiannelid).

Class Oligochaeta (Bristleworms)

This class includes the earthworms and many species of related freshwater worms that have many segments but few setae per segment, form a clitellum, but lack parapodia and a differentiated head region. They are generally monoecious, have permanent gonads, and development is direct. Mostly found in the soil and in freshwater habitats. Examples: *Lumbricus* and *Allolobophora* (earthworms); *Eisenia* (dung worm); *Tubifex* (sewage worm); *Aeolosoma; Chaetogaster.* The last three are aquatic genera and are common freshwater forms.

Class Hirudinea (Leeches)

This group includes the leeches, a relatively small class (about 500 species) of specialized annelids found mainly in freshwater habitats. A few species are also terrestrial. Leeches have dorsoventrally flattened bodies, anterior and posterior suckers, thick muscular bodies, no head or tentacles at their anterior ends, and lack parapodia and setae. The number of segments is limited (34) and a clitellum is formed during the reproductive season.

Three representatives of this phylum have been selected for special study to illustrate the principal features of the Annelida: the clamworm, an earthworm, and a leech.

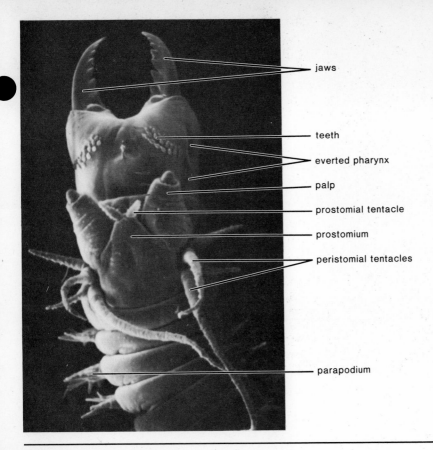

jaws

teeth

everted pharynx

palp

prostomial tentacle

prostomium

peristomial tentacles

parapodium

Fig. 12.2 *Nereis,* head region, dorsal view. (Scanning electron micrograph by Betsy Brown.)

Materials List

Living Specimens
    Earthworm
Preserved Specimens
    Nereis
    Leech
    Representative polychaetes (Demonstration)
    Polychaete tubes (Demonstration)
    Earthworm cocoons (Demonstration)
    Representative oligochaetes (Demonstration)
    Representative leeches (Demonstration)
    Leech, dissected to show internal anatomy (Demonstration)
Prepared Microscope Slides
    Earthworm, cross section
    Trochophore larva (Demonstration)
Miscellaneous Supplies
    Dissecting pans
    Pins
Audiovisual Materials
    Chart of earthworm anatomy

A Marine Annelid: *Nereis virens*

Class Polychaeta

The sandworm or clamworm, *Nereis virens* (figures 12.2 and 12.3), is a large marine annelid commonly found in many areas along the Atlantic coast of North America. *Nereis* usually lives in burrows in the sand or in mud bottoms of the shallow coastal waters during the day and emerges at night to search for food.

Select a preserved specimen of *Nereis* and observe the numerous body segments. Internally, the segments are separated by thin sheets of tissue called **septa.** The internal structure of *Nereis* is generally similar to that of the earthworm, which you will study later.

All the segments of *Nereis* except the first and the last bear a pair of lateral **parapodia** (see figure 12.4). Observe the demonstration slide of a cross section of *Nereis* to study the structure of these appendages. Note that the parapodia are **biramous** (two-branched), with a dorsal **notopodium** and a ventral **neuropodium.** Each main branch

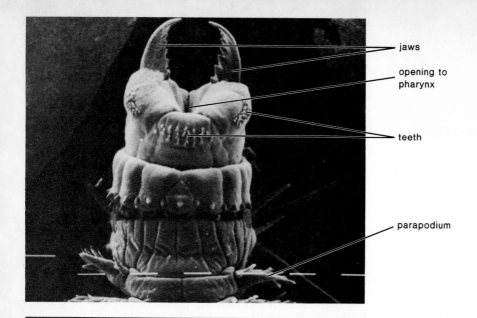

jaws

opening to
pharynx

teeth

parapodium

Fig. 12.3 *Nereis,* head region, ventral view. (Scanning electron micrograph by Betsy Brown.)

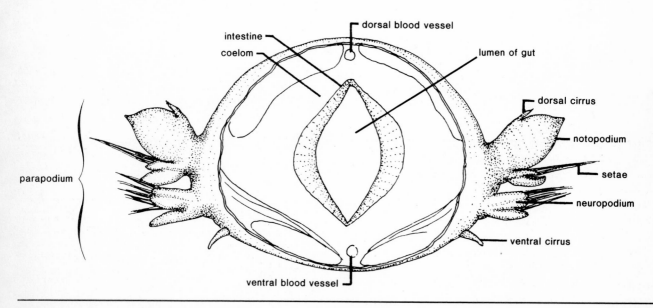

intestine

coelom

dorsal blood vessel

lumen of gut

dorsal cirrus

notopodium

setae

neuropodium

ventral cirrus

parapodium

ventral blood vessel

Fig. 12.4 *Nereis,* cross section.

bears bundles of bristles, or **setae,** and also has several smaller side branches. The parapodia play an important role both in locomotion (swimming and creeping along the sea bottom) and in respiration. What aspects of their structure appear to be advantageous for these functions?

Examine the well-developed head region made up of the **peristomium,** the first complete body segment that encircles the mouth, and the **prostomium,** a triangular mass of tissue located on top of the peristomium (figure 12.2). Several sensory structures are found on the prostomium, including two fleshy **palps** attached to the sides of the prostomium, two pairs of pigmented **eyes** on the dorsal

surface, and two **anterior tentacles.** Four pairs of **lateral tentacles** (or cirri) are also found attached to the peristomium.

During feeding the pharynx may be everted through the mouth exposing two powerful **jaws** and many small **teeth** on the wall of the pharynx (figures 12.2 and 12.3). The jaws aid in seizing prey, and the pharyngeal teeth help hold the prey as the pharynx is withdrawn through the mouth.

The sexes are separate in *Nereis,* as in most polychaete worms, although there are no permanent gonads. The gametes (eggs and sperm) are formed in the body

Chapter 12

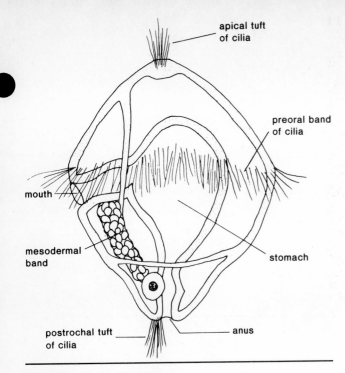

Labels on figure:
- apical tuft of cilia
- preoral band of cilia
- mouth
- mesodermal band
- stomach
- postrochal tuft of cilia
- anus

Fig. 12.5 Trochophore larva.

wall at certain times, released into the coelom, and released into the sea. There the eggs may be fertilized and develop into a **trochophore larva** (figure 12.5). Other larval stages may follow before metamorphosis into the adult worm.

Many species of polychaete worms show marked structural, physiological, and behavioral changes during the breeding season. Some species aggregate in large swarms and rise to the surface before releasing their gametes into the sea. This special behavior represents an important adaptation that serves to increase the likelihood of fertilization and the continuation of the species.

The trochophore larva formed by many marine polychaetes bears a close resemblance to the larvae of many marine molluscs. This similarity of trochophore larvae is one of the important pieces of evidence suggesting a close evolutionary relationship between these two phyla. Similarly, the resemblance of the trochophore larva of annelids and molluscs to the marine larvae of several other invertebrate phyla, plus other similarities in their patterns of embryonic and larval development, suggests other important evolutionary relationships with such groups as the flatworms and the arthropods.

Other Polychaetes

Polychaete worms are often abundant in the shallow coastal waters of the sea. This class includes many varied forms, some of which form permanent tubes that they secrete or build from various materials (often called sedentary polychaetes). Others do not form permanent tubes (called errant polychaetes). Observe the demonstrations

of various living and preserved polychaetes available in the laboratory . Three common and well-known species are illustrated in figures 12.6, 12.7, and 12.8.

Hydroides (figure 12.6) secretes a permanent calcareous tube on the surface of shells and rocks. The tube is open at the anterior end and tapers to the closed posterior end. While feeding, the worm extends a feathery crown of anterior appendages (extensions of the prostomium) from the open anterior end of its calcareous tube. Because of these feathery appendages, which are often brightly colored, *Hydroides* and other members of the Family Serpulidae are called feather duster or fan worms.

Diopatra, the plume worm (figure 12.7), is a polychaete common in intertidal areas on the Atlantic coast of the United States. The body of the plume worm is iridescent and bears several pairs of bright red (blood-filled) gills. *Diopatra* forms a vertical, parchmentlike tube which may reach three feet in length. Most of the tube is buried in the sediment, but 2–3 inches usually extend above the sediment. The tube is camouflaged by bits of shell, sea grass, and other bits of debris.

Arenicola, the lug worm (figure 12.8), is a burrowing polychaete which lives in a J-shaped tube lined with mucus. In its habits *Arenicola* resembles a seagoing earthworm. The worm burrows in shallow marine sediments by extending its proboscis into the sand or mud and retracting it filled with sand. The sand contains detritus and plankton on which the worm feeds as the sand passes through the digestive tract. *Arenicola* shows several external structural adaptations related to its peculiar burrowing habits, including reduced parapodia, small gills, a large bulbous proboscis, and the lack of tentacles and other head appendages.

Demonstrations

1. Representative species of Polychaeta, such as *Amphitrite, Aphrodite, Arenicola, Chaetopterus, Hydroides, Sabella, Diopatra,* and *Polygordius.*
2. Examples of different kinds of tubes formed by polychaetes.
3. Microscope slide of trochophore larva.

The Earthworm

Class Oligochaeta

Several features of earthworms appear to be more specialized than those of the polychaetes. Some of these features are apparent externally, such as the absence of parapodia and the lack of sensory structures in the head region. They are believed to be related to the burrowing habits and the underground existence of earthworms. Their internal anatomy shows fewer specializations, and earthworms, therefore, serve well as representatives of the general pattern of internal anatomy of annelids.

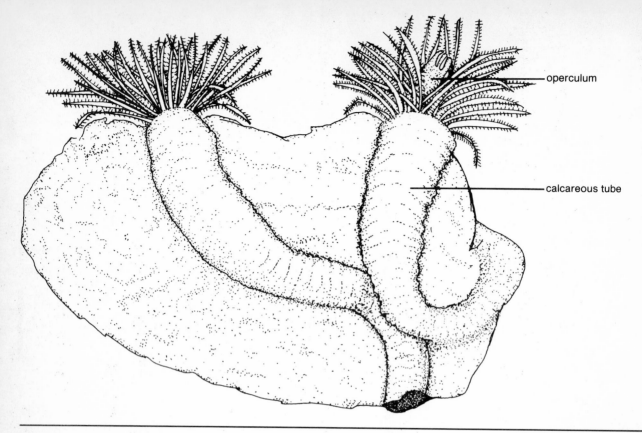

operculum

calcareous tube

Fig. 12.6 *Hydroides*, a fan worm.

External Anatomy

Obtain an anesthetized or freshly killed specimen of a large species of earthworm and examine some of the major external features. Preserved specimens may also be studied, but fresh specimens are much superior.

Observe the long, cylindrical body, which tapers at each end. The anterior end can be identified most readily by reference to the **clitellum**, a conspicuous swollen region including several segments located near the anterior end. The clitellum secretes a **cocoon** in which the fertilized eggs are incubated (figure 12.9). Note the absence of a distinct head and the lack of external sense organs at the anterior end.

The **mouth** is located in the first segment, and a small rounded projection, the **prostomium**, overhangs the mouth. The **anus** is located in the last segment.

Observe the darker and more rounded dorsal surface. Locate the **setae**, small chitinous bristles in each segment except the first and the last. To find the setae, rub your finger lightly back and forth along the sides of the worm. Observe how the setae protrude slightly from the body surface and catch on your finger. The setae are located in four distinct rows. Two pairs of setae are located in rows on the ventral surface, and two pairs are located on the sides (see figure 12.13). Toward which end of the worm do the setae point? What is the advantage of this in locomotion?

Note the iridescent **cuticle,** a protective covering which restricts water loss, that covers the surface of the body. The iridescence results from the many small striations in the surface of the cuticle. These striations cause the cuticle to act as a diffraction grating and produce the iridescence.

Also located on the surface of the body are several openings from the excretory and reproductive systems. A pair of small **excretory pores,** the nephridiopores, is found on the ventral surface of each segment except the first three or four and the last. Openings from the reproductive system include: a pair of **male pores** on the ventral surface of segment XV surrounded by swollen "lips," a small pair of **female pores** on the ventral surface of segment XIV, and openings to two pairs of **seminal receptacles** located laterally in the grooves between segments IX–X and X–XI. Notice that the segments are usually designated by Roman numerals. The use of a stereoscopic (dissecting) microscope will help in locating these tiny external openings to the excretory and reproductive systems of the earthworm.

(Note: The location of these pores and the internal reproductive organs differs in various species of earthworms. The traditional earthworm for dissection is *Lumbricus terrestris,* upon which this and most other textbook descriptions are based. Many of the earthworms supplied for dissection by American biological supply houses, however, are different species and therefore differ in some

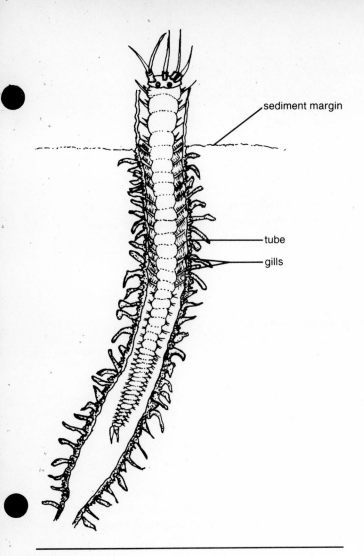

Fig. 12.7 *Diopatra*, the plume worm.

minor respects from textbook descriptions, particularly in respect to the location of the reproductive structures in particular segments.)

Internal Anatomy

Place your earthworm in a wax-bottom dissecting pan and carefully pin the specimen down, dorsal side up, with one pin through the prostomium and another near the posterior end. Study figure 12.13 and note the relative thickness of the body wall and the closeness of the underlying parts before you begin your dissection. Puncture the body wall slightly to one side of the dorsal midline with the tips of your scissors and make a longitudinal incision from behind the clitellum, continuing it anteriorly to the mouth. Cut to the side of the large dorsal blood vessel and avoid severing it. Take special care, also, to avoid damage to the brain, which is located near the mouth.

Figure 12.10 shows a dissected earthworm and its internal organs. Observe the large body cavity, the **coelom,** that is divided into many distinct compartments by the

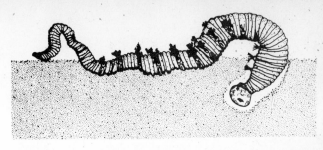

Fig. 12.8 *Arenicola*, the lug worm.

Fig. 12.9 Earthworm cocoons. (Courtesy Carolina Biological Supply Company.)

septa. The septa are membranous, transverse partitions that divide the body of the earthworm internally into many segments.

Carefully separate the body wall from the internal structures and pin down the body wall on each side. Place the first pins on opposite sides of segment XV, easily located because of the presence of the large male pores. Insert the pins through the tissue and into the wax bottom of the pan at a 45° angle to provide working space and a clear view of the internal organs. Place additional pins through each fifth segment, for example, segments V, X, XV, XX, XXV, XXX, etc., to serve as convenient landmarks to identify the location of various structures. Cover the specimen with about one-fourth inch of water to keep the internal organs moist and flexible.

Digestive System

The digestive tract is a straight tube extending from the mouth to the anus. Examine your specimen and locate the **mouth** at the anterior end. The mouth opens into the small **buccal cavity** (see figure 12.11). Immediately behind the buccal cavity is the muscular **pharynx.** The pharynx is attached to the body wall by numerous threadlike **dilator muscles** that can produce a sucking action to draw food materials through the mouth into the buccal cavity and pharynx. The tubular, thin-walled **esophagus** extends from segments VI to XIII and passes the food posteriorly to the **crop.**

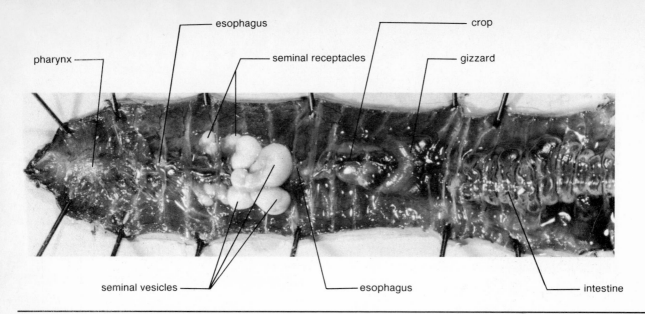

pharynx — esophagus — crop

— seminal receptacles — gizzard

seminal vesicles — — esophagus — intestine

Fig. 12.10   Earthworm, dissected, dorsal view. (Photograph by Indiana University Audiovisual Center.)

Associated with the lateral wall of the esophagus of earthworms are two or more pairs of **calciferous glands.** These glands develop as evaginations of the esophageal wall; they function as excretory rather than digestive organs. The calciferous glands serve to regulate the levels of calcium and carbonate ions (Ca^{++} and $CO_3^=$) and the pH of the blood.

Food is stored in the thin-walled, extensible crop before passing into the muscular **gizzard,** where it is ground into smaller pieces and passed to the **intestine** to be digested and absorbed. Undigested materials are passed to the **anus** for egestion.

Much of the intestine, as well as the dorsal blood vessel, is covered by a yellowish layer of **chlorogogue tissue.** The chlorogogue tissue does not appear to be directly involved in digestion, but it does store glycogen and lipids and has other functions analogous to the liver in vertebrate animals, including chemical defense of the body.

Circulatory System

The earthworm has a closed circulatory system, which contains red blood. The red color is due to the respiratory pigment, **hemoglobin,** dissolved in the **plasma,** or fluid portion, of the blood. In addition to the plasma, the blood also contains numerous colorless **blood cells,** or corpuscles.

Locate the large **dorsal blood vessel** lying just above the digestive tract, and also a pair of **parietal vessels** extending laterally from the dorsal blood vessel in each segment. In segments VII to XI, inclusive, the parietal vessels are enlarged and much more muscular. These five pairs of enlarged **aortic arches** are often called "hearts." Their contractions aid in propelling the blood through the circulatory system.

The yellowish tissue that covers the surface of the large blood vessels consists of **chlorogogue cells.** Some of the other major blood vessels include the **ventral vessel,** lying beneath the digestive tract; the **subneural vessel,** lying beneath the ventral nerve cord; and two **lateral neural vessels,** one on each side of the ventral nerve cord (see figure 12.13).

In an anesthetized worm, waves of muscular contraction can be readily observed along the dorsal vessel and the five pairs of "hearts." In which direction does the blood flow through these vessels? Valves in the dorsal vessel and in the hearts serve to prevent backflow and to aid in the efficient circulation of the blood.

Excretory Organs

Each segment, except the first three or four and the last, contains a pair of tubular excretory structures, the **nephridia.** Each nephridium is located partly in two adjacent segments. Consult figures 12.11 and 12.12. Locate the funnel-shaped opening into the coelom, the **nephrostome,** which projects through the septum into the next anterior segment. Locate also the **coiled tubular portion,** which empties through a ventral opening in the body wall, the **nephridiopore.**

The **calciferous glands,** formed as evaginations of the wall of the esophagus, also serve as excretory organs. They were described previously during our discussion of the digestive system.

Reproductive System

Cut across the digestive tract about a half-inch behind the gizzard and carefully free it with the associated blood vessels from the underlying structures up to the level of segment IV. The earthworm is **monoecious;** each individual has both male and female sex organs (see figure 12.12). Despite the monoecious condition, however, cross fertilization between different individuals is necessary for sexual reproduction in earthworms.

Chapter 12

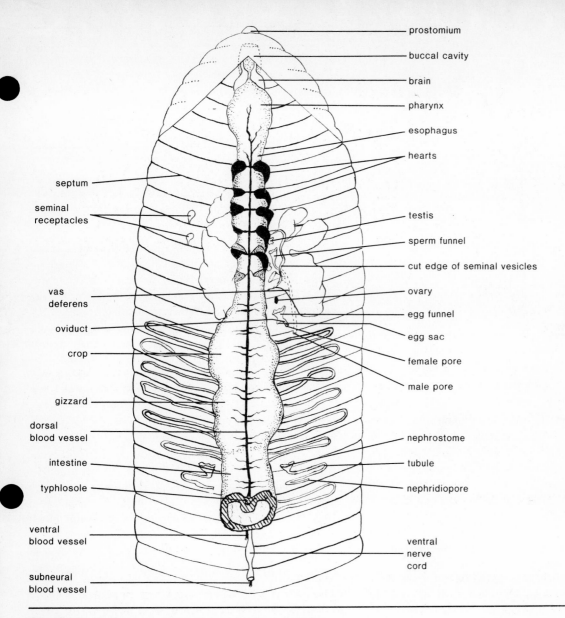

Fig. 12.11 Earthworm, principal internal organs, dorsal view.

Several of the male reproductive organs are larger and more conspicuous than the female organs. Locate the following male reproductive structures and note the segments in which they are located: (1) three pairs of large **seminal vesicles** (storage sacs in which the sperm mature and are stored until copulation); (2) two pairs of small **testes** embedded within the tissues of the seminal vesicles; (3) two pairs of small ciliated **sperm funnels** (one pair on each side), which connect via short tubules to (4) two **vasa deferentia** (sperm ducts), which lead to the male genital pores on segment XV.

The female reproductive structures are small and difficult to locate in most specimens. They include a pair of **ovaries** in segment XII, which discharge mature ova into the coelom. The ova are collected in two ciliated **egg funnels** and passed through the oviducts to the **female genital openings** on segment XIV. You should be able to locate two pairs of **seminal receptacles** in segments IX and

X, where sperm is received during copulation. Sperm stored in the seminal receptacles is later used in fertilization of eggs released into the cocoon.

Nervous System

The nervous system of the earthworm consists of the following principal components: (1) an anterior pair of large **suprapharyngeal ganglia** (the "brain"); (2) two **circumpharyngeal connectives**; (3) a pair of smaller **subpharyngeal ganglia**; and (4) a **ventral nerve cord** with an enlarged ganglion in each segment (see figure 12.12). Locate several nerves extending from the suprapharyngeal and subpharyngeal ganglia to the anterior segments. Three pairs of **lateral nerves**—two from the ganglion itself and one anterior to the ganglion—lead from the ventral nerve cord in each posterior segment.

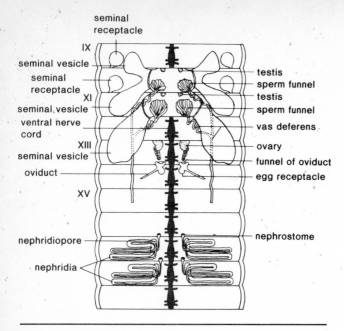

seminal
receptacle

IX

seminal vesicle

seminal
receptacle

XI

seminal vesicle

ventral nerve
cord

XIII

seminal vesicle

oviduct

XV

nephridiopore

nephridia

testis
sperm funnel
testis
sperm funnel
vas deferens
ovary
funnel of oviduct
egg receptacle

nephrostome

Fig. 12.12 Earthworm, reproductive and excretory systems.

Cross Sections

Obtain a slide with a stained cross section of an earthworm and study its histological structure (figure 12.13). Observe the body wall and its composition. Locate the outer **cuticle** and the underlying **hypodermis.** What is the nature of the cuticle and how is it formed? Beneath the hypodermis is a thin layer of **circular muscles** and a thicker layer of **longitudinal muscles.**

Observe the thin layer of flattened cells, the **peritoneum,** at the inner margin of the longitudinal muscle layer. Covering the outer surface of the intestine (and some of the larger blood vessels as noted earlier) is a specialized type of peritoneum, the **chlorogogue.** The cells of this layer are believed to function like the vertebrate liver in intermediary metabolism, in the removal of waste products, and also in the synthesis of hemoglobin, the red oxygen-carrying pigment in the blood. The body cavity of the earthworm is a true **coelom.** Why?

You may be able to observe within the coelom the **nephridia** lateral to the intestine. The nephridia in cross section may appear to be irregularly shaped since they are often cut through the coiled portions of the tubule.

The wall of the intestine consists of several layers of cells, including an outer **chlorogogue,** thin layers of **longitudinal muscle** and **circular muscle,** and a layer of **endodermal epithelium** lining the cavity, or lumen, of the gut. Dorsally observe the pronounced fold of tissue, the **typhlosole,** which extends into the lumen. What is the function of the typhlosole?

Observe also the **dorsal** and **ventral blood vessels** and the **ventral nerve cord.** On each side of the nerve cord is a **lateral neural blood vessel,** and beneath the nerve cord is the **subneural blood vessel.**

Demonstrations

1. Charts and models to illustrate earthworm anatomy.
2. Earthworm cocoons.
3. Behavior of living earthworms (locomotion, reactions to stimuli, etc.).
4. Examples of other Oligochaetes, such as *Tubifex, Aeolosoma, Enchytraeus, Stylaria, Chaetogaster.*

Leeches

Class Hirudinea

Although they are commonly called "bloodsuckers," many of the 500-plus species of leeches (figure 12.14) are actually scavengers or predators on small invertebrates, rather than ectoparasites that feed on the blood of some vertebrate host to which they periodically attach.

The large European leech *Hirudo medicinalis* described in most textbooks was used in the early days of medical practice for bloodletting, since it was once believed that many diseases and bodily disorders were caused by an accumulation of excess blood. For this reason this species is often called the "medicinal leech."

Most leeches live in freshwater streams, ponds, and lakes. There are also some marine species and a few terrestrial forms. Leeches are relatively large annelids, and the adults of most species reach about 2–5 cm in length. Among their distinctive features is the **lack of appendages and setae,** but the presence of large **anterior and posterior suckers,** a **fixed number of segments,** and male and female **copulatory organs.** The leeches most often available for laboratory study in the United States are large aquatic species of the genera *Haemopis, Malacobdella,* or *Placobdella.*

External Anatomy

Examine a preserved leech and observe the elongate, flattened body. The body is flattened dorsoventrally and usually tapers toward the anterior end. Note the absence of setae and lateral appendages. Locate the numerous rings or **annuli** that circle the body. Leeches are segmented both internally and externally; but, unlike the oligochaetes and polychaetes, leeches have a limited number of body segments (34). Also the external annuli are more numerous than the internal segments; various internal segments bear from 1–5 annuli each.

Identify the large **anterior sucker** and the smaller **posterior sucker.** The suckers serve both for attachment to the host and for locomotion. Leeches move in a looping fashion by attaching to the substrate by the anterior sucker, contracting the body, attaching by the posterior sucker, loosening the anterior sucker, and extending the body forward. Aquatic leeches can also swim by means of undulatory motions of the body resulting from rhythmic contractions of their muscular bodies.

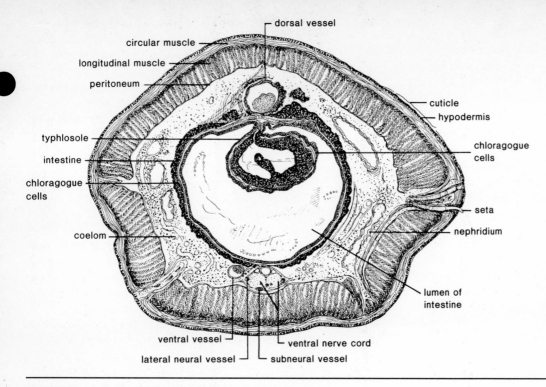

Fig. 12.13 Earthworm, cross section.

Labels: dorsal vessel, circular muscle, longitudinal muscle, peritoneum, cuticle, hypodermis, typhlosole, intestine, chloragogue cells, chloragogue cells, coelom, seta, nephridium, ventral vessel, lateral neural vessel, subneural vessel, ventral nerve cord, lumen of intestine

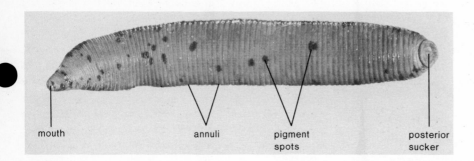

Fig. 12.14 Leech. (Courtesy Carolina Biological Supply Company.)

Labels: mouth, annuli, pigment spots, posterior sucker

Locate the **mouth** in the center of the anterior sucker. At the anterior margin of the posterior sucker find the **anus** on the middorsal surface.

Internal Anatomy

Study a demonstration of a dissected leech to observe its internal anatomy. The principal internal features include a complete digestive system with a **mouth** with teeth, a muscular **pharynx,** a large **crop** that serves to store blood in parasitic leeches, a small **stomach,** a narrow tubular **intestine,** and an enlarged **rectum** that opens dorsally via the **anus** located just anterior to the posterior sucker.

Leeches also have a closed circulatory system, several pairs of segmental nephridia that serve as excretory organs, and permanent gonads. The **reduced coelom** contains one pair of ovaries and several pairs of testes. Although leeches are monoecious (hermaphroditic), reproduction involves copulation and reciprocal fertilization between different individuals. The fertilized eggs are deposited in a cocoon secreted by the clitellum formed by an enlargement of segments 9–11. The eggs develop directly into miniature leeches; there are no larval stages.

Demonstrations

1. Preserved specimens of several different leeches, such as *Haemopis, Malacobdella, Placobdella,* and *Hirudo.*
2. Dissected leech to illustrate internal anatomy.

Key Terms

Biramous having two branches or rami; as the parapodium of a polychaete worm with a dorsal notopodium and a ventral neuropodium.

Clitellum a thickened, glandular portion of certain midbody segments of many oligochaetes and leeches; characteristic of sexually mature individuals. Forms a cocoon in which eggs are deposited and incubated.

Coelom a central body cavity fully lined by peritoneum, an epithelial lining of mesodermal origin. It forms the principal body cavity of annelids and most higher phyla, although in some groups the coelom is secondarily reduced.

Nephridium a tubular excretory organ found in annelids and other invertebrate phyla; may function both in osmoregulation and in the removal of nitrogenous and other body wastes. Plural: nephridia.

Parapodium a lateral appendage typical of polychaete worms; always occurs in pairs on opposite sides of the body; usually with many setae arranged in bundles. Plural: parapodia.

Segmentation having a body consisting of many similar units or subdivisions arranged along the anterior-posterior axis. Each unit is called a segment, somite, or metamere. Also called metamerism.

Septum a thin wall of tissue separating two adjacent segments or masses of tissue. Plural: septa.

Seta a chitinous bristle or rod secreted by certain epidermal cells of oligochaetes and polychaetes. Setae are also found in certain arthropods and other invertebrates. Plural: setae.

Trochophore larva a marine larval form characteristic of many polychaete worms and certain other invertebrate groups; with a pear-shaped body, one or more bands of cilia around the body, tufts of cilia at the anterior and posterior ends, and a complete digestive tract. Similar larvae are formed by molluscs and several other invertebrate phyla.

Chapter 12

13
Arthropoda

Objectives

After completing the laboratory work for this chapter, you should be able to complete the following tasks:

1. List the four subphyla of arthropods. Briefly characterize each subphylum and cite examples.

2. Identify the principal external features of the horseshoe crab, *Limulus;* briefly describe its feeding and locomotion.

3. Demonstrate the chief external features of a spider on a preserved specimen.

4. Compare the basic morphological organization of a spider with that of the horseshoe crab, a crayfish, and a grasshopper.

5. Identify the chief external features of a crayfish. Compare and contrast the morphology of the crayfish with that of the horseshoe crab. Tell which animal is the more primitive and explain why.

6. Discuss the concept of serial homology and give examples from the appendages of a crayfish.

7. Describe the respiratory system of a crayfish and tell how the appendages are involved in respiration.

8. Describe the circulatory system of a crayfish and compare it with that of an earthworm.

9. Describe the eyes of a crayfish and compare them with the eyes of a spider.

10. Identify the main morphological features of *Daphnia* and explain how this animal is adapted for its planktonic life.

11. Describe the feeding mechanism of *Daphnia* and identify the chief structures involved.

12. Identify the principal external features of the grasshopper *Romalea* and tell how they illustrate the basic organization of the insects.

13. List and demonstrate the main segments of a grasshopper leg. Explain the function of each part.

14. Describe the special adaptations of the appendages of the honeybee and demonstrate them on a specimen. Compare the basic morphology of a grasshopper and the honeybee. Tell which insect is more primitive and explain why.

15. Describe the principal types of insect development and identify the stages in complete metamorphosis.

Introduction

The arthropods represent the largest and most successful of the animal phyla. More than 1,000,000 species of arthropods have now been described, and members of this phylum dwell in almost every type of habitat. In addition to the many kinds of aquatic arthropods, there are also many terrestrial species and numerous species of flying insects—the only invertebrates that have evolved a capability for flight. The body of arthropods is covered externally with a chitinous exoskeleton and, in varying degrees, is segmented both internally and externally. The jointed appendages are among the most apparent and versatile features of the arthropods and have been adapted for walking, swimming, jumping, feeding, reproduction, sound production, sensory perception, and defense.

Among the important distinguishing characteristics of the Phylum Arthropoda are the following: (1) a hardened, chitinous exoskeleton; (2) paired, jointed appendages, often modified for various functions; (3) segmented body divided into two or three functional regions, such as head and trunk, cephalothorax and abdomen, prosoma and opisthosoma, or head, thorax, and abdomen; (4) reduced coelom; (5) complete digestive tract with mouthparts developed from modified anterior appendages; (6) an open venous system but a closed arterial system with a dorsal heart, arteries, and open spaces (sinuses) in the tissues that serve to collect blood prior to return to the heart; (7) a highly organized brain and a nervous system with paired ventral ganglia and nerve cords; and (8) complex sense organs and behavior.

Fig. 13.1 Fossil trilobite, portion head surface broken away.

In this exercise you will have an opportunity to study the organization of representatives of three of the four major groups (subphyla) of arthropods: the **chelicerates** (horseshoe crab, spider); the **crustaceans** (crayfish, water flea); and the **uniramians** (grasshopper, honeybee).

Classification

The classification of the arthropods is complicated because the phylum is so large and diverse. We can present only a brief summary of some of the major classes here. Recent studies have led to some changes in the classification of the phylum. Most current authorities recognize four major groups of arthropods—the trilobites, the chelicerates, the crustaceans, and the uniramians.

Subphylum Trilobita (Trilobitomorpha)

The trilobites (figure 13.1). Primitive marine arthropods with a flattened, ovoid body divided into three parts, an anterior head (cephalon), a middle thorax, and a posterior pygidium ("tail"). Two longitudinal grooves divide the body into three lobes. One pair of simple antennae on the head; all other appendages biramous and similar. All extinct.

Subphylum Chelicerata

The chelicerates. Body with distinct prosoma and opisthosoma, or prosoma and opisthosoma broadly fused. Antennae absent. Compound eyes absent, simple eyes varying from four pairs to none. Six pairs of appendages, including one pair of chelicerae (specialized pincerlike appendages modified for feeding), but no mandibles.

Class Merostomata (Water Scorpions and Horseshoe Crabs)

Marine animals with abdominal gills and one pair of lateral compound eyes. Examples: *Eurypterus* (water scorpion), *Limulus* (*Xiphosura*) (horseshoe crab).

Class Pycnogonida (Sea Spiders)

Small, marine, spiderlike chelicerates; well-developed cephalothorax and reduced abdomen. Example: *Pycnogonum*.

Class Arachnida (Scorpions, Spiders, Ticks, and Mites)

Mainly terrestrial animals, although a few inhabit aquatic areas, with six pairs of appendages. From the anterior, the first pair of appendages are chelicerae with claw or fang; the second pair are pedipalps (variously modified for sensory perception, sperm transfer, grasping); the remaining four pairs are walking legs. Head and thoracic areas fused to form cephalothorax (=prosoma). Prosoma and opisthosoma distinct and narrowly connected by pedicel, or else broadly fused. Examples: *Centruroides* (scorpion), *Argiope* (spider), *Dermacentor* (tick), *Trombicula* (mite).

Subphylum Crustacea

The crustaceans. Mainly aquatic animals with cephalothorax (fused head and thorax) usually covered by a dorsal carapace; hardened chitinous exoskeleton; biramous appendages. Head with two pairs of antennae, one pair of mandibles, and two pairs of maxillae (accessory mouthparts).

Class Branchiopoda (Branchiopods)

Small crustaceans with trunk appendages flattened and leaflike, often modified for filter-feeding. Commonly live in temporary ponds or pools, mainly freshwater. Examples: *Daphnia* (water flea), *Artemia* (brine shrimp).

Class Cirripedia (Barnacles)

Adults sessile and modified for sessile or parasitic existence; carapace covered with calcareous plates; six pairs of thoracic appendages (cirri) modified for feeding. All marine. Examples: *Balanus* (acorn barnacle), *Lepas* (goose barnacle).

Class Malacostraca

Large crustaceans with trunk consisting of fourteen segments plus a posterior telson, thorax with eight segments and abdomen with six segments. All segments typically bear appendages, first antennae, walking legs, and anterior abdominal appendages biramous, first one, two, or three pairs of thoracic appendages modified as maxillipeds. Examples: *Homarus* (lobster), *Penaeus* (shrimp), *Procambarus* (crayfish), *Callinectes* (blue crab).

Subphylum Uniramia (Uniramians)

Mainly terrestrial animals with uniramous (unbranched) appendages; head with one pair of antennae, one pair of mandibles, and one or two pairs of maxillae. Respiration via tracheal tubules and spiracles; excretion by Malpighian tubules.

Class Diplopoda (Millipedes)

Cylindrical body; one pair of short antennae; two pairs of walking legs per abdominal segment. Example: *Narceus* (*Spirobolus*).

Class Chilopoda (Centipedes)

Elongate, flattened body with many segments; one pair of antennae; abdominal segments with one pair of walking legs. Examples: *Lithobius, Scolopendra.*

Class Pauropoda (Pauropods)

Tiny cylindrical body; one pair of triramous (three-branched) antennae; no eyes; abdomen with nine or ten pairs of walking legs. Example: *Pauropus.*

Class Symphyla (Garden Centipede)

Mostly minute, slender, elongate bodies with fourteen segments; one pair of long antennae; one pair of spinnerets on last segment; twelve pairs of walking legs. No eyes. Example: *Scutigerella.*

Class Insecta (Insects)

Body with three regions: head, thorax, and abdomen; one pair of antennae; six pairs of walking legs and usually two pairs of wings. Examples: *Romalea* (grasshopper), *Musca* (housefly), *Apis* (honeybee), *Pulex* (flea).

Materials List

Living Specimens
  Crayfish (*Procambarus*, etc.)
  Daphnia
  Honeybee (*Apis*)
  Spider (*Argiope* or similar)
Preserved Specimens
  Limulus
  Romalea
  Argiope
  Crayfish, female with eggs (Demonstration)
  Representative crustaceans (Demonstration)
  Queen and drone bees (Demonstration)
  Honeybee, larva and pupa (Demonstration)
Prepared Microscope Slides
  Crayfish, compound eye (Demonstration)
  Crayfish, gill, cross section (Demonstration)
  Honeybee, mouthparts (Demonstration)
  Honeybee, spiracle (Demonstration)
  Insect cuticle (Demonstration)
  Argiope, walking leg

Plastic Mounts
  Argiope
  Peripatus
  Lepas
  Balanus
  Crayfish appendages
  Millipede
  Centipede
  Honeybee
  Insect life cycle

Subphylum Chelicerata

The Horseshoe Crab: *Limulus*

The horseshoe crab, *Limulus* (Xiphosura) *polyphemus* (figure 13.2), is common in the shallow waters and sandy shores of the Atlantic coast of North America from Canada to Mexico. It is one of the few surviving members of an ancient group of chelicerate arthropods some 200 million years old.

Study a preserved or dried specimen of *Limulus* and observe the tough, leathery **carapace** that covers the **cephalothorax** (fused head and thorax). Behind the horseshoe-shaped cephalothorax is a tapering, hexagonal **abdomen** and a long posterior **telson.** On the dorsal surface of the carapace, identify the two lateral **compound eyes** and the two **simple eyes** located on opposite sides of a small anterior spine. Along the sides of the abdomen find six pairs of **movable spines.**

On the ventral surface of the cephalothorax (figure 13.3), identify the **seven pairs of appendages**: an anterior pair of **chelicerae** located in front of the mouth, four pairs of **chelate** (pincer) **legs,** a longer sixth pair of appendages with specialized tips for cleaning the gills and pushing the carapace through the sea bottom sediments, and a small pair of spiny **chilaria.** On the abdomen find the six pairs of broad, flat appendages, one pair per abdominal segment. The first pair of plates forms the **genital operculum,** which bears a pair of genital pores through which the gametes are shed. Behind the second to sixth pair of abdominal appendages are the many thin, leaflike, highly vascularized folds of the **book gills.** When submerged, these abdominal plates beat almost constantly to provide a continuous flow of seawater over the respiratory surfaces of the book gills.

The **mouth** is located ventrally near the center of the base of the walking legs. The spiny **gnathobases** at the base of the legs shred the food and push it toward the mouth.

Limulus feeds on worms, molluscs, and other small invertebrates in shallow ocean sediments. The crab plows through the sand by arching its back and pushing against the sand with the sixth pair of appendages and the telson. Awkward as it may sound, this crude form of locomotion and feeding has been effective for some 200 million years. *Limulus* is also able to swim short distances by vigorous beating of the abdominal platelike appendages.

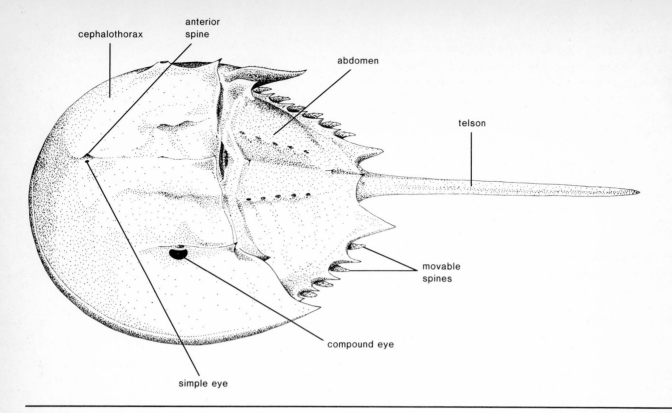

Fig. 13.2 Horseshoe crab, *Limulus,* dorsolateral view.

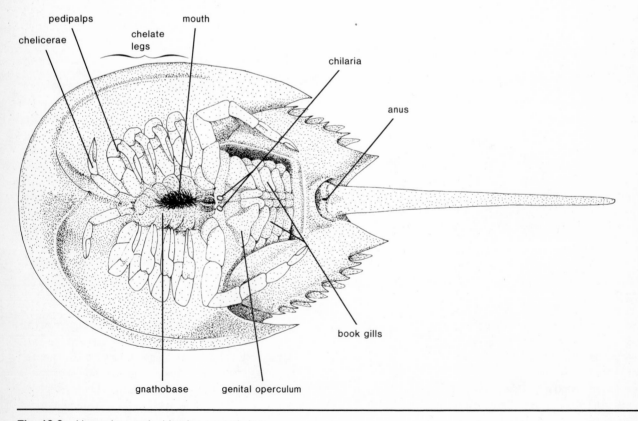

Fig. 13.3 Horseshoe crab, *Limulus,* ventral view.

The Spider: *Argiope*

Spiders, belonging to the Order Araneida (=Araneae), are an ancient group, with some fossils dating back over 300 million years. With two exceptions, Antarctica and the oceans below the intertidal zone, spiders can be found throughout the world. Of an estimated world total of 50,000 species, about 30,000 have been named and described. Sometimes spiders are very abundant; grassy fields may contain over two million spiders per hectare. All spiders are carnivorous, and thus are important for controlling insect pests. The common garden spider serves well as an example of this important group of chelicerate arthropods.

Obtain a preserved or plastic embedded specimen of the garden spider *Argiope* (formerly called *Miranda*) and study it under your dissecting microscope. The anterior region from which the legs project is called the **prosoma** (=cephalothorax) and is composed of the head and thorax fused together. The posterior part is called the **opisthosoma** (=abdomen). The two regions are connected by a narrow **pedicel.** The pedicel is seen most easily by gently bending the spider to expose the area between the overlapping opisthosoma and the prosoma (figures 13.4b and 13.4c). If your specimen is of a genus other than *Argiope,* the shape and size may vary, but the anatomical parts will be similar.

Observe that when spiders die, their legs close over the ventral side. The reason for this is that spiders do not have extensor muscles for their legs; they have only flexor muscles. Living spiders extend their legs by using blood as a hydraulic fluid.

Externally, the body of a spider is covered by a **chitinous exoskeleton,** which consists of numerous hardened plates called **sclerites** (for example, carapace, sternum). The sclerites are separated by sutures or thin membranous areas. The structure of the body wall is the same as that described for the grasshopper (p. 191). The sclerites provide a number of functions: body rigidity, muscle attachment sites, prevention from desiccation, and protection from predators.

Male and female spiders are usually very different in appearance from one another. The females may grow to be ten or more times larger than the males and may be very differently marked and colored. Such difference between the male and female is called **sexual dimorphism.**

The head region of the spider bears the eyes and mouthparts (figure 13.4a). Most spiders have eight eyes (four pairs), but some have fewer or even none at all. Garden spiders have eight simple eyes called **ocelli,** visible at the anterior end of the carapace. Vision plays a relatively minor role in the behavior of most spiders; they rely more on tactile and chemical senses than vision to locate their prey. Eyes often aid in detecting subtle changes in light intensity and motion; but, with the exception of hunting spiders (such as the wolf, crab, and jumping spiders), the eyes of spiders do not form actual images.

The area immediately below the eyes and down to the edge of the carapace is called the **clypeus.** Extending downward from below the clypeus is a pair of **chelicerae.** The chelicerae consist of a thick basal segment called the **paturon** and a small distal segment called the **fang** (figure 13.5). The fang articulates with the paturon. It is with the fang that a spider bites and poisons its prey.

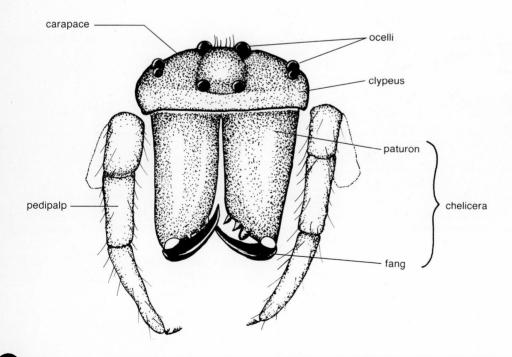

Fig. 13.4a   Garden spider, *Argiope,* anterior view of head.

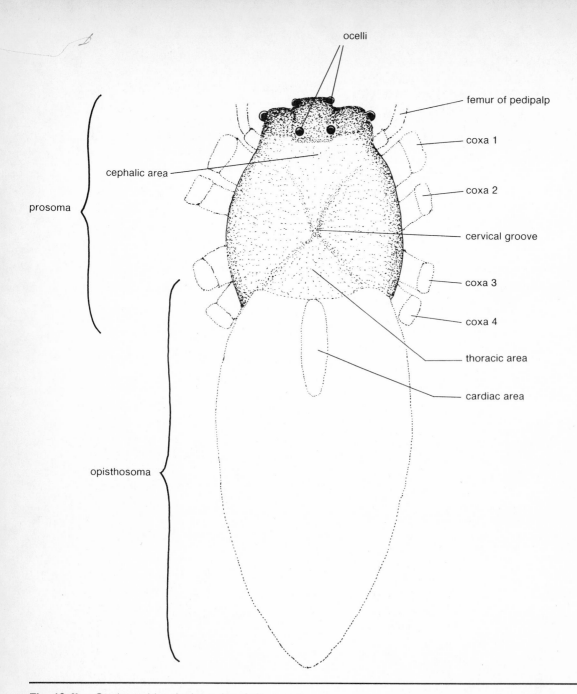

ocelli

femur of pedipalp

coxa 1

coxa 2

cephalic area

cervical groove

prosoma

coxa 3

coxa 4

thoracic area

cardiac area

opisthosoma

Fig. 13.4b Garden spider, *Argiope,* dorsal view, most appendages removed.

Lateral to and slightly behind the chelicerae is a pair of **pedipalps** (=palps) that resemble small legs. (If your specimen is a male, the distal end of the pedipalp is enlarged to form the copulatory organ.) Now, turn the spider ventral side up and gently pry the legs away from the underside of the spider. Pedipalps consist of only six segments (figures 13.6b and 13.6c). The basal segment (coxa) of the pedipalp is expanded to form the **maxilla** or **endite** (=gnathobase). The endite often has a brush of hairs at its distal end called the **scopula.** The scopula aids in sponging up fluids from the prey as the spider feeds. The endites are manipulated to squeeze the fluids from the prey.

The **labium** (see figure 13.4c) is located between the two endites and immediately anterior to the sternum. The labium is sternite No. 2, or the ventral plate of the second fused segment.

The mouth (not visible) is located between the area formed by the endites, labium, and chelicerae. Note that there are no mandibles with which to chew the food. Spiders can eat only liquid foods. When spiders "eat" their prey, they first suck out the blood. Then, through the wound, they pour in digestive juices (the chief enzymes are proteases and lipases) that convert the prey's inner tissues to a soup. The soup is then sucked out by the spider, leaving the hollow remains of the prey.

Chapter 13

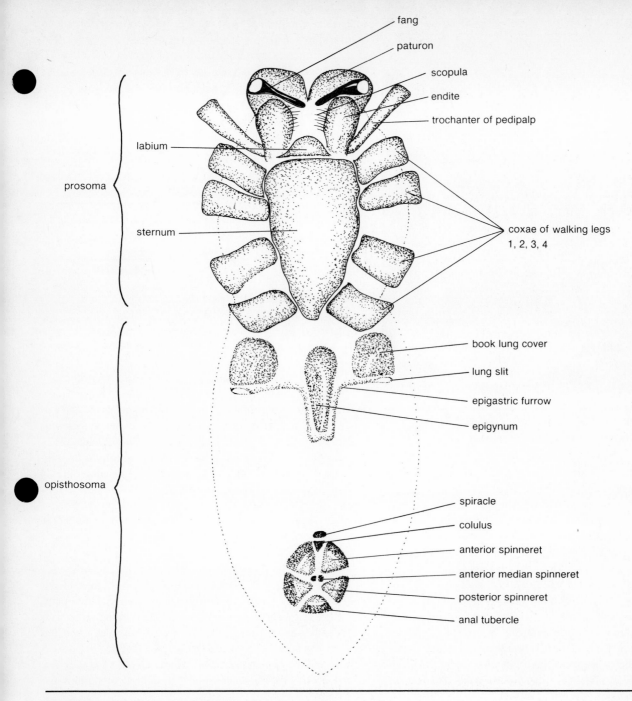

Fig. 13.4c Garden spider, *Argiope,* ventral view, most appendages removed.

The **sternum** is the large plate on the ventral surface of the prosoma. It is surrounded laterally by the coxae of the four pairs of walking legs. The sternum represents the fused segments of sternites Nos. 3–6.

The pedicel connects the prosoma to the opisthosoma, and represents the seventh body segment. Except in the very primitive spiders (Suborder Liphistiomorpha), none of the original opisthosomal segmentation is evident in spiders. From histological studies, we know that the opisthosoma in spiders represents segments 7 to 17, and that the spinnerets come from segments 9 and 10, and the anus from segment 18.

On the ventral, anterior portion of the opisthosoma, find the distinct groove that runs from one side to the other (figure 13.4c). This is the **epigastric furrow.** At each end of the epigastric furrow is a **lung slit,** the entrance to a **book lung.** Most of the air that larger spiders breathe enters through the lung slits. On the central, anterior edge of the epigastric furrow are the **gonopores.** If your specimen is a female, there will be a large, intricate, sclerotized plate extending posteriorly, the **epigynum.** If your specimen is a male, a small plate may or may not be visible. Male spiders do not have a penis. Instead, the palpal

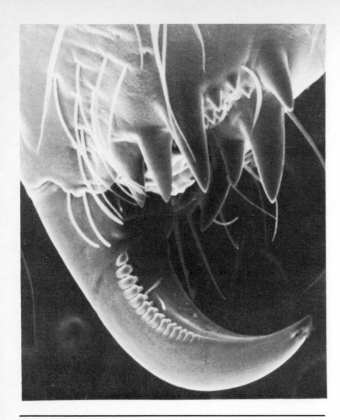

Fig. 13.5 Fang of spider. (Scanning electron micrograph courtesy of North Carolina State University.)

organ of the pedipalp is enlarged and serves as the **copulatory organ** (penis analogue) (figure 13.6c).

Near the ventral, posterior end of the opisthosoma locate the three pairs of **spinnerets** (see figure 13.4c). Two pairs, the anteriors and posteriors, are larger and distinct; the anterior median pair is smaller and tucked in between the larger pairs. The single **anal tubercle** is located at the central posterior edge of the spinneret cluster. Between the anterior bases of the anterior spinnerets find the small organ called the **colulus.** It is considered to be a small, vestigial spinneret. Immediately anterior to the spinnerets, in a groove, is an extremely small **spiracle,** which serves as the entrance to the tracheal system. As the spiracle is all but invisible, it is not suggested that you look for it.

Turn the spider over and observe the dorsal side. On the central, slightly posterior part of the carapace, is a depression that represents the **cervical groove** (see figure 13.4b). Aside from marking the boundary between the cephalic and thoracic regions, it is also an apodeme, serving as a muscle attachment site on the inside of the carapace.

Manipulate the spider gently so that the pedicel can be seen from above. On the dorsal surface of the pedicel is a small pair of plates called the **lorum** (may be difficult to see).

There are no external, dorsal features worthy of note on the opisthosoma of spiders. However, in the central, dorsal anterior region is the cardiac area, beneath which lies the longitudinal **heart** that pumps blood anteriorly. Pale markings on the opisthosoma result from the deposition of guanine, an excretory product of spiders.

Examine a walking leg from your specimen and in a prepared microscope slide. Identify the parts as shown in figure 13.6a. Compare the segments of a pedipalp with those of a walking leg (figures 13.6b and 13.6c). How do these appendages illustrate adaptation? Which segments of the pedipalps show the highest degree of specialization?

Demonstrations

1. Living *Argiope* or other spider in terrarium.
2. Plastic mount of *Argiope.*

Subphylum Crustacea, Class Branchiopoda

A Water Flea: *Daphnia*

The water flea, *Daphnia* (figure 13.7), is a common microscopic crustacean found in many bodies of fresh water. Its structure is simpler than that of the crayfish, and because of its transparency and small size, living *Daphnia* can be studied easily in the laboratory. *Daphnia* and other members of the Class Cladocera of the Subphylum Crustacea are popularly known as "water fleas" because of their characteristic swimming movements. These animals constitute an important element in the diet of many fish and other, larger, aquatic animals.

External Anatomy

Daphnia swims by rapid movements of its two large **antennae** and feeds on microscopic food particles filtered from the water by means of complex movements of five pairs of thoracic appendages, which bear many setae. Food particles removed from the water are collected in a **median ventral groove** located at the base of the legs and are passed forward to the mouthparts where larger particles are macerated by the sclerotized (hardened) **mandibles** before passing into the mouth.

The body of *Daphnia* is laterally compressed and not obviously segmented. It consists of three main regions—an anterior head, a large thorax, and a smaller postabdomen. A true abdomen is absent in the Cladocera. The thorax is covered by a chitinous **carapace,** which has a bivalved appearance, but which is fused dorsally and open ventrally. The compact head is fully enclosed in the exoskeleton (that is, it does not open ventrally) and is bent downward. Under certain conditions, some Cladocera exhibit seasonal changes in the morphology of the head by forming a dorsal elongation or "helmet." (Certain other parts of the body may be modified also.) This kind of morphological variation is called **cyclomorphosis,** and there is evidence that it is influenced by temperature, turbulence of the water, heredity, and other factors.

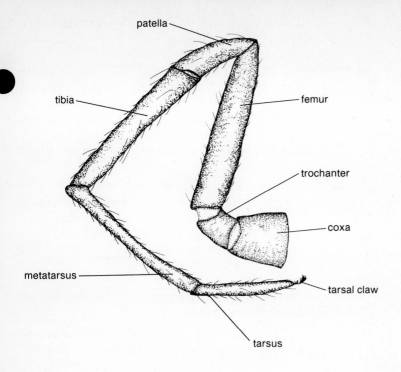

patella

tibia

femur

trochanter

coxa

metatarsus

tarsal claw

tarsus

Fig. 13.6a Garden spider, *Argiope,* walking leg.

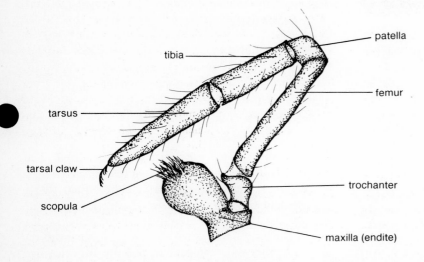

patella

tibia

tarsus

femur

tarsal claw

scopula

trochanter

maxilla (endite)

Fig. 13.6b Garden spider, *Argiope,* pedipalp of female.

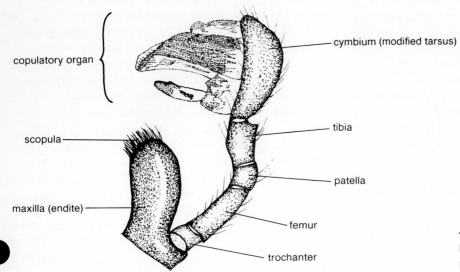

copulatory organ

cymbium (modified tarsus)

scopula

tibia

patella

maxilla (endite)

femur

trochanter

Fig. 13.6c Garden spider, *Argiope,* pedipalp of male.

Arthropoda

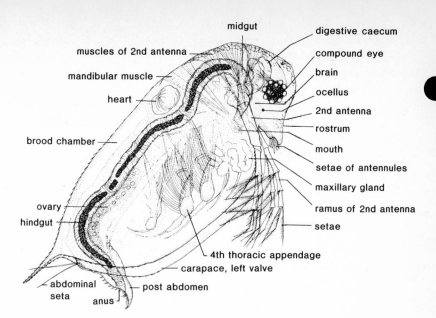

Fig. 13.7 *Daphnia*, lateral view.

The most conspicuous structure on the head is the single large compound eye with a central mass of pigmented granules surrounded by numerous hyaline lenses. A single small ocellus (simple eye) lies posterior to the compound eye.

Locate the first and second antennae. The first antennae, or antennules, are small and unsegmented, and are located on the ventral side of the head near the posterior margin. They bear chemical sense receptors (specialized setae) and are primarily sensory in function. The tapering projection of the head between the antennules is the **rostrum.**

The second antennae are very large structures attached laterally near the posterior margin of the head. Each second antenna consists of a stout, unjointed **basal segment** and segmented **dorsal** and **ventral rami.** Each ramus consists of several segments and bears numerous **plumose** (featherlike) **setae.** The second antennae are moved by a set of strong muscles that originate on the dorsal side of the carapace near the junction of the head and the thorax. Ventrally, near the same junction, the mouth may be found. It is surrounded by six mouthparts: (1) a dorsal median **labrum** (upper lip); (2) a pair of stout **mandibles;** (3) two small pointed **maxillae,** which are used to push food between the mandibles; and (4) a ventral median **labium** (lower lip).

The **thorax** consists of six segments, and all but the fifth segment bear paired appendages. These five pairs of lobed, leaflike (foliaceous) appendages are basically biramous, but this condition is masked by their high degree of specialization. All of the thoracic appendages bear many hairs and setae used to strain food particles from the water.

As previously stated, a true abdomen is absent in *Daphnia,* but there is a definite **postabdomen** located at the posterior end of the body. It is usually held bent forward under the thorax so that the dorsal side of the postabdomen is directed downward. The postabdomen bears

no limbs; but it has two rows of spines, and it terminates in two long abdominal setae. It is equipped with an effective musculature and appears to function mainly in cleaning debris from the thoracic legs, although it may also aid locomotion.

Internal Anatomy

The complicated musculature of *Daphnia* tends to obscure some of the smaller anatomical details, but the major elements of most of the internal systems can be seen in a whole specimen. Study as many as possible of the following features on your specimen.

1. **Digestive system**—Relatively unspecialized, and consisting of the **mouth, foregut** (esophagus) with a cuticular lining, **midgut** (stomach-intestine), and a **hindgut** (rectum), the latter also with a cuticular lining. Digestion takes place principally in the midgut region. Located anteriorly (in the head region above the compound eyes) are two digestive caeca, which probably secrete digestive enzymes.

2. **Circulatory system**—A simple oval **heart** lies behind the head and dorsal to the gut. Blood enters through two lateral **ostia** and leaves via an anterior opening. There are no definite blood vessels, and the blood is roughly channeled through the large central **haemocoel** by a series of thin mesenteries.

3. **Respiratory system**—None. Exchange of gases occurs through the body wall, especially through that portion lining the inner surfaces of the valves, and it occurs, also, through the surfaces of the legs.

4. **Excretory system**—Two looped **maxillary glands** located near the anterior end of the valves are thought to be excretory in function.

5. **Nervous system**—Double **ventral nerve cord,** few **ganglia,** paired **nerves,** and a **brain** located between the foregut and the compound eyes.

6. **Reproductive system**—Two elongated **ovaries** lie lateral to the midgut in the female *Daphnia*. Posteriorly, each ovary is connected to the **brood chamber** (a cavity dorsal to the body proper and located between the valves of the carapace; closed posteriorly by dorsal processes of the postabdomen) by a thin **oviduct.**

In the male, two **testes** occupy the same location but are smaller in size and are continued posteriorly as **sperm ducts,** which follow the gut and open on the postabdomen near the anus. In some species of Cladocera, a part of the postabdomen of the male is modified to form a copulatory organ.

Under favorable conditions, a female *Daphnia* lays thin-shelled, diploid eggs, which contain little yolk and develop parthenogenetically. Under adverse conditions, however, the females produce thick-shelled, resistant eggs with large yolk reserves. The latter type of eggs are haploid and require fertilization by a male for their development. Both types of eggs are deposited in the brood chamber.

Male *Daphnia* appear in natural populations only at certain times of the year and often constitute a small proportion of the total population. The specific factors responsible for the production of males are not known, despite much investigation, but it appears that male production can be induced by a combination of environmental factors: crowding, decreasing food supply, and water temperature. These environmental factors appear to affect the metabolism of the female in such a way that parthenogenetic male eggs rather than parthenogenetic female eggs are released into the brood chamber.

There are no larval stages in the development of the Cladocera (a situation not at all typical of Crustacea!), but four different periods can be recognized in the life history of a cladoceran: egg, juvenile, adolescent, and adult.

Parthenogenetic eggs develop in the brood chamber into a stage already having the adult form before their release (first instar young). Subsequent development continues through a series of molts in which the carapace and other chitinous portions are lost and the animal increases in size. The average life span is said to cover about seventeen instars (instar = intermolt stage).

Resistant eggs are retained in the brood chamber until the next molt. A portion of the brood chamber surrounding them is modified into a thick-walled protective case called an **ephippium.** At the next molt of the parent, the ephippium and its contained eggs are shed.

Subphylum Crustacea, Class Malacostraca

The Crayfish: *Procambarus*

The crayfish is a large aquatic arthropod, which effectively illustrates many of the basic characteristics of the phylum. Large southern crayfish (*Procambarus*) or western crayfish (*Astacus*) are especially suitable for laboratory study because of their large size, but members of other common genera, such as *Cambarus* and *Orconectes,* can also be used. The American lobster *Homarus* is very similar in structure to the crayfish and may also be used when available.

External Anatomy

Obtain an anesthetized or freshly killed crayfish and study the general organization of the body (figure 13.8). Observe that the body is divided into two major regions: an anterior **cephalothorax,** which is made up of a fused head and thorax, and a posterior **abdomen.** Extending anteriorly from the cephalothorax between the compound eyes is a pointed **rostrum.** Find the **cervical groove,** which represents the line of fusion between the head and the thorax. The hard outer covering (exoskeleton) of the cephalothorax is the carapace; note that the carapace and the chitinous exoskeleton on the dorsal side of the abdomen are hardened by the deposition of mineral salts (except in newly molted specimens). The crayfish body is made up of a total of nineteen segments. Segments I–XIII make up the cephalothorax (five in the head and eight in the thorax), and segments XIV–XIX constitute the abdomen.

Observe the two large **compound eyes** on movable eyestalks, two large **antennae** (sometimes called second antennae), and two smaller **antennules** (sometimes called first antennae). The pointed rostrum projects forward between the eyes. The **mouth** is ventral, surrounded and largely concealed by several pairs of modified appendages that serve as mouthparts, and will be studied later. Locate the five pairs of large **walking legs** and five pairs of small abdominal appendages or **swimmerets.** The first pair of walking legs bear large pincers or chelae and thus they are called **chelipeds** ("pincer legs"). How many of the other pairs of walking legs also bear chelae? On the last abdominal segment is a pair of large flattened lateral appendages, the **uropods.** Find the **anus** on the ventral side of the **telson,** a medial extension of the last abdominal segment (not an appendage).

Crayfish are dioecious, and each sex exhibits distinctive secondary sexual characteristics. Females have a broad abdomen, rudimentary appendages (swimmerets) on the anterior abdominal segments, a **seminal receptacle** on the ventral surface of the exoskeleton between the

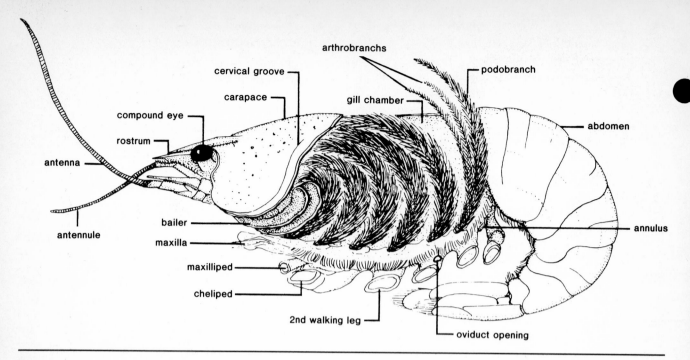

Fig. 13.8 Crayfish, lateral view, portion of carapace removed to show gills.

fourth and fifth pairs of walking legs, and two female sex openings located at the bases of the third pair of walking legs. Males have a narrower abdomen, enlarged swimmerets (**gonopods**) on the first abdominal segments (modified for the transfer of sperm to the seminal receptacle of the female), and openings of the vasa deferentia (male sex ducts) at the base of each fifth walking leg. Determine the sex of the specimen you are examining and observe also specimens of the opposite sex distributed around the laboratory.

Appendages

In early ancestors of the arthropods, the appendages of all body segments were biramous and similar on all segments. During the evolution of arthropods, various segments and the attached appendages have become adapted to serve in sensory perception, defense, swimming, walking, grasping, sperm transfer, and so forth. Since the body segments and appendages arise similarly during embryonic development, they are said to be homologous structures. **Homology** is an important principle in zoology, and the study of structural homologies has provided much valuable evidence about the evolutionary history and relationships of animals.

The paired appendages of the crayfish provide an excellent illustration of **serial homology,** the adaptation of a longitudinal series of originally similar organs to carry on different functions. The appendages of primitive arthropods were **biramous** (two-branched). This is the simplest type of arthropod appendage and is most nearly approximated in the crayfish by the swimmerets of the abdominal segments. Remove an appendage from the third

abdominal segment by cutting it free near its attachment, and observe that it consists of a basal portion, the **protopodite,** and two terminal branches, an outer **exopodite** and an inner **endopodite.**

Carefully remove the appendages from one side of the body, starting with the first walking leg (cheliped) and working forward and then backward. Remove each appendage close to its attachment to the body and arrange it in order on a sheet of paper as in figure 13.9. Study each appendage with the aid of figures 13.9 and 13.10 and the descriptions provided below.

1. **Antennule** (first antenna)—Base of three segments, and two long, many-jointed filaments. This appendage is probably not truly biramous. The slender inner branch arises from the base of the outer thicker filament and is, therefore, probably not homologous with an endopodite. The two basal segments are homologous with the protopodite. Within the basal segment of each antennule is a **statocyst,** a saclike sense organ that opens dorsally via a small pore.

2. **Antenna** (second antenna)—Protopodite of the two segments; endopodite long, many-jointed; exopodite a short scale.

3. **Mandible**—Basal segment of protopodite greatly enlarged as functional jaw; second segment small and forming the base of the palpus; endopodite, the two small distal segments of the palpus; exopodite lost.

   (Note: The metastoma, or bifurcated "lower lip" between the mandible and the first maxilla, is a leaflike blade that fits closely over the convex surface of the mandible. It is not a true appendage.)

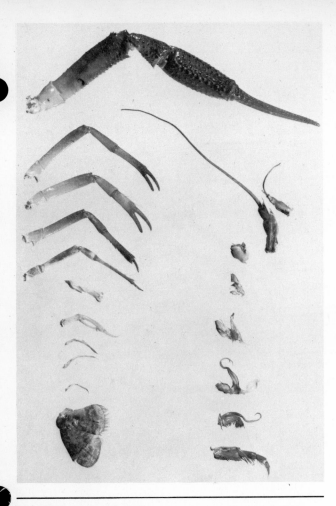

Fig. 13.9 Crayfish appendages. (Photograph by Indiana University Audiovisual Center.)

4. **First maxilla**—Protopodite of two flat, leaflike segments; endopodite of two segments, the distal one narrow and pointed; exopodite lost.

5. **Second maxilla**—Protopodite of two flat, bilobed segments; endopodite, a small pointed segment; exopodite forming part of a large elongated plate, the **bailer.**

6. **First maxilliped**—Protopodite of two flat segments extending inward; endopodite and exopodite are both present.

7. **Second and third maxillipeds**—Protopodite of two segments; endopodite of five distinct segments; exopodite relatively thin.

8. Five walking legs or **pereiopods**—Protopodite of two segments; endopodite of five joints as in the second and third maxillipeds; exopodite lost. In the large chelipeds (first walking legs), the second segment of the protopodite and the first segments of the endopodite are fused together.

9. **First and second swimmerets**—The first is rudimentary in the female; in the male, the protopodite and the endopodite are fused, and the entire structure is modified for transferring spermatozoa to the seminal receptacle of the female; exopodite is lost.

10. **Third to fifth** (second to fifth in the female) **abdominal appendages** (swimmerets)—Protopodite of two segments; endopodite and exopodite flat and filamentous.

11. **Sixth abdominal appendage** (uropod)—Greatly enlarged and modified to form (with the telson) the tail fan, which is used for swimming backward.

Respiratory System

Locate the gills within the **branchial chambers** at each side of the carapace. The lateral flaps of the carapace, covering the gills, are termed **branchiostegites.** Remove the left branchiostegite carefully by cutting away the carapace with your scissors to expose the gills; make your cut carefully to avoid damage to the underlying gills. The gills are feathery projections of the body wall, which contain blood channels. Water is pumped through the gill chamber by the action of the **bailer,** a paddlelike projection of one of the oral appendages (the second maxilla). Oxygen dissolved in the water diffuses across the thin walls of the gills and combines with the blood in the gills. Carbon dioxide diffuses across the walls of the gills in the opposite direction.

Observe that the gills occur in distinct longitudinal rows. How many rows of gills are there in the specimen? The outer row is attached to the base of certain of the appendages. Which ones? These outer gills are the **podobranchs** ("foot gills"). How many podobranchs do you find in the specimen? Separate the gills carefully with your probe or dissecting needle, and locate the inner row(s) of gills. These inner gills are the **arthrobranchs** ("joint gills") and are attached to the chitinous membrane, which joins the appendages to the thorax (see figure 13.8). How many rows of arthrobranchs do you find in the specimen?

Remove a gill with your scissors by cutting it free near its point of attachment and place it in a watch glass filled with water. Observe the numerous gill filaments arranged along a central axis. Study also a demonstration slide of a crayfish gill and note the afferent and efferent blood vessels in the central axis.

Internal Anatomy

Remove the remainder of the carapace up to the rostrum by cutting forward from the rear of the carapace up to a level just behind the eyes. Leave the eyes, rostrum, and adjacent anterior portion of the carapace intact. With your scalpel, carefully separate the hard carapace from the thin, soft underlying layer of **hypodermis** as you cut away the carapace. The hypodermis is the tissue that secretes the carapace and other parts of the exoskeleton. Next, remove the remaining gills on the left side by cutting them near their points of attachment, and cut away the chitinous membrane underlying the gills to expose the internal organs as in figure 13.11.

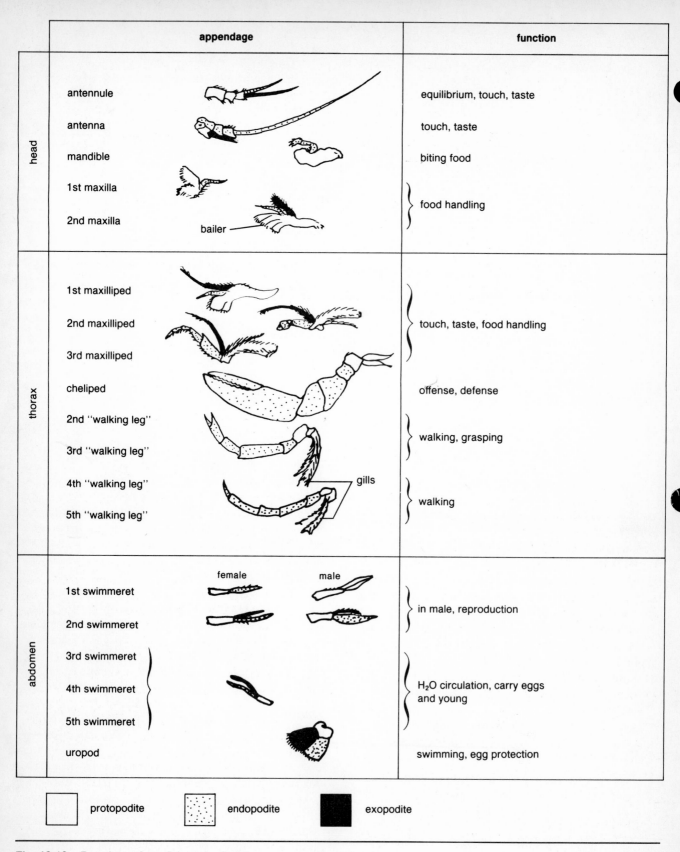

Fig. 13.10 Functions of crayfish appendages.

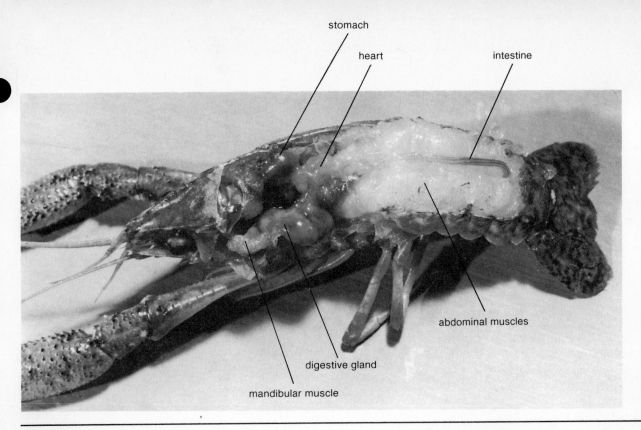

stomach

heart

intestine

abdominal muscles

digestive gland

mandibular muscle

Fig. 13.11 Crayfish, dissected, dorsolateral view.
(Photograph by Indiana University Audiovisual Center.)

Muscular System. Locate first some of the principal muscles of the crayfish. These include the two longitudinal bands of **extensor muscles** that run along the dorsal side of the thorax and extend back along the dorsal surface of the abdomen. What is their function? Find the large **flexor muscles** in the abdomen lying below the extensor muscles. These large muscles serve to bend or flex the abdomen and hence provide the force for the quick backward thrust of the animal when it is alarmed.

The **gastric muscles** attach the stomach to the inner wall of the carapace, and two large **mandibular** muscles **originate** (fixed end of the muscle) on the carapace and **insert** (movable end of the muscle) on the mandibles. Contraction of these muscles causes the crushing or grinding action of the mandibles. Each of the other appendages is also equipped with a musculature; **flexor muscles** serve to draw the appendage closer to the body or to its point of attachment, and **extensor muscles** serve to extend, or straighten out, the appendage.

Circulatory System. Locate the small membranous **heart** just posterior to the stomach (see figure 13.11). If you are dissecting a preserved specimen, you may find the heart filled with colored latex.

Identify the **pericardial sinus** (the space in which the heart lies), the **ostia** (openings in the wall of the heart), and the delicate **arteries** attached to the heart. Because of their small size, the arteries are often difficult to locate.

Carefully search among the internal organs and try to locate the following arteries and determine which structures they supply with blood: (1) **ophthalmic artery,** (2) two **antennary arteries,** (3) two **hepatic arteries,** (4) **dorsal abdominal artery,** and (5) **sternal artery.** The circulatory system of the crayfish is an open system; blood is pumped from the heart through the major arteries that branch into smaller arteries and supply blood to the various organs of the body. From the arteries, the blood flows into open spaces, or **sinuses,** within and between the organs and then into a large **sternal sinus** along the floor of the thorax. Channels from the sternal sinus lead to the gills where the blood is oxygenated before returning to the pericardial sinus and into the heart via the ostia.

Crayfish blood consists of a nearly **colorless plasma** containing a dissolved respiratory pigment **hemocyanin** (a complex copper-containing protein) and numerous amoeboid blood cells.

Female Reproductive System. Locate the paired **ovaries** lying ventral and lateral to the pericardial sinus (figure 13.12) and trace the delicate **oviducts** to their openings at the base of the third walking legs. Are there **eggs** in the ovaries? What color are they? Locate again the **seminal receptacle** where sperm is deposited during copulation. Observe the demonstration of a female "in berry" with clusters of eggs attached to the bristles on the swimmerets.

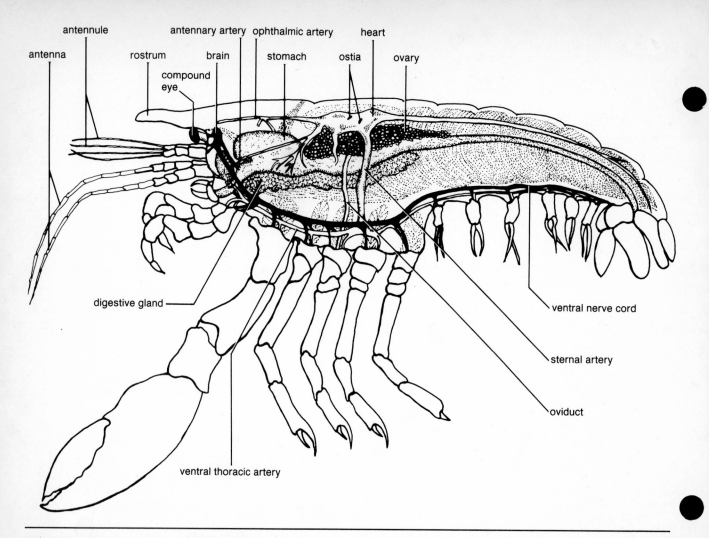

antenna · antennule · rostrum · compound eye · brain · antennary artery · ophthalmic artery · stomach · heart · ostia · ovary · digestive gland · ventral nerve cord · sternal artery · oviduct · ventral thoracic artery

Fig. 13.12 Crayfish, internal anatomy, female.

Male Reproductive System. Find the paired **testes** beneath the pericardial cavity, and trace the **vasa deferentia** to their openings. Where are the male openings located? Compare the first abdominal appendages of the male specimen to those of a female specimen.

Digestive System. The **mouth** is obscured by the several oral appendages and can be located by moving these appendages aside with your forceps. Locate, behind the mouth, the tubular **esophagus** leading to the **stomach** just behind the rostrum. The stomach is divided into a large anterior **cardiac chamber** and a smaller posterior **pyloric chamber.** Cut open the stomach and study the **gastric mill,** a grinding apparatus consisting of three chitinous teeth. Observe in the pyloric chamber the **chitinous bristles,** which serve as a strainer to prevent large chunks of food from passing into the intestine. You may also find calcareous **gastroliths** attached to the walls of the stomach. These calcareous bodies serve as stores of calcium salts, and they appear to play an important role in the calcification of the exoskeleton after molting. Locate the large digestive gland,

or **hepatopancreas.** What is its function? Trace the **intestine,** behind the **stomach,** into the abdomen. Find the blind, saclike **intestinal caecum** in the abdomen, and the terminal **anus.**

Excretory System. The excretory structures of the crayfish are two large **green glands** (also called antennal glands) located ventrally in the head region near the base of the antennae. These glands are usually dark red or brown (not green!) in preserved specimens. Find the two **excretory pores** at the bases of the antennae through which the wastes are discharged to the exterior.

Nervous System and Sense Organs. Carefully remove the organs of the reproductive, digestive, and circulatory systems in the head and thoracic regions to expose the ventral surface of the cephalothorax. Remove also the muscles and intestine from the abdominal region, carefully cut away the skeletal plates that cover the anterior part of the nervous system, and locate the **ventral nerve cord.** Observe the **segmental ganglia** and their **paired lateral nerves.** Trace

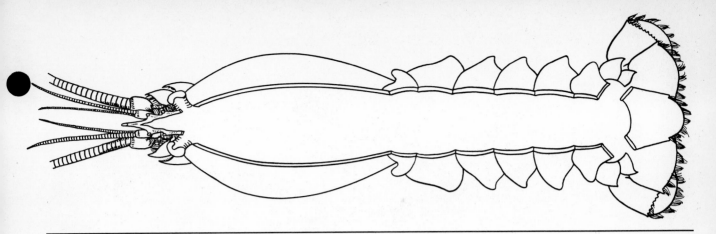

Fig. 13.13 Drawing of nervous and excretory systems of crayfish.

the nerve cord forward to the thorax and carefully cut away the skeleton that surrounds the nerve cord in this region. Follow the nerve cord anteriorly and locate the **supraesophageal ganglion** (the "brain"), the **subesophageal ganglion,** and the **circumesophageal connectives.** Find the several pairs of nerves leading from the brain to the eyes, antennules, antennae, and mouthparts.

The most prominent sense organs of the crayfish are the **compound eyes** and the **statocysts,** both located previously. The crayfish also has numerous **sensory hairs** found on various appendages that are sensitive to touch. Special **chemoreceptors** on the antennae, antennules, and mouthparts also provide senses of taste and smell.

The statocysts are a pair of small sacs located in the basal segments of the antennules. They have a cuticular lining bearing sensory hairs. A sand grain within the sac, the **statolith,** stimulates the sensory hairs and thus provides a sense of equilibrium. The lining of the statocysts and the statoliths are lost and replaced at each molt.

Demonstrations

1. Compound eye (microscope slide).
2. Section of crayfish gill (microscope slide).
3. Female crayfish with eggs attached to swimmerets.
4. Live crayfish in aquaria for study of behavior.
5. Preserved specimens of several other types of freshwater and marine Crustacea.

The large compound eyes are made up of many individual units, the **ommatidia.** Remove one of the compound eyes and examine its surface under your stereoscopic microscope. The surface of the eye is covered with a transparent **cornea** secreted by underlying cells. Observe that the cornea is divided into many sections, or **facets.** Each facet of the cornea represents the outer end of an ommatidium. The compound eye forms a mosaic impression of the surroundings; also because of its structure, the compound eye is well adapted for detecting movements.

Draw the principal structures of the nervous and excretory systems in the outline of a crayfish provided in figure 13.13.

Subphylum Uniramia, Class Insecta

A Grasshopper: *Romalea*

Insects constitute the largest, most diverse, and most widespread class of arthropods, and indeed of all the multicellular animals. More than 900,000 species of insects have been described, and several thousand new species are added to the list each year. The insects are easily distinguished from the other classes of arthropods by a combination of the following characteristics of adult insects: (1) body divided into **three distinct regions** (head, thorax, and abdomen); (2) **one pair of antennae;** and (3) **three pairs of legs** and, usually, two pairs of wings.

Although the insects are an exceedingly diverse group of running, jumping, flying, swimming, and creeping organisms, the fundamental organization of the insect body is effectively illustrated by the grasshopper. Grasshoppers are members of a relatively unspecialized order of insects (Order Orthoptera), along with the katydids, locusts, and crickets. Specimens of the large, black, lubber grasshopper, *Romalea microptera* (figure 13.14), are usually provided for laboratory study, but the following information is generally applicable, also, to other species of grasshoppers.

Externally, the body of a grasshopper is covered by a **chitinous exoskeleton** consisting of numerous hardened plates, or **sclerites,** separated by sutures, or thin membranous areas. The latter provide the flexibility necessary for movements of the body segments and the appendages. The body wall is made up of three principal layers—an outer **cuticle;** an underlying **hypodermis,** which secretes the cuticle; and a thin, noncellular **basement membrane.** The cuticle is a complex structure consisting of three layers: a very thin outer epicuticle, an exocuticle, and an innermost

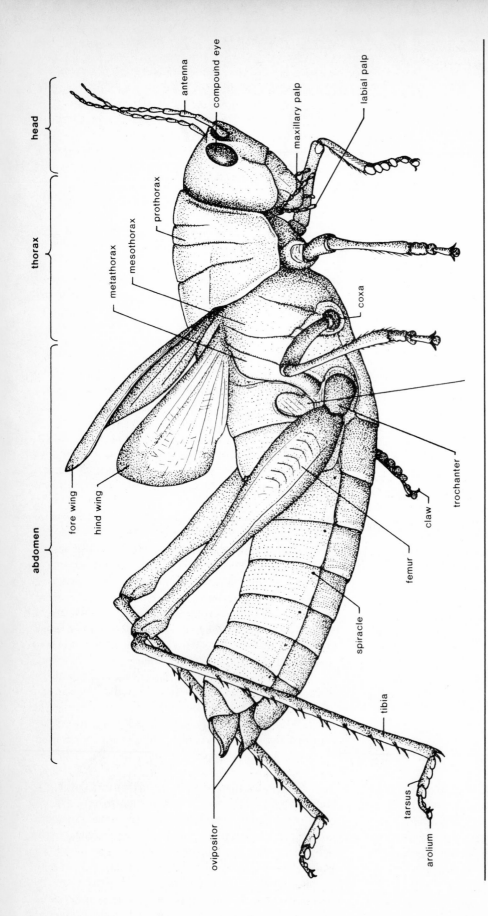

Fig. 13.14 Grasshopper, *Romalea*, lateral view, female.

head

thorax

abdomen

antenna

compound eye

maxillary palp

labial palp

prothorax

mesothorax

metathorax

coxa

fore wing

hind wing

femur

claw

trochanter

spiracle

tibia

tarsus

arolium

ovipositor

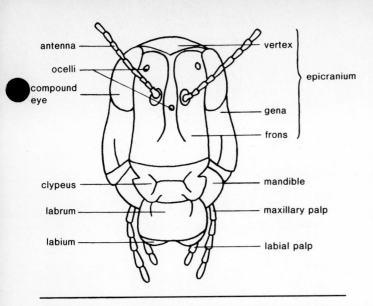

Fig. 13.15 Grasshopper, head, anterior view.

Labels on figure: antenna, ocelli, compound eye, clypeus, labrum, labium, vertex, epicranium, gena, frons, mandible, maxillary palp, labial palp

endocuticle lying directly upon the hypodermis. The endocuticle and exocuticle are composed largely of chitin, a nitrogenous polysaccharide, while the epicuticle is non-chitinous and consists chiefly of a network of fibrous proteins impregnated with wax. Study the demonstration provided in the laboratory, showing the microscopic structure of insect cuticle.

The head of the grasshopper (figure 13.15) bears the eyes, antennae, and mouthparts. Two large **compound eyes** occupy prominent sites on opposite sides of the head, and three **ocelli** (simple eyes) form an inverted triangle on the front of the head between the compound eyes. The two jointed **antennae** arising anterior to the compound eyes serve as organs of touch, smell, and taste. The **mouthparts** of a grasshopper are relatively unspecialized in comparison with those of most other kinds of insects and are of the **chewing type.** They consist of seven parts: a dorsal **labrum** or upper lip; a pair of sclerotized (hardened) **mandibles;** a pair of **maxillae,** which manipulate the food; a broad median **labium** or lower lip; and, in the center of these mouthparts, a cylindrical **hypopharynx** or tongue at the base of which the **salivary glands** empty. The labium and the maxillae bear sensory palps, which project ventrally. The mouthparts can be most easily observed and studied by removing them carefully from the head with your forceps and studying them under your stereoscopic microscope.

The thorax, like the head and abdomen, consists of several fused segments, illustrating the multisegmented ancestry of the insects. In the thorax, there are three segments, a large anterior **prothorax,** followed by two smaller segments, the **mesothorax** and the **metathorax.** Each segment bears one pair of legs. Also, the mesothorax and the metathorax each bear a pair of wings.

The legs are clearly jointed and consist of five main segments: a short **coxa,** which is attached to the ventral body wall; a small **trochanter** fused with a stout **femur;** a

slender, spiny **tibia;** and a distal **tarsus** bearing two terminal claws. Between the claws is a fleshy pad, the **arolium,** which provides a grip on smooth surfaces. All three pairs of legs are used by the grasshopper for walking and climbing, but the third pair (the metathoracic legs) are specially modified for leaping. They are equipped with an enlarged femur containing strong voluntary muscles and with an elongated tibia to provide additional leverage. Another modification of the metathoracic legs can be seen in male specimens of *Romalea* and its close relatives. The inner surface of the femur bears a row of small spines and is rubbed against the lower edge of the front wing to produce the sounds that comprise the characteristic song of the males. Certain other species of grasshoppers produce their song by rubbing together their fore and hind wings. This type of sound production is called **stridulation.**

The long, narrow **fore wings** are borne on the mesothorax and serve as a covering for the hind wings when at rest. Observe the numerous **wing veins** in the thin, membranous **hind wings.** Note the difference in texture and pigmentation between the fore and the hind wings. The grasshopper wing develops as a saclike outgrowth of the body wall; this outgrowth later flattens to form a thin double membrane, and the two opposing membranes fuse together enclosing tracheae, nerves, and blood sinuses. The hollow veins are formed by thickening of the cuticle around the **tracheae** (air tubes) or along the sinuses. The pattern of wing venation in insects is very precise and characteristic, and therefore plays an important role in classification.

The cylindrical abdomen of *Romalea* consists of eleven segments, the last three of which are reduced in size and modified either for copulation or for egg laying. Along the sides of certain of the thoracic and abdominal segments are ten pairs of tiny respiratory openings, the **spiracles,** which open into the system of **tracheal tubules.** The first abdominal segment also bears on each side a large, oval **tympanic membrane,** which serves as an organ of hearing. There is one pair of **sensory cerci** on the eleventh segment.

Internal Anatomy

Living or freshly killed specimens are most suitable for the study of internal anatomy, although well-preserved specimens can also be used with some success. The internal tissues and organs are difficult to preserve, however, because of the chitinous exoskeleton of the grasshopper, which limits penetration of the preservative fluids. Study figure 13.16 for orientation before starting your dissection.

Remove the wings, and starting at the posterior end of the body, cut through the body wall along each side just above the spiracles. Remove the freed dorsal wall and pin back the sides of the remaining body wall in your dissecting pan. Cover your dissection with water to keep the internal organs moist and pliable.

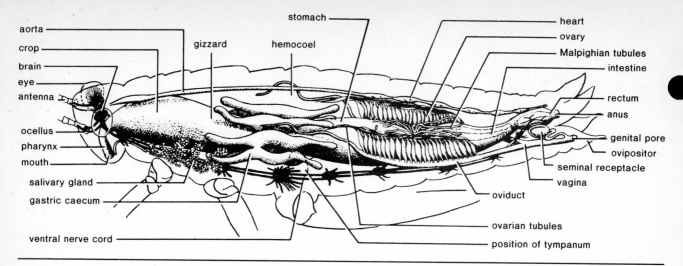

Fig. 13.16 Grasshopper, internal anatomy, female.

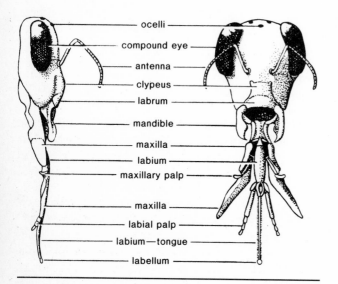

Fig. 13.17 Honeybee, head, anterior view.

Observe the **haemocoel,** the tubular **heart** with **ostia,** and the various parts of the digestive tract. The digestive tract of insects consists of three main divisions: **foregut, midgut,** and **hindgut.** These divisions are often functionally subdivided for specific functions in particular insects.

The **Malpighian tubules** are outgrowths of the digestive tract originating at the junction of the midgut and hindgut. They serve for excretion. Other internal organs include the **tracheal tubules** and the **gonads.** Is your specimen a male or female? Along the ventral surface find the **salivary glands** and the **nerve cord.** Note the **segmental ganglia** and the numerous **lateral nerves.** Observe the powerful muscles of the thorax associated with the legs and wings.

The Honeybee: *Apis mellifera*

The grasshopper was studied as an example of a relatively simple or generalized insect. Many insects, however, have become highly specialized and exhibit numerous adaptations for life in particular habitats or under special conditions. The honeybee is an example of such a specialized insect. Bees are social insects, and they live in permanent colonies organized into three **castes.** Each caste plays a distinct role in the colony and exhibits certain distinctive morphological features. The **queen** lays the eggs; the **drones** (males) serve only to fertilize the eggs; and **workers** (sterile females) carry on all the other functions of the colony, including constructing and protecting the hive, gathering food, caring for the queen, and rearing the young.

Obtain a freshly killed worker for study and examine the specimen for the general characteristics of the phylum, class, and order. Compare the general organization of the bee with that of the grasshopper. How does the surface of the exoskeleton differ from that of the grasshopper? Note the **antennae,** the **compound eyes,** and the three **ocelli** on the head. Locate the **chewing** and the **sucking mouthparts.**

Mouthparts

Remove the head and place it on a microscope slide face-up (figure 13.17). Press on the face and spread the mouthparts with your dissecting needles. Identify the two chewing structures, the **mandibles.** The sucking mouthparts consist of the **labium** or tongue (a long, middle process with a spoonlike structure at the tip), a pair of **labial palps** (one on each side of the labium), and a pair of broad, bladelike structures on the outside, the **maxillae.** Determine how these parts fit together to form a sucking tube. Compare these special sucking mouthparts with the corresponding mouthparts in the grasshopper.

Wings

Examine the wings of the bee and compare them with those of the grasshopper. Study the interlock mechanism, which links the two wings on each side together for flight. Note, also, the network of wing veins, which serves to strengthen the wings.

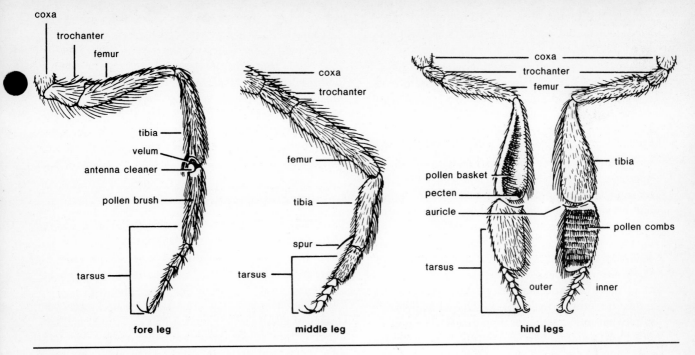

Fig. 13.18 Honeybee, leg adaptations of worker.

Legs of the Worker Bee

Consult figure 13.18 and identify the segments of each of the thoracic legs. Starting with the segment next to the body, find the **coxa, trochanter, femur, tibia,** and **tarsus** (foot). The latter is composed of several segments; observe the **tarsal claws** on the terminal segments. The median pad between the claws is the **pulvillus,** which secretes a sticky substance that enables the bee to cling to smooth surfaces. Each of the three legs of the worker exhibits some highly specialized structures; these special adaptations are not present on the appendages of the queen and the drones.

Note on the **first leg** the flattened movable spine, or **velum,** at the distal end of the tibia. This spine, with a notch in the proximal end of the first tarsus, forms the **antenna cleaner.** Note on the opposite side of the velum a number of curved bristles, which form the **pollen brush,** and at the distal end of the tibia of the **middle leg,** the long **spur,** which is used in picking and transferring wax for comb building.

The third, or **metathoracic leg,** possesses the greatest degree of specialization. Note the **pollen basket,** a concave depression surrounded by incurving hairs on the external side of the expanded tibia. Note, also, the **pecten,** a row of spines at the end of the tibia, and the **auricle,** a structure opposite the pecten, on the end of the first tarsus. These two adaptations are used to convey pollen from the pollen combs of the opposite leg to the pollen basket. Find the **pollen combs,** consisting of nine rows of stiff bristles on the inner side of the tarsus. The pollen combs also aid in the removal of wax scales. Observe the wax scales secreted by the wax glands on the ventral side of the abdominal segments.

Abdomen

Count the abdominal segments and compare with the abdomen of the grasshopper. Try to locate the **spiracles.** Observe the sting at the rear of the abdomen. Pull it out slightly with your forceps and note that it consists of a dorsal hollow **sheath** and two **darts.** Observe the **barbs** on the tips of the darts. Which way do they point? What effect would this have on removal of the sting from a wound?

Demonstrations

1. Preserved queen and drone bees for comparison with the worker bee.
2. Microscopic mounts of mouthparts.
3. Microscope slide showing spiracle.
4. Life cycle stages (larvae, pupae).
5. Microscope slide with section of cuticle.

Insect Metamorphosis

Many insects undergo marked changes of body form during their postembryonic life so that the immature stages bear little resemblance to the adult stages. These form changes are called **metamorphosis.** Insects differ considerably in the pattern of their postembryonic development, and we can distinguish four general types of postembryonic development.

1. **No metamorphosis.** Some primitive insects (**ametabolous insects**) exhibit no metamorphosis but hatch directly from the egg in a form resembling miniature adults. Examples of this type of development

are provided by the silverfish, or bristletails (Order Thysanura), and the springtails (Order Collembola).

2. **Incomplete or gradual metamorphosis.** Young grasshoppers emerge from the egg generally resembling the adult but with a disproportionately large head and without wings. The emergent form is the **nymph,** and transformation to the adult form involves a series of five molts during which the wings are gradually developed and the head-body proportions approach the adult condition. Insects that exhibit incomplete metamorphosis are called **hemimetabolous insects.** In addition to grasshoppers, this type of metamorphosis is exhibited by several other groups of insects including the cicadas, walking sticks, mantises, cockroaches, termites, and the true bugs (Order Hemiptera).

Some aquatic hemimetabolous insects have immature forms (naiads) that generally resemble the adults but are adapted for aquatic life. The naiads often have external gills, and they lack wings. They, too, undergo a series of molts and gradually assume the adult form. Dragonflies (Order Odonata), stone flies (Order Plecoptera), and mayflies (Order Ephemeroptera) are examples.

3. **Complete metamorphosis.** Most insects do undergo striking changes in form during their postembryonic life. The characteristic series of stages in complete metamorphosis is: **egg, larva, pupa,** and **adult.** The **egg** hatches into a segmented, wormlike larva. The **larva** feeds, grows, and undergoes several molts, and after the final molt, transforms into a pupa. The **pupa** is a stationary, nonfeeding stage. Although, externally, the pupa appears to be inactive, internally, radical changes are taking place. Most of the larval tissues are resorbed, and the adult organs are newly formed from special groups of embryonic cells called **imaginal discs.** These groups of undifferentiated cells are retained through the larval period and do not resume their development until the pupal stage. When the adult is fully formed within the pupal case, the case is ruptured and the **adult** emerges. Insects demonstrating complete metamorphosis include the beetles (Order Coleoptera), the flies (Order Diptera), the bees and wasps (Order Hymenoptera), and the butterflies and moths (Order Lepidoptera).

Study the demonstration materials on insect metamorphosis provided in the laboratory. The grasshopper affords an excellent illustration of incomplete metamorphosis. Observe the eggs and several nymph stages, and compare each of them with the adult. Note especially the changes in the head-body proportions and the gradual development of the wings.

Study complete metamorphosis in a beetle (such as the June beetle, the Mexican bean beetle, or the meal worm *Tenebrio*) and in the silkworm moth (*Bombyx*), or a similar moth. Observe each of the principal stages—the **egg,** the wormlike **larva,** the **pupa,** and the **adult.** On figure 13.19, draw the series of stages for one of the insects studied.

Phylum Onychophora

The onychophorans have often been cited as an important evolutionary link between the annelids and arthropods. Thus, they should be mentioned here. Recent studies, however, have resulted in a reevaluation of the probable evolutionary significance of the onychophorans.

The onychophorans are a small group of about 80 species of small (1.5 to 15 cm. long) wormlike animals found in tropical rain forests and other damp or wet habitats in tropical and subtropical areas. Distribution is widespread but scattered and discontinuous. *Peripatus* is the best-known genus. The caterpillarlike onychophorans exhibit several morphological characteristics similar to those of the annelids and the arthropods. In addition, however, recent investigators have found that they also have several unique features.

The Onychophora are an ancient group that arose in the Paleozoic Era and apparently have changed little since the Cambrian Period. The current belief of most authorities is that long ago the onychophora evolved separately from the annelids and the arthropods, and that perhaps the three phyla evolved independently from some common ancestral stock.

Examine a preserved or plastic-embedded specimen of *Peripatus* and observe its general appearance. The body has a velvety surface that is covered by a thin, chitinous **cuticle.** The body is divided into an anterior **head** and a posterior **trunk. Segmentation** is not apparent from the dorsal side, but note the numerous paired, segmentally arrange legs. The short, stumpy legs bear terminal **claws.**

Study the anterior **head** with one pair of segmented **antennae,** one pair of **simple eyes** (ocelli), and a pair of **oral papillae** just anterior to the **ventral mouth.** Inside the mouth is a pair of sclerotized, hooklike **mandibles** used to hold and to tear prey.

Among the annelid characteristics shown by onychophorans are the muscular body wall, segmental nephridia with ciliated openings, and nonjointed legs. Arthropodlike features include the chitinous cuticle, a reduced coelom, an open circulatory system with a large interior hemocoel, and jaws formed from modified appendages.

Fig. 13.19   Drawing of insect metamorphosis stages.

Key Terms

Biramous appendage   a two-branched appendage characteristic of the crustaceans and certain other invertebrate groups. Consists of a basal portion (protopodite) and two distal branches or rami. The medial branch is the endopodite and the lateral branch is the exopodite.

Book gills   respiratory organs consisting of many thin, highly vascularized sheets of tissue. Found in *Limulus*.

Cephalothorax   fused head and thorax regions typical of the Crustacea and the Arachnida.

Chelicera   pincerlike first appendages characteristic of the Chelicerata. Plural: chelicerae.

Cheliped   one of the first paired walking legs in the crayfish, lobster, and many other large crustacea. Equipped with a terminal pincerlike claw called a chela.

Chitin   a complex structural carbohydrate that forms a chief component of the exoskeleton of many arthropods and other invertebrates.

Compound eye   a type of image-forming photoreceptor with many individual units, the ommatidia. Typical of insects but also found in many crustaceans.

Endopodite   the medial branch (ramus) of the distal portion of a biramous appendage.

Exopodite   the lateral branch (ramus) of the distal portion of a biramous appendage.

Exoskeleton   an external supporting structure or support for the body. Found in arthropods and many other kinds of animals.

Extensor muscle   a muscle that extends or straightens out an appendage or other body part.

Flexor muscle   a muscle that flexes or bends an appendage or other body part toward the body.

Green glands   type of excretory organ found in the crayfish and many other crustacea. Located near the base of the antennae; also called antennal glands.

Hepatopancreas   spongy internal organ of many arthropods, which secretes digestive enzymes, absorbs food, and stores food reserves. Formed as an outgrowth of the midgut.

Larva   an independent, immature stage of an animal morphologically dissimilar to the adult.

Metamorphosis   a change from one form to another in the developmental history of an animal.

Nymph   an immature stage in the life history of insects exhibiting incomplete metamorphosis. Morphologically similar to the adult.

Ocellus   a simple eye found in many invertebrates. Plural: ocelli.

Protopodite   the basal portion of a biramous appendage in arthropods.

Pupa   a dormant stage following the larva stage and preceding the adult stage in insects with complete metamorphosis.

Serial homology   the adaptation of a longitudinal series of originally similar organs to perform different functions, as in the appendages of the crayfish and many other arthropods.

Sexual dimorphism   significant structural differences between the male and female members of a species.

Statocyst   organ of equilibrium found in certain invertebrates, including crustaceans.

Statolith   a sand grain or calcareous granule within a statocyst.

14
Echinodermata

Objectives

After you have completed the laboratory work for this chapter, you should be able to perform the following tasks:

1. List and briefly characterize the five classes of living echinoderms.
2. Identify the principal external features of a starfish. Explain the function of the madreporite, pedicellariae, dermal branchiae, and tube feet.
3. Identify the parts of the digestive tract of a starfish and explain the function of each part.
4. Identify the parts of the water vascular system of a starfish and give the function of each part.
5. Identify the main external features of a sea urchin.
6. Compare the basic organization of a starfish, a sea urchin, and a sea cucumber.

Introduction

The Phylum Echinodermata is a group of spiny-skinned marine animals, which includes the starfish or sea stars, sea urchins, sand dollars, sea cucumbers, brittle stars, and sea lilies (figure 14.1). Some 6,000 species of living echinoderms and more than 20,000 fossil species have been described. Echinoderms are well preserved in the fossil record of ancient times and flourished during several past eras.

Adult echinoderms are easily recognized by their obvious five-part (pentamerous) radial symmetry and calcareous endoskeleton made up of many small plates, which may be separate (as in the sea cucumbers) or fused into a rigid framework or test (as in a sea urchin or sand dollar). In many of the echinoderms, these mesodermally derived skeletal plates bear protruding spines. Echinoderms have a large coelom lined with a ciliated peritoneum, a portion of which forms into the unique water vascular system during embryonic development. The water vascular system is of great importance to the echinoderms and plays a vital role in their locomotion, attachment, respiration, food handling, and sensory perception.

The echinoderms are of special interest because of their close evolutionary relationship to the Phylum Chordata. Strong embryological evidence suggests a close link between the echinoderms and the chordates. Their mesodermal endoskeleton, the formation of the coelom from outpockets of the archenteron (primitive gut) in the embryo, the formation of the anus from the embryonic blastopore, and the radial indeterminate cleavage of the early embryo are all chordatelike characteristics.

Classification

Some disagreement exists among specialists on the classification of the Phylum Echinodermata, but most recent workers have adopted the following classification of living echinoderms, with three subphyla and five classes represented among the present North American fauna.

Subphylum Echinozoa

Nonsessile animals with rounded or disc-shaped body without arms.

Class Echinoidea

Sea urchins, heart urchins, and sand dollars. Bottom-dwelling (benthic) echinoderms with a rigid test of fused skeletal plates. Body covered with movable spines. Five rows of tube feet (bearing suckers) around the test. Examples: *Arbacia*, *Lytechinus*, *Strongylocentrotus* (all sea urchins); *Echinarachnius*, *Mellita*, and *Dendraster* (sand dollars).

Class Holothuroidea

Sea cucumbers. Soft-bodied animals having a flexible body wall with many tiny embedded calcareous ossicles; no

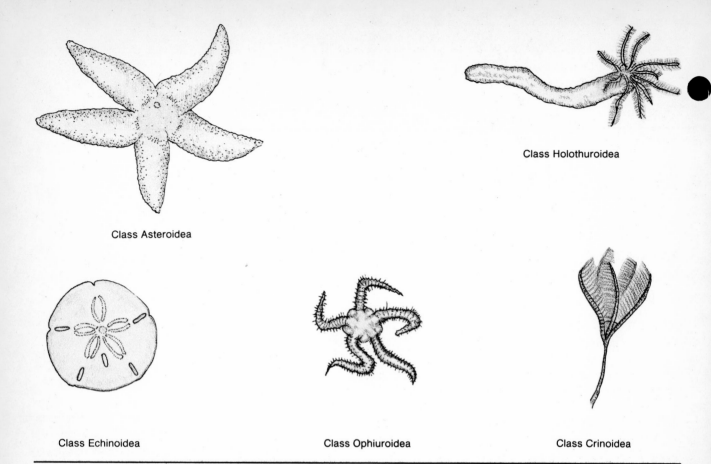

Class Asteroidea

Class Holothuroidea

Class Echinoidea

Class Ophiuroidea

Class Crinoidea

Fig. 14.1 Major classes of living echinoderms.

spines or arms. Body elongated in the oral-aboral axis to a wormlike form; a circle of tentacles surround the mouth. Usually with five or more series of tube feet extending longitudinally. Examples: *Thyrone, Cucumaria, Stichopus.*

Subphylum Crinozoa

Spherical or cup-shaped body with branching arms; attached during all or part of life by a stem; oral surface directed upward.

Class Crinoidea

Sea lilies, basket stars. Flowerlike echinoderms with a central calyx and five (or multiples of five) branching arms. Ciliated ambulacral groove along the oral surface of the arms serves for food gathering. Some species attach to the sea bottom by a stalk, others free-swimming. Examples: *Cerocrinus* (crinoid), *Hathrometra* (=*Antedon*) (feather star).

Subphylum Asterozoa

Nonsessile, star-shaped echinoderms, with radial arms and oral surface directed downward.

Class Asteroidea

Starfish or sea stars. Animals with a star-shaped body, typically with five arms and a flexible skeleton of many calcareous plates. The arms are continuous with and not clearly distinct from the central disc. With ambulacral grooves on the oral side of each arm; tube feet bearing suckers. Examples: *Asterias, Pisaster, Astropecten.*

Class Ophiuroidea

Brittle stars. Star-shaped echinoderms with arms distinct from the central disc; no ambulacral grooves along the arms; tube feet absent or reduced to sensory organs. Examples: *Amphipholis, Ophiothrix* and *Ophioderma* (brittle stars); and *Gorgonocephalus* (Pacific basket star).

The Common Starfish: *Asterias*

Class Asteroidea

Asterias forbesi is a common starfish (see figure 14.2) found along the Atlantic coast of the United States. Obtain a living or preserved specimen of *Asterias* and observe its radial symmetry. The radial symmetry of adult echinoderms appears to be a secondary condition, since

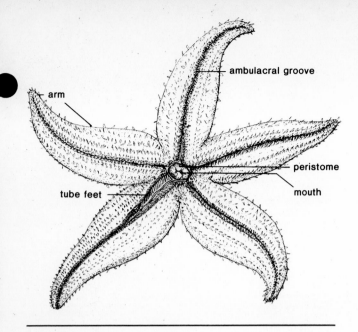

Fig. 14.2 Starfish, oral view.

echinoderm larvae are bilaterally symmetrical. Observe a demonstration slide of the bipinnaria larva of the starfish. In what ways does the radial symmetry of the echinoderms differ from that of the coelenterates?

Materials List

Preserved Specimens
  Asterias
  Arbacia
  Cucumaria or *Thyone*
  Aristotle's Lantern (Demonstration)
  Starfish, dried to show ossicles (Demonstration)
  Representative echinoderms (Demonstration)
Prepared Microscope Slides
  Starfish, pedicellaria (Demonstration)
  Starfish, dermal branchiae (Demonstration)
  Starfish, bipinnaria larva (Demonstration)
  Starfish, metamorphosis (Demonstration)

External Anatomy

Specimens preserved for laboratory study often are injected with a colored material to facilitate the study of the water vascular system, including the external tube feet. Keep your specimen wet by adding a little water to the dissecting pan. Note the **central disc** on the upper, or **aboral,** side; the five rays or **arms;** and the **madreporite,** a light-colored circular area near the edge of the disc at the junction of two rays. The madreporite is a calcareous disc that serves as the entrance to the water vascular system. The two arms adjacent to the madreporite comprise the **bivium;** the other three arms make up the **trivium.** The arm opposite the madreporite is referred to as the **anterior** arm.

Fig. 14.3 Drawing of pedicellariae from starfish.

Note the many spines scattered over the surface of the arms and the central disc. Among the spines there are also many small pincerlike structures located near the bases of the spines (see demonstration slide). These structures are the **pedicellariae.**

Carefully scrape a small area of the arm with your scalpel, add the scrapings to a drop of water, and observe the material under the low power of your microscope. Can you identify any pedicellariae among the scrapings? From their structure, what do you think the function of the pedicellariae might be? Draw one of the pedicellariae in figure 14.3. Also among the spines are numerous **papulae** (dermal branchiae), soft, hollow projections of the body wall used for respiration (see demonstration slide). Observe the skeletal calcareous plates, or **ossicles,** on a dried starfish specimen.

Note on the oral side of the starfish: the **mouth** guarded by specialized **oral spines** and surrounded by a soft membrane, the **peristome;** the five **ambulacral grooves,** extending from the mouth and along the middle of each arm; and the numerous **tube feet** in the grooves, extending from the water vascular system. There is also a pigmented **eyespot** and a **sensory tentacle** at the tip of each ray, but these latter structures are usually difficult to observe in preserved specimens.

Draw an enlarged view of the central disc and one arm of a starfish (in figure 14.4). Show an oral view and

Fig. 14.4 Drawing of central disc and one arm of starfish, oral view.

include the following structures in your drawing: **mouth, peristome, oral spines, ambulacral groove,** and **tube feet.** Label each structure.

Internal Anatomy

Carefully examine figure 14.5 for orientation before starting your dissection. Cut off about one-half inch from the tip of the anterior arm. Then carefully cut along each side of the anterior arm to its junction with the central disc along each side and across the top of the arm at the margin of the disc. Observe the hard, calcareous plates, or ossicles, of the skeleton as you cut. Remove the portion of the body wall covering the aboral surface of the arm and examine the large **coelom** within the arm, which contains the internal organs. The coelom is lined with a ciliated **peritoneum,** as mentioned previously, and is normally filled with coelomic fluid. This fluid carries oxygen and absorbed food to various parts of the body. Next, remove the portion of the body wall covering the aboral side of the central disc, but leave the madreporite in place by carefully cutting around it.

Digestive System

The starfish has a complete digestive system. The mouth of the starfish is on the aboral surface and opens into a short esophagus. The esophagus leads to a large, two-chambered stomach that fills most of the central disc. The

large, thin-walled cardiac chamber lies below (oral to) and a smaller, thick-walled chamber lies above (aboral to) the cardiac chamber.

Emptying into the pyloric stomach through five pyloric ducts are five pairs of large, branched pyloric caecae (digestive glands). The pyloric caecae fill most of the interior space in the arms. Also leading from the pyloric stomach is a small, short intestine that empties through the aboral anus.

The cardiac chamber can be everted through the mouth to envelop its prey. Most starfish are carnivores and feed on many kinds of invertebrates, including clams, oysters, gastropods, crustaceans, polychaetes, other echinoderms, and small fish. Prey is often engulfed whole, including shell or exoskeleton, and the undigestible parts are later regurgitated. Typically, food is digested in the cardiac chamber of the stomach. Digestive enzymes are secreted principally by the pyloric caecae, which also play a major role in absorption and storage of food materials.

Water Vascular System

The water vascular system (figure 14.6) is a unique feature of the echinoderms. It develops embryologically as a specialized part of the coelom and is lined with a **ciliated epithelium.** The water vascular system has several important functions, including locomotion, feeding, respiration, excretion, and sensory perception. The canals of this system are filled with a water vascular fluid that has a

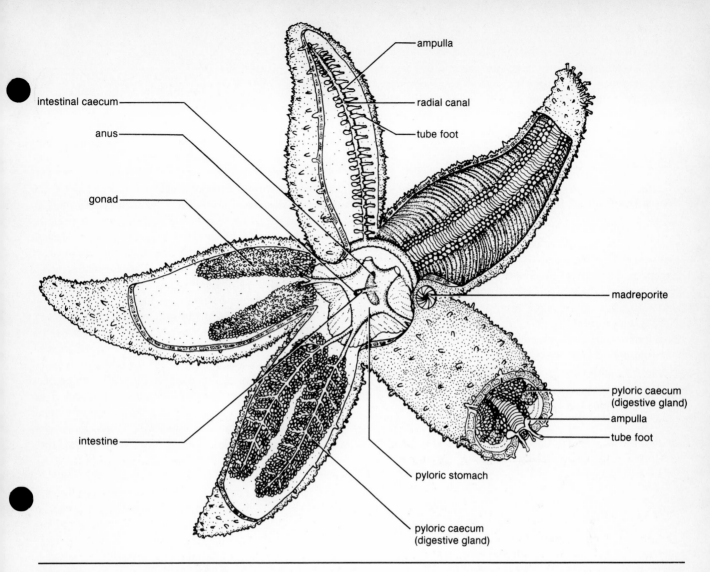

Fig. 14.5 Starfish, internal anatomy.

chemical composition resembling seawater plus some soluble proteins and an elevated level of potassium ions. Suspended in the water vascular fluid are numerous **coelomocytes,** amoeboid cells that serve in defense, excretion, and perhaps in several other roles as well.

Starfish have a well-developed water vascular system. Among its principal parts is the external **madreporite,** seen previously on the aboral surface, which marks the entrance to a short **stone canal** (see figure 14.6). The madreporite is covered by the thin ciliated epidermis of the body surface, and is believed to function as a pressure regulator for the water vascular system rather than as a simple filter, as has sometimes been assumed. The stone canal gets its name from the calcareous deposits surrounding it. The stone canal leads to the **circular canal,** which encircles the mouth. Five **radial canals** lead from the circular canal along the top of the **ambulacral groove** leading into each arm. Many short **lateral canals** connect the radial canals with each pair of **tube feet.** Each tube foot consists of a bulblike **ampulla** attached to each suckerlike tube foot.

Within each lateral canal is a valve that closes to allow local pressure changes in the individual tube feet. Contraction of the circular muscles in the wall of the ampulla increases the internal pressure and extends the foot. Contraction of longitudinal muscles in the wall of the tube foot withdraws the foot. These pressure changes within the water vascular system plus nervous control from the central nervous system allow coordinated movements of the tube feet and, thus, coordinated locomotion of a starfish.

Nervous System

The starfish has a very simple nervous system. A circular **nerve ring** surrounds the mouth, and a **radial nerve** extends from the nerve ring into each arm. A simple light-sensitive **eyespot** at the tip of each arm is the only differentiated sense organ, but sensory cells are scattered throughout the epidermis.

Respiration and Excretion

Respiration is carried on by the **papulae** (dermal branchiae) extending from the body surface and by the tube feet. Excretion of nitrogenous wastes is accomplished mainly by diffusion of ammonia and urea from the respiratory surfaces and the body wall. Coelomocytes engulf insoluble particulate matter and carry these wastes to the surface where they are discharged. Echinoderms have no differentiated excretory organs.

Reproductive System

The reproductive system consists of a pair of **gonads** located near the base of each ray (see figure 14.5). The sexes are separate in sea stars, but are difficult to distinguish except by microscopic examination of the gonad contents. Take a small amount of tissue from one of the gonads and macerate it in a drop of water on a clean microscopic slide. Can you identify any large oocytes (eggs) or any flagellated sperm? From each gonad a duct leads to an external pore on the aboral surface. Eggs and sperm are released by mature sea stars into the seawater, where fertilization occurs, and the zygotes develop into free-swimming **bipinnaria larvae.**

A Sea Urchin

Class Echinoidea

Sea urchins are bottom-dwelling (benthic) echinoderms with a **rigid endoskeleton** consisting of many fused plates or ossicles. They lack arms and typically are globular or ovoid in shape, and their bodies are covered with movable **spines** (figure 14.7). Among the spines are many thin-walled **dermal branchiae** (gills), which serve in respiration, and numerous **pedicellariae,** which serve to keep the test clean of debris and fouling organisms.

Study a living or preserved specimen of *Arbacia* or a similar sea urchin. Identify the rigid **test, spines, dermal branchiae,** and **pedicellariae.** Locate also the five series of **tube feet** along five symmetrically placed **ambulacral regions** around the globular body. In a living specimen the tube feet can extend beyond the spines.

On the oral side of the urchin (flattened surface), find the **mouth** surrounded by five **calcareous teeth** and a circular oral membrane, the **peristome.** The teeth are connected internally to a complex chewing apparatus, called **Aristotle's Lantern,** which anchors the teeth and enables them to scrape algae from submerged rocks. Urchins are principally herbivorous, and many species feed on algae from the seafloor; they are sometimes called the grazers

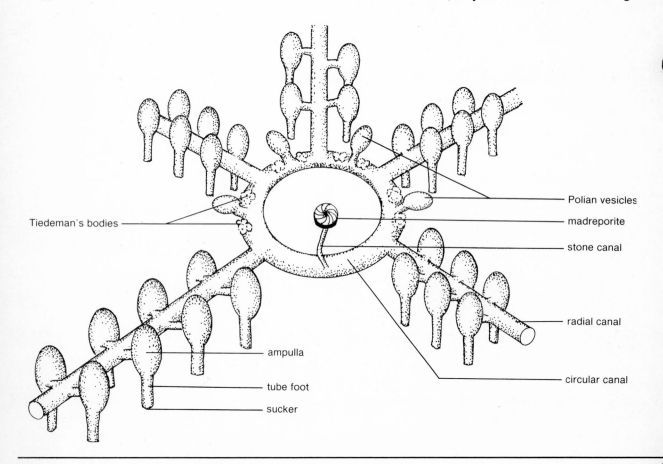

Polian vesicles
madreporite
stone canal
radial canal
circular canal
Tiedeman's bodies
ampulla
tube foot
sucker

Fig. 14.6 Diagram of the water vascular system of a starfish.

of the sea. They live mostly along rocky shores. Their relatives, the sand dollars and heart urchins, are adapted for life on soft sea bottoms in which they burrow in search of food and shelter.

Some Common Echinoids

Arbacia punctulata—purple sea urchin found along the Atlantic coast from Cape Cod to Florida.

Strongylocentrotus drobachiensis—green sea urchin found in colder waters north of Cape Cod along the western Atlantic, in northern Europe, and from Washington north to Alaska on our Pacific coast.

Strongylocentrotus purpuratus—the Pacific purple sea urchin found on surf-swept rocks from California to Alaska.

Fig. 14.7 Sea urchin, *Arbacia*. (Courtesy Carolina Biological Supply Company.)

Echinarachnius parma—Northern Atlantic sand dollar found on sandy areas on the sea bottom from Labrador to New Jersey, and also in the Northern Pacific from British Columbia across to Japan.

Mellita quinquiesperforata—Keyhole sand dollar abundant from Cape Hatteras to the Caribbean.

A Sea Cucumber

Sea cucumbers are soft-bodied echinoderms with many tiny plates or ossicles embedded in their **leathery body wall.** They are principally filter feeders, collecting plankton with their tentacles, or deposit feeders, collecting food materials mixed with the bottom sediments.

Study a living or preserved *Cucumaria* (figure 14.8) or *Thyone* and identify the **oral tentacles,** the external **tube feet,** and the **leathery body wall.**

Some Common Holothuroidea

Thyone briareus—burrowing sea cucumber found in shallow waters on soft, sandy, or muddy sea bottoms from Cape Cod to the Gulf of Mexico. Tube feet scattered over body surface.

Cucumaria frondosa—deeper water, North Atlantic species in which the tube feet are arranged along five distinct ambulacral areas.

Leptosynapta tenuis—long, wormlike sea cucumber, modified for burrowing in soft bottom sediments. Lacks regular tube feet.

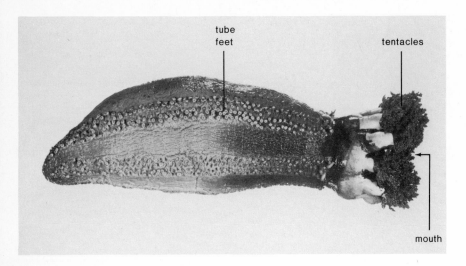

tube feet

tentacles

mouth

Fig. 14.8 Sea cucumber, *Cucumaria*. (Courtesy Carolina Biological Supply Company.)

1. Pedicellariae (microscope slide).
2. Dermal branchiae (microscope slide).
3. Bipinnaria larva (microscope slide).
4. Starfish metamorphosis (microscope slide or color transparency series).
5. Preserved or dried Aristotle's Lantern from sea urchin.
6. Dried starfish to show ossicles.
7. Preserved specimens representing other types of echinoderms, such as crinoids, basket stars, and brittle stars.

Key Terms

Pedicellariae small, pincerlike structures found on the surface of many starfish (Class Asteroidea), sea urchins (Class Echinoidea), and brittle stars (Class Ophiuroidea). Serve to keep the body surface free from debris and fouling organisms.

Radial symmetry type of body organization in which all parts of the body are arranged symmetrically around a central axis.

Water vascular system a unique organ system found only in the Phylum Echinodermata. Functions in feeding, attachment, locomotion, respiration, and sensory perception. Formed embryologically as a specialized portion of the coelom.

15
Chordata

Objectives

After completing the laboratory work in this chapter, you should be able to perform the following tasks:

1. Describe the life cycle of a tunicate (Subphylum Urochordata) and identify the principal features of a tunicate tadpole larva. Explain the morphological significance of the larva.

2. Identify the main organs in an adult tunicate and discuss the adaptations of the adult for its sessile mode of life.

3. Explain the feeding mechanism of a tunicate and identify the principal structures involved.

4. Identify the main organs in a lancelet and discuss the significance of its basic chordate features.

5. Compare the organization of a tunicate and a lancelet and explain their main similarities and differences.

Introduction

The Phylum Chordata includes a remarkable range of forms, varying from relatively simple marine animals to highly specialized birds and mammals. Despite the wide range of diversity in the phylum, however, all members exhibit three fundamental chordate characteristics during some stage in their life histories. These distinctive chordate features are (1) a dorsal hollow nerve cord; (2) a notochord, a dorsal, elastic supporting rod; and (3) paired pharyngeal gill slits. Some writers add a fourth characteristic, a post-anal tail. Nonetheless, the chordates comprise a large, diverse, and distinct phylum of animals that have successfully populated the land, the waters, and the air. This large and important phylum is commonly divided into three subphyla.

Classification

Group Acrania (Protochordata)

Chordates without a cranium or braincase.

Subphylum Urochordata (Tunicates or Sea Squirts)

Animals with a well-developed notochord and dorsal nerve cord in the free-swimming larva; specialized adults, sessile or planktonic, and lacking a notochord and dorsal nerve cord. Examples: *Molgula* (sea grape), *Styela, Amaroucium* (sea pork).

Subphylum Cephalochordata (Lancelets)

Elongate, fishlike chordates with a persistent notochord and dorsal nerve cord. Example: *Branchiostoma (Amphioxus)*.

Group Craniata

Chordates with a cranium enclosing the brain and sense organs of the head; exhibit substantial cephalization.

Subphylum Vertebrata (Vertebrates)

Chordates with a backbone, skull, brain, and kidneys.

Class Agnatha (Lampreys). Fishlike vertebrates without jaws or appendages. Example: *Petromyzon*.

Class Elasmobranchiomorphii (Chondrichthyes) (Sharks and Rays). Cartilaginous fishes, usually with external placoid scales. Examples: *Squalus* (shark), *Raja* (skate), *Chimaera* (ratfish).

Class Osteichthyes (Bony Fishes). Fishes with skeleton primarily of bone, usually with swim bladder or lungs. Examples: *Amia* (bowfin), *Perca* (perch), *Lepistoseus* (gar), *Micropterus* (bass).

Class Amphibia (Salamanders, Frogs, and Toads). Four-limbed vertebrates (tetrapods) with a soft, moist skin; three-chambered heart; eggs enclosed in gelatinous covering; fertilization and development usually restricted to fresh water. Adults usually aquatic or semiaquatic. Examples: *Rana* (frog), *Bufo* (toad), *Ambystoma* (salamander).

Class Reptilia (Lizards, Snakes, Turtles, and Alligators). Four-limbed vertebrates with a dry, cornified skin; eggs enclosed in a protective shell resistant to drying (amniote egg); four-chambered heart. Examples: *Anolis* (lizard), *Chrysemys* (turtle), *Crotalus* (rattlesnake), *Alligator.*

Class Aves (Birds). Winged vertebrates with feathers and constant high body temperature. Examples: *Sturnus* (starling), *Cyanocitta* (blue jay), *Corvus* (crow).

Class Mammalia (Mammals). Warm-blooded vertebrates; body covered with hair; young nourished by milk produced by female. Examples: *Homo* (human), *Sus* (pig), *Equus* (horse), *Canis* (dog).

In this chapter we will study representatives of the first two subphyla, the urochordates and the cephalochordates.

Materials List

Preserved Specimens
  Molgula
  Branchiostoma
  Representative tunicates
Prepared Microscope Slides
  Amaroucium, tadpole larva
  Branchiostoma, whole mount
  Branchiostoma, cross section

Subphylum Urochordata, Tunicates or Sea Squirts

Urochordates are interesting, important, and unusual animals. They are interesting and important because they represent the simplest living chordates. They are unusual because adult urochordates became adapted to a special mode of life as sedentary, filter-feeding, marine animals, and in the process have lost two of the three basic chordate features. Larval chordates, however, have retained all three chordate features and clearly establish the urochordates as legitimate ancestors of the higher chordates, including fishes, birds, mammals, and humans.

The Tunicate Larva

The larval stages of tunicates are of special importance because they clearly exhibit the three fundamental chordate characteristics: tubular nerve cord, notochord, and pharyngeal gill slits. The first two of these features are lost in the adult tunicate, presumably because of the specialization of the adult due to its sessile mode of life.

Tunicate larvae are often called **tadpole larvae** because of their superficial resemblance to the tadpole larvae of amphibians. Study a prepared microscope slide with a whole mount of the larva of *Amaroucium,* or a similar tunicate larva, and observe its characteristic form (see figure 15.1). Note the **muscular tail** and thickened body. At the anterior end of the body find the adhesive papillae by means of which the larva attaches prior to metamorphosing into a sessile adult. Near the dorsal surface is a **sensory vesicle** with two conspicuously pigmented sense organs—a light-sensory **ocellus,** or eye-spot, and a **statolith,** which serves as a balancing organ.

Anterior to the darkly pigmented sensory vesicle, locate the **incurrent siphon** through which water is pumped into the pharynx. The **excurrent siphon** is found slightly posterior to the sensory vesicle, near the attachment of the tail. Locate also the **gill slits** in the wall of the pharynx.

The **notochord,** the **dorsal nerve cord,** and conspicuous **muscle bands** are present in the tail. At the time of metamorphosis, the larva settles to the bottom, attaches by means of a secretion of glands in the adhesive papillae, and the tail is partly resorbed. The notochord, nerve cord, and muscles are broken down, and several new organs appear in the body of the tunicate to complete the transformation into the adult. The life cycle of a tunicate is illustrated in figure 15.1.

The Adult Tunicate

Adult tunicates may be solitary or colonial, sessile or planktonic. The most familiar ones, however, are the sessile forms found attached to rocks, jetties, and pilings in shallow seas.

The typical structure of an adult tunicate is well illustrated by *Molgula,* the sea grape, found in many places along the Atlantic coast. Study the demonstration of a partially dissected specimen of *Molgula* and note the tough, fibrous outer covering called the tunic (see figure 15.2). This outer covering is unusual because it is composed partly of cellulose, a carbohydrate normally produced only by plants.

Projecting from the globular body are two siphons: the incurrent siphon is anterior, and its opening is divided into six lobes. The more posterior excurrent siphon has a square opening. Locate internally the large pharynx with numerous gill slits. Ventral to the pharynx are the short esophagus, the bulbous stomach, and the beginning of the intestine. The intestine continues dorsally and ends at the anus just below the excurrent siphon.

Molgula, like most other tunicates, is a specialized filter-feeder. Water is drawn in through the incurrent siphon by the action of cilia lining the internal chambers. The water is drawn into the large pharynx, and minute

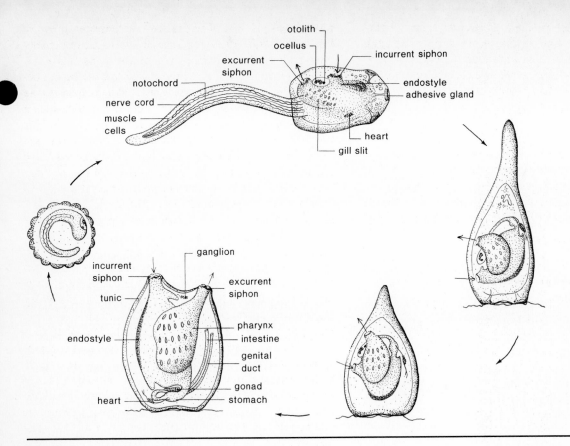

Fig. 15.1 Tunicate, life cycle.

organisms suspended in the water (plankton) are trapped in the mucus on the pharyngeal walls as the water passes out through the many gill slits.

Special ciliary currents collect the food within the pharynx and pass it into the esophagus and on to the stomach. Water leaving the pharynx passes into the atrium, a large cavity surrounding the pharynx, and from the atrium the water flows out through the excurrent siphon.

The endostyle, a ciliated groove located on one side of the large pharynx, is of special importance. The endostyle secretes a stream of mucus that is carried along by the motion of the cilia to form a sheet that traps small food particles from the incoming seawater. The mucus sheet with its captured food particles is then passed to the esophagus and into the stomach and through the digestive system. The endostyle has been found to be the forerunner of the vertebrate thyroid gland.

The circulatory system in *Molgula,* as in most tunicates, consists of a ventral heart located near the stomach. Attached to the heart are two large vessels that carry blood to other organs. The tunicate heart is unique in that it provides two-way propulsion. The heart first pumps blood in one direction, then reverses and pumps blood in the opposite direction.

In the adult tunicate the dorsal hollow nerve cord and the notochord are absent. Their loss is generally believed to represent a specialization of the adult tunicate for its peculiar sessile mode of life.

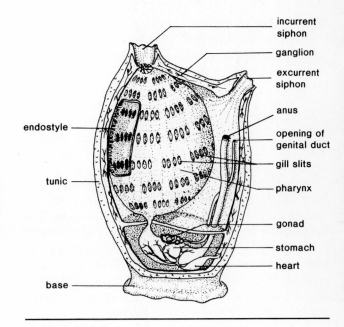

Fig. 15.2 Tunicate, adult.

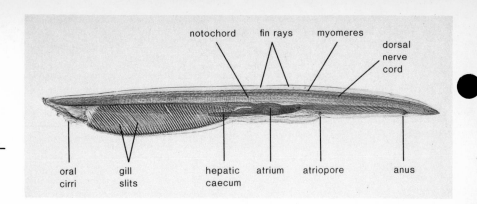

Fig. 15.3 Lancelet, whole mount. (Courtesy Carolina Biological Supply Company.)

A Lancelet: *Branchiostoma*

Subphylum Cephalochordata

The Subphylum Cephalochordata consists of a few species of small fishlike animals called lancelets (see figure 15.3), which live in shallow seas in many parts of the world. In certain areas, the lancelets are sufficiently abundant that they are used as human food. The common American lancelets are traditionally called *Amphioxus* (formerly a generic name), although most specialists now use the generic name *Branchiostoma*.

Internal Anatomy

Study first a preserved specimen and observe its general form and external features. Then obtain a prepared microscope slide with a stained whole mount of a lancelet to study its internal anatomy. Mature lancelets are usually between two and three inches in length, but smaller, immature forms are normally used in making microscopic whole mounts.

In the preserved specimen, note the slender, elongate shape of the animal, the absence of a distinct head, and the lack of paired fins or limbs. Handle the specimen with care and do not dissect it. Return it intact to the proper container when you have completed your study.

Refer to figure 15.4 and identify the anterior **rostrum** and the **oral hood** bordered by a fringe of ciliated **oral cirri** enclosing the large **vestibule.** The **mouth** is an opening in a membrane, the **velum,** located at the rear of the vestibule. Surrounding the mouth are several **velar tentacles.** How many? Cilia arranged in bands along the walls of the vestibule form the "wheel organ," which generates water currents and carries seawater containing suspended food organisms through the mouth.

Behind the mouth is a large pharynx with many diagonal **gill slits** on each side. Between the gill slits are **gill bars,** each supported by a thin cartilaginous rod. The pharynx plays an important role both in feeding and in respiration, but its primary function is in feeding. The respiratory function is clearly secondary. Posterior to the pharynx is a straight intestine, which ends at the subterminal anus. A slender pocket, the **midgut caecum** (believed by some workers to be homologous with the vertebrate liver), opens on the ventral side of the intestine near the junction of the pharynx and intestine and extends forward.

Locate the **dorsal, caudal** (tail), and **ventral fins.** Note the short **fin rays** composed of connective tissue within the fins. Observe the **atriopore,** a midventral opening located anterior to the ventral fin. Water taken into the pharynx passes out through the gill slits into the **atrium** (see figure 15.6) and out of the atrium via the **atriopore** (see figure 15.4).

Lancelets, like the tunicates studied earlier, are filter-feeders. Seawater containing planktonic organisms is drawn by ciliary currents through the mouth into the pharynx where food particles are trapped in mucous secretions. The mucus, containing trapped food particles, is swept posteriorly to the intestine, and the water passes out of the pharynx through the lateral gill slits. Respiratory exchange also occurs as the water passes through the gill slits and past the gill bars, which contain blood vessels.

Note the conspicuous V-shaped structures along each side of the body; these are the muscle segments, or **myomeres.** Contraction of these muscles produces a lateral bending of the body, which aids the lancelet in swimming and burrowing in the bottom sediments where it commonly dwells.

Locate the dorsal **notochord,** which extends longitudinally just dorsal to the pharynx and intestine. The notochord is a thin, cartilaginous rod surrounded by a sheet of connective tissue. The contraction of the myomeres against the rigidity of the notochord produces the lateral swimming movements of the body that propel the lancelet forward. Find the **dorsal nerve cord** just ventral to the dorsal fin and dorsal to the notochord. At its anterior end is a slight enlargement, the **cerebral vesicle,** a very primitive sort of "brain."

The circulatory system of *Branchiostoma* consists of a network of elastic vessels similar to that found in the higher chordates, but it lacks a distinct heart. It is difficult to study the details of the circulatory system except by the dissection of specially prepared specimens, but portions of the circulatory system can be observed in the microscopic cross sections to be studied later. Figure 15.5 shows the principal blood vessels and the general pattern of circulation in a lancelet.

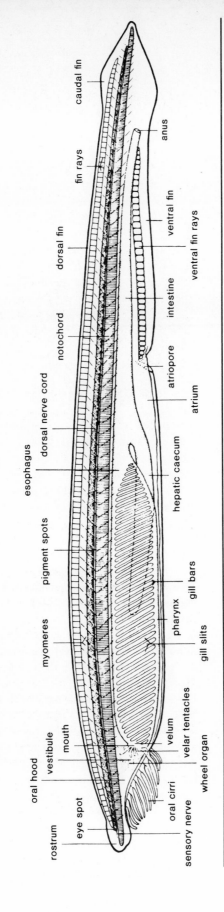

Fig. 15.4 Lancelet, internal organs.

Labels for Fig. 15.4:
caudal fin, fin rays, anus, ventral fin, dorsal fin, ventral fin rays, intestine, notochord, atriopore, dorsal nerve cord, atrium, esophagus, pigment spots, hepatic caecum, myomeres, gill bars, pharynx, gill slits, oral hood, vestibule, mouth, velum, velar tentacles, wheel organ, oral cirri, sensory nerve, rostrum, eye spot

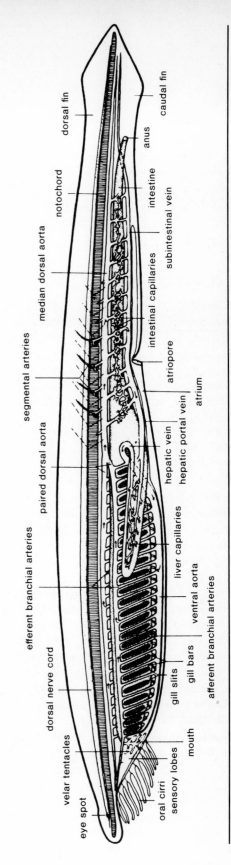

Fig. 15.5 Lancelet, circulatory system.

Labels for Fig. 15.5:
dorsal fin, caudal fin, notochord, anus, median dorsal aorta, intestine, subintestinal vein, segmental arteries, intestinal capillaries, paired dorsal aorta, atriopore, atrium, efferent branchial arteries, hepatic vein, hepatic portal vein, dorsal nerve cord, liver capillaries, velar tentacles, ventral aorta, gill slits, gill bars, afferent branchial arteries, oral cirri, sensory lobes, mouth, eye spot

215

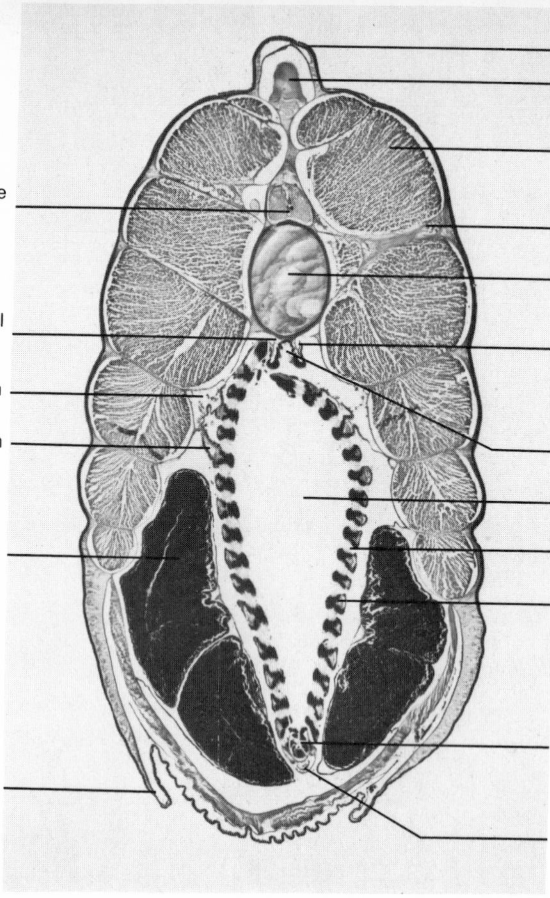
dorsal fin
fin ray
myomere
dorsal nerve cord
myoseptum
notochord
dorsal aorta
dorsal aorta
coelom
nephridium
epibranchial groove
pharynx
testis
gill slit
gill bar
endostyle
metapleural fold
ventral aorta

Fig. 15.6 Lancelet, male, cross section through atrium. (Courtesy Carolina Biological Supply Company.)

Blood from the digestive tract is collected by the **subintestinal vein,** which leads to the **hepatic portal vein,** which, in turn, carries the blood to the midgut caecum. The **hepatic vein** leaves the liver and leads to the **ventral aorta** below the pharynx. Numerous **afferent branchial arteries** (each with a contractile bulb at its base) branch from the ventral aorta and carry the blood upward to the gill bars where it is oxygenated. Pulsations of the ventral aorta and of the enlargements at the bases of the afferent branchial arteries appear to aid in pumping the blood through the system. From the gills, the blood is transported by the **efferent branchial arteries** and is collected in the **paired dorsal aortas** (right and left) above the gills. The two dorsal aortas (see figure 15.6) join posteriorly to form a single **median dorsal aorta** just behind the pharynx. This latter vessel carries oxygenated blood posteriorly to the body tissues and the intestine to complete the circuit.

Cross Sections

Prepared microscopic cross sections can help greatly to supplement your observations on the preserved specimens and whole mounts and to improve your understanding of the anatomy of amphioxus. Study several cross sections from different regions of the body and attempt to identify as many of the internal and external structures as possible. The following description should aid you in identifying various structures in the cross section. It is based upon a cross section through the pharyngeal region as illustrated in figure 15.6.

Observe the two-layered **skin,** with an outer **epidermis** consisting of a single layer of columnar epithelium, and an underlying **dermis,** a thin layer of gelatinous connective tissue. Find the **dorsal fin** and **fin ray,** the **myomeres** bounded by the myosepta of connective tissue, the **notochord,** and the **dorsal nerve cord** beneath it. Locate the **central canal** within the dorsal nerve cord. Observe also the nerve cells and fibers of the dorsal nerve cord and the **spinal nerves** (not present in every section because of differences in the plane of sectioning).

Observe the laterally compressed pharynx and the numerous **gill slits** between the **gill bars.** The ventral ciliated groove is the **endostyle,** or hypobranchial groove, and the dorsal ciliated groove is the **epipharyngeal groove.** Both play an important role in the trapping and transporting of food particles. There is some evidence that the endostyle may be homologous with the vertebrate thyroid gland.

The chamber around the pharynx is the **atrium.** What tissue layer forms the lining of the atrium? What tissue layer lines the coelom? Note the **gonads** extending

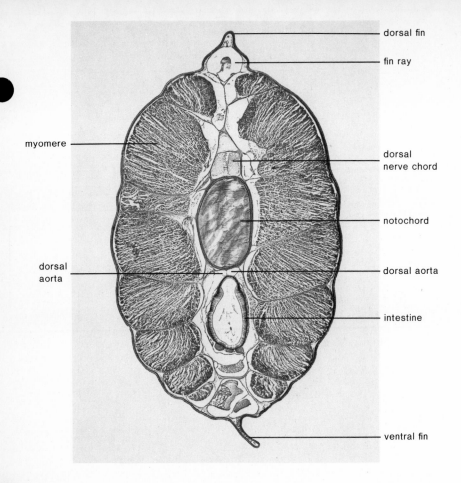

dorsal fin

fin ray

myomere

dorsal
nerve chord

notochord

dorsal
aorta

dorsal aorta

intestine

ventral fin

Fig. 15.7 Lancelet, cross section through intestinal region. (Courtesy Carolina Biological Supply Company.)

into the atrium and the section of the midgut caecum also extending into it. The paired cavities above and lateral to the pharynx are portions of the **coelom,** as is the small cavity found below the endostyle. Locate the **blood vessels** in your section, including the **paired dorsal aortae** above the pharynx and the **single median ventral aorta** below it. Find, also, the **hepatic vein** or veins closely associated with the midgut caecum. The **nephridia** are small ciliated ducts which connect the dorsal portions of the coelom with the atrium. The two **metapleural folds** at the two sides on the ventral surface of the body should be apparent.

Compare the structures observed in the cross section through the pharyngeal region with a cross section through a more posterior region, as illustrated in figure 15.7.

Demonstrations

1. Preserved lancelets.
2. Examples of solitary and colonial tunicates.

Key Terms

Endostyle a ciliated groove located on the ventral surface of the pharynx in tunicates and lancelets. Functions in the capture of food particles. Possibly homologous with the thyroid gland of vertebrates.

Gill slits paired openings in the lateral walls of the pharynx in urochordates and cephalochordates. Important in food capture and secondarily in respiration.

Myomeres segmented muscle blocks arranged longitudinally along the dorsal portion of the lancelet body. Contraction of the myomeres, combined with the relatively stiff notochord, causes the lancelet body to flex, and produces effective swimming and burrowing movements.

Notochord a stiff supporting cartilaginous rod of mesodermal origin found in the tunicate tadpole and in adult cephalochordates. Absent in adult tunicates.

Tadpole larva characteristic larval form of the tunicates with a notochord, dorsal tubular nerve cord, and paired gill slits. Superficially resembles an amphibian tadpole larva.

16
Shark Anatomy

Objectives

After you have completed the laboratory work for this chapter, you should be able to perform the following tasks:

1. Identify the principal external features of the dogfish shark.
2. Locate the pelvic and pectoral girdles of the shark and explain their function.
3. Locate and identify the parts of the digestive system of the shark and explain the function of each part.
4. Locate the parts of the male and female reproductive systems of the shark and give the function of each part.
5. Locate the principal arteries and veins of the shark and explain the pattern of blood circulation.
6. Demonstrate the parts of the heart and explain the function of each part.
7. Describe the hepatic and renal portal systems and trace their paths in a specimen.
8. Describe the pattern of branchial circulation in a shark, explain its importance, and point out the chief blood vessels involved on a specimen.
9. Identify the main parts of the brain and explain the function of each.
10. Locate the eleven pairs of cranial nerves on a specimen and list their names.

The Dogfish Shark: *Squalus acanthias*

Sharks, skates, rays, and chimaeras are primitive fishes with a cartilaginous endoskeleton, biting jaws, paired appendages, and a tough, leathery skin that is usually covered with placoid scales. These fishes belong to the Class Elasmobranchiomorphii (Chondrichthyes) of the Subphylum Vertebrata and are commonly called elasmobranchs (because of their exposed gill openings) or cartilaginous fishes (because of the nature of their skeleton). Most of the members of the group are marine, although a few species have secondarily invaded fresh waters. Paleontological studies have revealed that this group evolved from ancestors with bony skeletons and which inhabited fresh waters. Therefore, both the cartilaginous skeleton and the marine habitat of the elasmobranchs must be regarded as specialized rather than primitive characteristics. Nonetheless, despite certain specialized features, the elasmobranchs clearly illustrate many basic vertebrate characteristics, and the shark has long been studied in zoology laboratories for this reason.

The spiny dogfish, *Squalus acanthias*, is a small shark (figure 16.1) common in shallow coastal waters on our Atlantic and Gulf coasts. A similar small shark from the Pacific coast, sometimes called *Squalus suckleyi*, which some authorities consider to belong to the same species as the Atlantic form, is also commonly used for laboratory study. Mature specimens usually range from three to four feet in length, although smaller, immature specimens are normally used for laboratory study. Formalin-preserved specimens with the arterial and/or venous systems injected with latex are most satisfactory for dissection.

Materials List

Preserved Specimens
  Squalus acanthias
  Squalus, pregnant female with embryos in uterus (Demonstration)
  Squalus, skeleton (Demonstration)
  Bony fish skeleton (Demonstration)
  Representative elasmobranchs (Demonstration)
Prepared Microscope Slides
  Shark, placoid scales (Demonstration)
  Shark, retina (Demonstration)

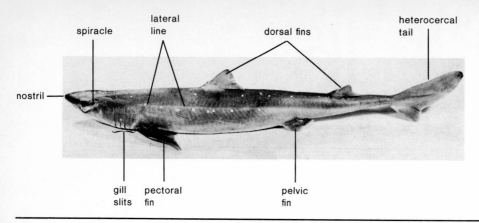

Fig. 16.1 Shark, lateral view. (Courtesy Carolina Biological Supply Company.)

External Anatomy

Obtain a preserved dogfish shark and study the general form of the body; note the broad, flat **head,** the tapered **trunk,** and the laterally compressed **tail.** The surface of the body is covered with many tiny **placoid scales.** Rub your finger lightly over the surface of the skin and feel the sandpaperlike texture caused by the minute spines borne on each scale. In which direction do the spines point?

Locate the paired **pectoral** (anterior) and **pelvic** (posterior) **fins.** In male sharks, the pelvic fins become enlarged and their inner borders are modified to form long, rodlike **claspers,** which aid in mating. Note also the two median **dorsal fins,** each with a sharp spine at its anterior edge, and the large **caudal** (tail) **fin.** The caudal fin consists of two lobes, a larger **dorsal lobe** and a smaller **ventral lobe.** This type of asymmetrical caudal fin with unequal dorsal and ventral lobes constitutes a **heterocercal tail.** Note how the end of the vertebral column curves upward into the dorsal lobe of the caudal fin.

On the head find the ventral, curved, slitlike **mouth,** and anterior to the mouth locate the two **nostrils.** Inside the mouth note the numerous pointed **teeth,** believed to be evolutionarily derived from placoid scales. The **eyes** are located on the sides of the head. Note the rudimentary **eyelids.**

Slightly behind and above the eyes find the two **spiracles,** openings for water intake. The spiracles lead into the pharynx and represent a modified first pair of **gill slits.** Behind the mouth and near the ventral surface find the five pairs of **external gill slits,** a fundamental chordate characteristic. Internally, the gill slits open into the **pharynx.**

Locate also the two light-colored **lateral lines** running posteriorly from the region of each spiracle to the tail on each side of the shark and the **cloaca opening** located ventrally between the two pelvic fins. The lateral lines of the shark represent a special kind of sensory system found only in fishes and in some larval amphibians. It appears to function in the perception of water movements and current changes and thus to aid in orientation and locomotion.

Internal Anatomy

Take care in your dissection to avoid damage to structures important for subsequent study. Do not tear, cut, or pierce parts until you are sure that you know what you are doing. Never cut and discard parts until you are sure of their identity and know that you will have no further use for them.

Be conservative in your cutting; often you will find that structures can be separated neatly and distinctly by teasing them free with a blunt instrument (a blunt probe, back of a scalpel blade, or even the handle of a scalpel) rather than by cutting. When cutting is necessary, make clean incisions with sharp instruments; dull scissors or a dull scalpel will tend to tear rather than to cut tissues and will lead to unsatisfactory results. Read the directions thoughtfully and follow them carefully.

Coelom and Visceral Organs

Place your specimen on its back and carefully locate with your fingers the cartilaginous **pectoral** and **pelvic girdles,** which support the pectoral and pelvic fins, respectively (see figure 16.2). Also using the same figure as a guide, estimate the relative thickness of the body wall in your specimen. Carefully make an incision through the body wall along the midventral line from the pectoral girdle backward through the pelvic girdle, cutting to one side of the cloaca and ending your incision at a point just posterior to it.

If the interior of the body cavity appears to be oily, rinse it out carefully with **cold** water. Take care not to disturb the position of the internal organs during the washing. Now make two transverse incisions through the body wall, one to the rear of the pectoral fins and one anterior to the pelvic fins, each about two inches in length. Pin or tie back

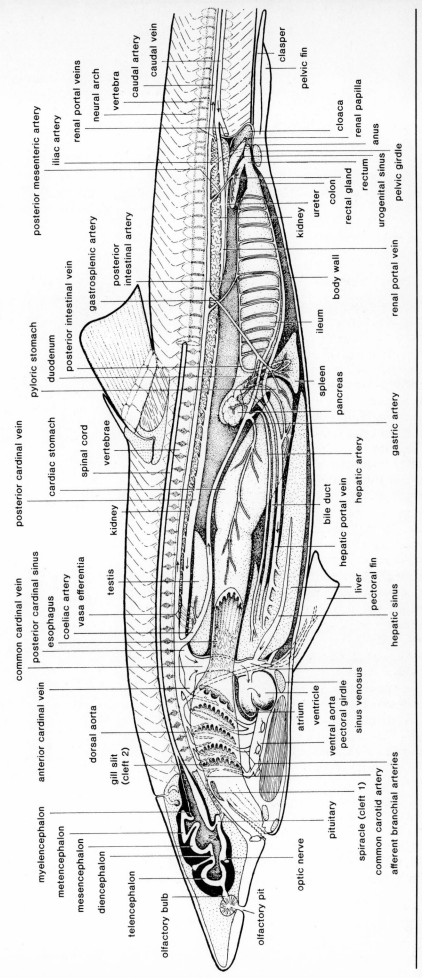

Fig. 16.2 Shark, sagittal section.

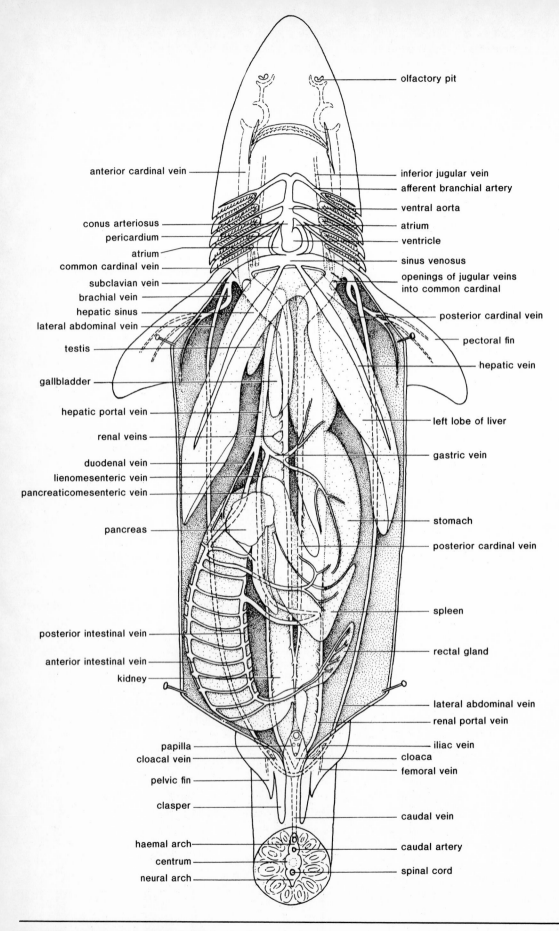

olfactory pit

anterior cardinal vein

inferior jugular vein
afferent branchial artery
ventral aorta

conus arteriosus
pericardium
atrium
common cardinal vein

atrium
ventricle
sinus venosus

subclavian vein
brachial vein
hepatic sinus
lateral abdominal vein

openings of jugular veins
into common cardinal

posterior cardinal vein

testis

pectoral fin

hepatic vein

gallbladder

hepatic portal vein

left lobe of liver

renal veins

gastric vein

duodenal vein
lienomesenteric vein
pancreaticomesenteric vein

pancreas

stomach

posterior cardinal vein

spleen

posterior intestinal vein

anterior intestinal vein

rectal gland

kidney

lateral abdominal vein
renal portal vein

papilla
cloacal vein

iliac vein
cloaca
femoral vein

pelvic fin

clasper

caudal vein

haemal arch

caudal artery

centrum

neural arch

spinal cord

Fig. 16.3 Shark, dissected, ventral view.

the flaps of tissue to expose the large coelomic cavity containing the visceral organs. Carefully study figures 16.2 and 16.3 and identify the various structures visible within the coelom. Note the location, relative size, shape, color, and texture of each structure. Attempt during your subsequent study to relate each structure with its principal function or functions.

The coelom of the shark is divided into two portions, the **pericardial cavity,** found anterior to the pectoral girdle, and the **pleuroperitoneal cavity,** found posterior to the pectoral girdle. These two portions of the coelom are separated by a thin partition, the **transverse septum.**

The smooth lining tissue of the coelom is the **peritoneum,** which also covers the surface of the various organs suspended within the coelom. Dorsally, the peritoneum is continued as a double epithelial membrane, the **dorsal mesentery** (see figure 16.4), which supports the digestive tract within the coelom.

Locate the large esophagus anterior to the J-shaped **stomach.** The stomach is divided into a larger anterior **cardiac** region and a smaller posterior **pyloric** region (see figure 16.5). The pyloric region lies beyond a sharp bend in the stomach and constitutes the lower portion of the J. The constriction between the pyloric region of the stomach and the small intestine is the pylorus. A sphincter muscle in this region controls the movement of food from the stomach into the **small intestine.**

The anterior segment of the small intestine is the **duodenum,** a short and narrow portion. The posterior, longer, segment of the small intestine is the **ileum.** Find the long **bile duct** extending from the liver to the duodenum. Within the ileum is the spiral valve, which serves to increase the intestinal surface area for digestion. The spiral valve and other structures inside the digestive tract will be examined later during your study of the shark.

Posterior to the small intestine the digestive tract is continued by a short and narrow **colon,** which connects with a short **rectum.** Attached dorsally at the junction of the colon and rectum is a blind pocket called the **rectal gland** (see figure 16.4), which plays an important role in maintaining the proper salt balance in the blood of sharks. The rectal gland secretes a fluid consisting mainly of a concentrated solution of sodium chloride. The rectum discharges into the cloaca, a common chamber in which the ducts of the digestive and urogenital systems also terminate. Technically, the opening from the rectum into the cloaca is the **anus,** and the opening from the cloaca to the exterior is the **cloaca opening** (also called the vent).

Also associated with the digestive system are two large digestive glands, the **liver** and the **pancreas.** The liver consists of three lobes, two long lobes on the **right** and **left** sides and a shorter **median** lobe. Find the thin-walled, greenish **gallbladder** along the margin of the median lobe and the common bile duct leading from the anterior end of the gallbladder to the small intestine. The **pancreas** of the shark consists of two distinct parts—a round, flattened **ventral lobe** attached to the surface of the duodenum and a long, narrow **dorsal lobe** lying between the pyloric portion of the stomach and the duodenum.

Another large organ, the **spleen,** is also found attached to the stomach, although it is a part of the circulatory system rather than the digestive system. Locate the dark, triangular spleen closely applied around the outer curvature of the stomach.

The Urogenital System

The reproductive system is poorly developed in the immature dogfish sharks usually provided for laboratory study. Therefore, it is advisable for you to supplement your observations of the reproductive system, particularly of the female shark, by viewing demonstrations of more mature specimens. Consult your instructor for information on the demonstrations available.

Examine figure 16.6, which shows the urogenital systems of male and female sharks. **Do not remove** or **cut** any of the structures shown, but simply spread the abdominal organs apart to facilitate your observation. Note the long, flat **kidneys** closely applied to the body wall. The kidneys are actually **retroperitoneal;** that is, they are located **between** the coelom and the body wall and not suspended into the peritoneal cavity as are the other abdominal organs.

If you have a male specimen, observe the coiled **archinephric ducts** as shown in figure 16.4 (also sometimes called the pronephric, mesonephric, opisthonephric, or Wolffian ducts; and sometimes incorrectly called ureters), which lie on the ventral surface of the kidneys and extend posteriorly to the cloaca. These ducts carry both urine from the kidneys and sperm from the testes. Locate the paired **testes,** a pair of elongated organs dorsal to the anterior end of the liver. Leading from each of the testes and emptying into the mesonephric duct are several small efferent ductules (**vasa efferentia**). Within the cloaca, the right and left mesonephric ducts join and empty through a common **urogenital pore.** The urogenital pore is located at the tip of the **urogenital papilla,** a small, fleshy projection from the dorsal wall of the cloaca.

If you have a female specimen, locate the **ovaries,** a pair of oblong, lobed bodies situated near the dorsal body wall above the anterior portion of the liver. From the ovaries, a pair of slender **oviducts** extend posteriorly along the length of the body cavity. The two oviducts originate as a common duct with a single opening, the **ostium tubae,** located anterior to the liver and ventral to the esophagus. (Search carefully in this area; the ostium tubae is often difficult to find.) From the ostium tubae the **oviducts** loop anteriorly and laterally over the anterior end of the liver and pass back posteriorly along the dorsal body wall. Ripe eggs discharged from the ovaries enter the body cavity, pass through the ostium tubae (see arrows in figure 16.6),

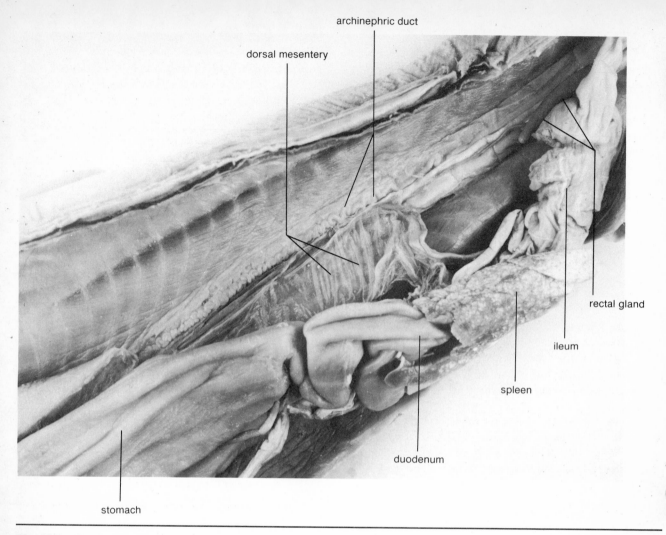

archinephric duct

dorsal mesentery

rectal gland

ileum

spleen

duodenum

stomach

Fig. 16.4 Shark, dissected, showing dorsal mesentery and abdominal organs of posterior region. (Photograph by Carol Majors.)

and enter the oviducts, where they may be fertilized. Most of the development of the embryos takes place in the **uterus,** an enlarged portion of each oviduct. The uteri open into the cloaca. Note also the **archinephric ducts** (you may require assistance from your instructor since these ducts are sometimes difficult to find in immature specimens) of the female shark lying along the ventral surface of the kidney. Trace one of the mesonephric ducts posteriorly to its entrance into the cloaca. Inside the cloaca, the mesonephric ducts empty through a **urinary pore** located on a **urinary papilla.** How do the male and female urogenital systems differ in this respect? Locate a student in your class who has a specimen of the opposite sex and compare the male and female reproductive systems or (alternatively) study the demonstration materials provided by your instructor.

Dogfish sharks (*Squalus acanthias*) give birth to living young, unlike most other sharks, which are egg-laying (**oviparous**). *Squalus acanthias* is **ovoviviparous** since more-or-less typical eggs are produced but are retained

within the reproductive system of the female until hatching. Many of the mature female sharks used for laboratory study are pregnant because of the unusually long gestation period of the dogfish shark, which ranges from twenty to twenty-four months. Higher mammals that form a true placenta and give birth to living young without forming a shelled egg are **viviparous.**

The Vascular System

Before you attempt to make a detailed study of the vascular system, you should study the general plan of the circulatory system and the direction of blood flow from the heart through the arteries to the capillaries of various organs and back to the heart by the veins.

In certain cases, blood from the capillary beds in the tissues does not return directly to the heart but is forced through an intervening bed of capillaries enroute back to the heart. The veins connecting the two beds of capillaries are called **portal veins.** Two **portal systems** exist in the shark, the **hepatic portal system** and the **renal portal**

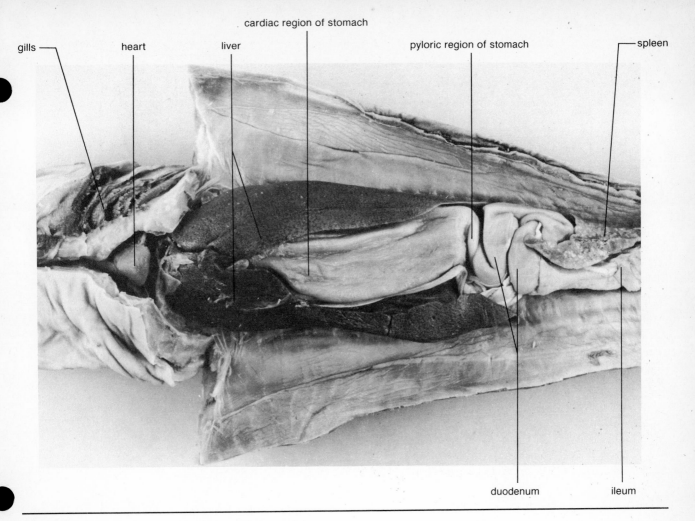

gills — heart — liver — cardiac region of stomach — pyloric region of stomach — spleen

duodenum — ileum

Fig. 16.5 Shark, dissected, ventral view. (Photograph by Carol Majors.)

system. Study figures 16.2 and 16.7 for this preliminary survey and note the following principal structures: (1) the **heart,** which pumps blood to the gills where it is oxygenated and then flows to the dorsal aorta for distribution to various parts of the body; (2) the **hepatic portal system,** which returns blood chiefly from the digestive system to the liver and thence to the heart via the hepatic vein and sinus venosus; (3) the **renal portal system,** which returns blood from the posterior portion of the body to the kidneys, from which it goes to the heart via the postcardinal sinuses and the sinus venosus; and (4) the **anterior cardinal veins,** which collect blood from the head.

From your study of the circulatory system you should learn the following: (1) the names of the principal parts of the system; (2) the direction of blood flow in each part; (3) the organs served by the major blood vessels (both arteries and veins), and (4) the gains and losses from the blood as it flows through the capillaries of the various organs, particularly in regard to oxygen, carbon dioxide, nutrients, and nitrogenous wastes.

The Arterial System

The arteries of the shark can be conveniently studied in three groups: (1) the **visceral arteries,** consisting principally of the dorsal aorta and its branches, (2) the **afferent branchial arteries** and their branches, which carry blood to the gills, and (3) the **efferent branchial arteries,** which carry blood away from the gills and connect with the dorsal aorta.

Visceral Arteries. Spread apart the organs and carefully separate the blood vessels from the mesenteries as necessary to locate the arteries. Study the principal arteries in your specimen, using figure 16.2 as a guide. Locate first the large median **dorsal aorta,** visible through the peritoneal lining of the dorsal wall of the coelom. Posteriorly, the dorsal aorta is continued as the caudal artery, best seen in cross sections of the tail (see figure 16.3). Within the pleuroperitoneal cavity the dorsal aorta gives off several arteries, some of which are paired and some of which are

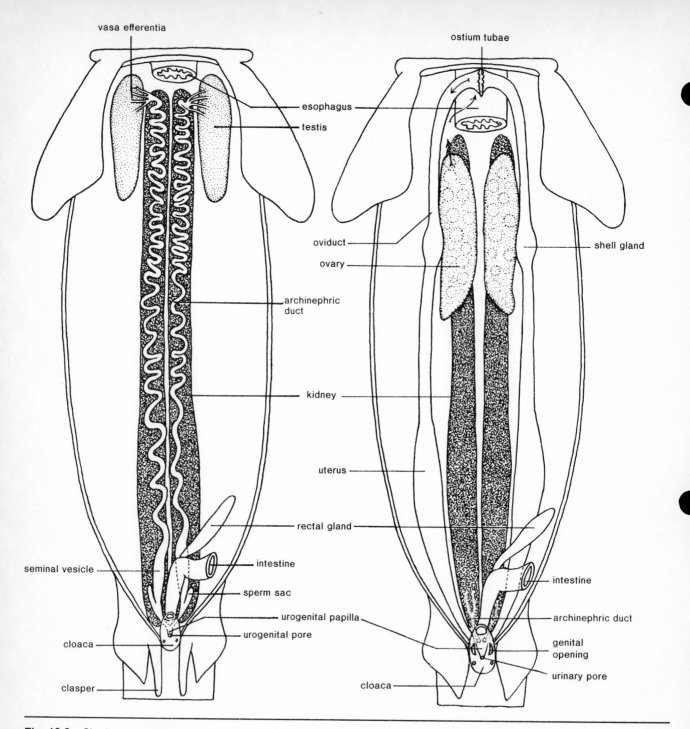

Fig. 16.6 Shark, urogenital system.

unpaired. The principal branches of the dorsal aorta that can be found within this cavity are listed below (anterior to posterior):

Coeliac artery—arises just posterior to the transverse septum and gives off branches to the gonads, esophagus, stomach, liver, and pancreas.

Posterior intestinal artery—arises from the aorta behind the stomach and near the posterior end of the mesentery and supplies one side of the ileum and part of the spiral valve.

Gastrosplenic (lienogastric) artery—arises just behind the posterior intestinal (in some specimens the two arteries arise from the aorta as a single vessel and then split) and supplies blood to the spleen, stomach, and a portion of the pancreas.

Posterior mesenteric artery—arises from the aorta near the anterior end of the rectal gland, which it supplies.

Iliac arteries (2)—a pair of large arteries arising just anterior to the cloaca. Supply the pelvic fin and the posterior portion of the body wall.

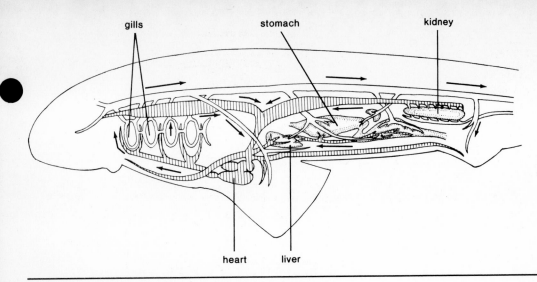

Fig. 16.7 Shark, diagram of circulatory system. The heart, venous system, and the afferent branchial arteries are shaded with vertical lines; the efferent branchial arteries are shown in white. (After Romer.)

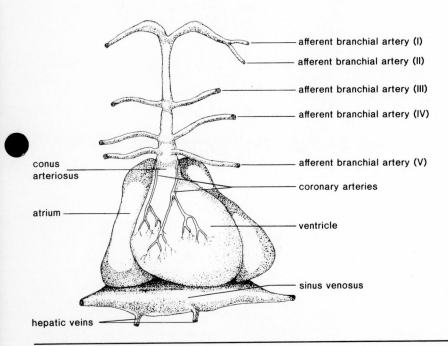

Fig. 16.8 Shark heart, ventral view. (After Wischnitzer.)

The Heart, Afferent Branchial Arteries, and Gills. Slice off the skin and muscles from the ventral surface of the head posterior to the mouth. Carefully cut away the muscles just anterior to the pectoral girdle until you reach the membrane enclosing the pericardial cavity. Cut through the membrane to expose the heart within its cavity, and carefully cut through the pectoral girdle and remove a section of the pectoral girdle (about one inch wide) to expose further the heart. Consult figures 16.8 and 16.9 and identify the parts of the heart and the surrounding blood vessels. Note that the shark has a **two-chambered heart**

with a thick-walled, muscular **ventricle** and a thin-walled **atrium** (auricle). Identify also the **sinus venosus,** a flattened, thin-walled sac closely applied to the posterior surface of the ventricle and lying between the ventricle and the transverse septum. Lift up the posterior end of the heart to facilitate your viewing of the sinus venosus. Note that the transverse septum separates the **pericardial cavity** from the **pleuroperitoneal cavity,** thus dividing the **coelom** into two distinct parts. Locate the muscular **conus arteriosus** extending anteriorly from the ventricle.

Blood enters the heart through the sinus venosus, passes into the atrium, thence to the muscular ventricle,

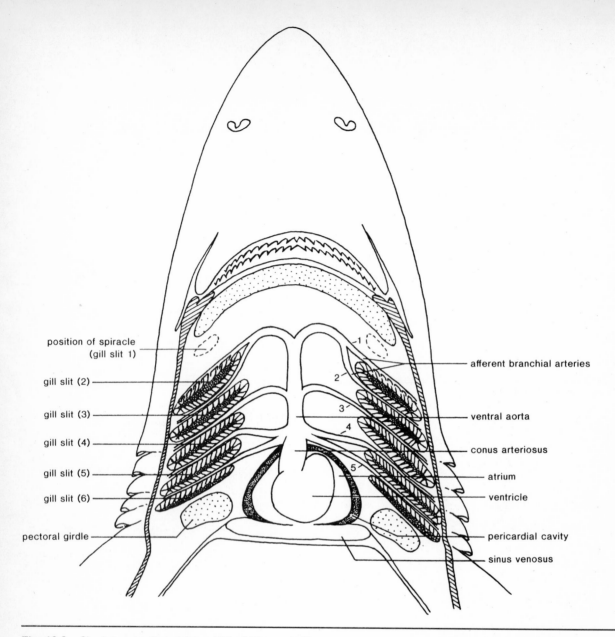

Fig. 16.9 Shark heart, afferent branchial arteries and gills.

Labels on figure:
- position of spiracle (gill slit 1)
- gill slit (2)
- gill slit (3)
- gill slit (4)
- gill slit (5)
- gill slit (6)
- pectoral girdle
- afferent branchial arteries
- ventral aorta
- conus arteriosus
- atrium
- ventricle
- pericardial cavity
- sinus venosus

and is pumped out through the conus arteriosus into the ventral aorta (uninjected). The ventral aorta extends anteriorly from the heart and gives rise to five pairs of **afferent branchial arteries.** Trace the ventral aorta forward, and on one side locate the afferent branchial arteries that carry blood to the gills for oxygenation. Cut away the lower portion of one of the gills and note the **cartilaginous bars** within the **gill arches,** which support the numerous **gill filaments.** Find also the **gill-rakers,** cartilage-supported, fingerlike projections from the gill arches that guard the internal gill slits and prevent large food particles from entering the **gill chamber.** Count the number of gills and gill slits on your specimen. The **spiracle,** although nonfunctional as a gill in the shark, is customarily still designated as the first gill slit (see figure 16.2). Note the many small branches of the afferent branchial arteries, which carry blood into the gill filaments where gas exchange occurs.

Efferent Branchial, Subclavian, and Carotid Arteries.
Cut through the angle of the jaw on the left side of the head and continue the incision posteriorly through the pharynx, esophagus, and the ventral body wall to a point just beyond the pectoral girdle. Make a similar cut from the angle of the jaw on the right side through the pectoral girdle. Fold back the lower jaw and the parts immediately behind it to expose the roof of the mouth and pharynx. Carefully remove the membrane from the roof of the mouth and pharynx to expose the **efferent branchial arteries.** Study figure 16.10 and continue your dissection until you have exposed all of the efferent branchial arteries and other major blood vessels in this region. Identify in your specimen each of the blood vessels shown in figure 16.10. Note that each efferent branchial artery is formed by the union of two smaller arteries, a **pretrematic artery** and a **posttrematic artery,** which collect blood from

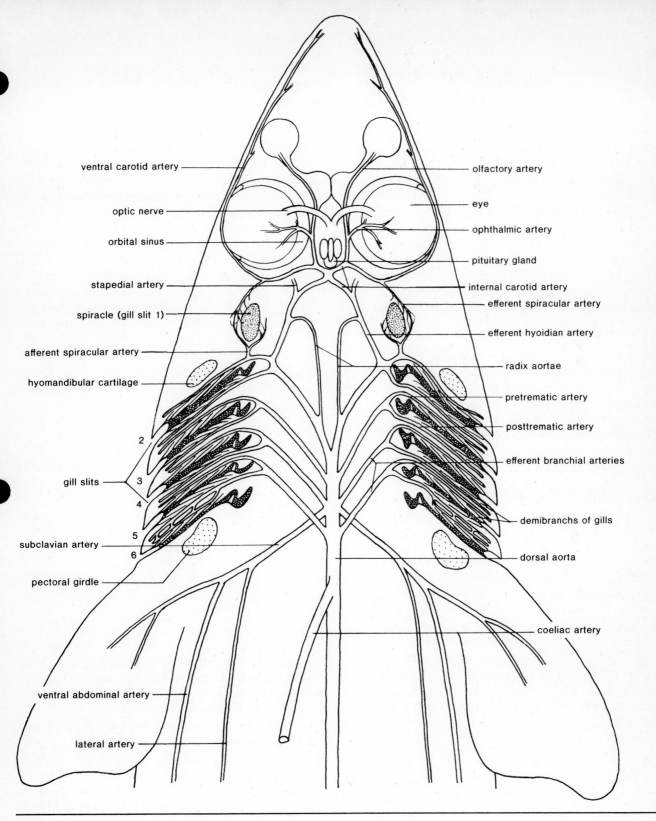

Fig. 16.10 Shark, efferent branchial arteries, carotid, and subclavian arteries.

Labels (left side, top to bottom): ventral carotid artery, optic nerve, orbital sinus, stapedial artery, spiracle (gill slit 1), afferent spiracular artery, hyomandibular cartilage, gill slits, subclavian artery, pectoral girdle, ventral abdominal artery, lateral artery

Gill slit numbers: 2, 3, 4, 5, 6

Labels (right side, top to bottom): olfactory artery, eye, ophthalmic artery, pituitary gland, internal carotid artery, efferent spiracular artery, efferent hyoidian artery, radix aortae, pretrematic artery, posttrematic artery, efferent branchial arteries, demibranchs of gills, dorsal aorta, coeliac artery

Shark Anatomy

the demibranchs ("half-gill") on the anterior and posterior sides of each gill slit. The pre- and posttrematic arteries are joined at both the dorsal and ventral ends of the gill slit, thus forming an **arterial collecting loop** (see figure 16.10), which encircles the gill slit.

Find also the **dorsal aorta** and the paired **subclavian arteries** running toward the two lateral pectoral fins. Arising from each subclavian artery, there are two smaller arteries that carry blood posteriorly. These branches are the **ventral abdominal arteries** (larger and more lateral in location) and the **lateral arteries** (smaller and more medial in location).

Between the gills and the eyes, several other important arteries can also be found, including the **hyoidean epibranchial arteries** (or **efferent hyoidean arteries**), which arise from the arterial collecting loops around the first gill slits on each side and extend diagonally across the roof of the mouth. The hyoidean epibranchials form two branches, the **temporal arteries** and the **internal carotid arteries.**

The Venous System

The venous circulation (figures 16.2 and 16.7) of the shark includes three systems: (1) the **hepatic portal system,** which carries blood from the digestive tract to the liver; (2) the **renal portal system,** which carries blood from the tail region to the kidneys; and (3) the **systemic veins,** which collect blood from the other tissues and organs of the body. All three venous systems empty into the two large common cardinal veins and thence into the sinus venosus.

Hepatic Portal System. Study figures 16.2, 16.3, and 16.7 and identify on your specimen the **hepatic portal vein** extending forward from the stomach and intestine toward the liver. Note that the hepatic portal vein runs along the bile duct and enters the liver. Inside the liver it branches several times and terminates in several capillary beds.

Posteriorly, the hepatic portal vein is formed by the junction of three branches: (1) the **gastric vein** from the stomach (left branch); (2) the **lienomesenteric vein** with branches from the intestine and spleen (central branch); and (3) the **pancreaticomesenteric vein** with branches from the intestine, stomach, and pancreas (right branch).

Some of the food materials absorbed from the stomach and intestine are removed and stored in the liver before the blood is returned to the heart and circulated to the rest of the body.

Renal Portal System. The caudal vein leads from the tail region forward into the trunk. Near the level of the cloacal opening the caudal vein divides into two **renal portal veins,** which pass lateral to the kidneys. The renal portal veins give off many small **afferent renal veins,** which empty into the kidneys.

Systemic veins. The systemic veins collect blood from the various organs and return it to the heart. The systemic veins are more difficult to study than the arteries because most of them lack definite walls and appear as more or less open tissue spaces or sinuses. The **posterior cardinal veins** run along the sides of the dorsal aorta and carry blood forward to the large **posterior cardinal sinus.** The posterior cardinal veins collect blood from the kidneys via many small **efferent renal veins,** which drain the renal sinuses within the kidneys. In addition to the efferent renal veins, the posterior cardinal veins also collect blood from many small, segmentally arranged **parietal veins,** which collect blood from the muscles of the body wall.

Other important systemic veins include the two large **hepatic veins,** which drain the liver and carry blood to the sinus venosus, the **lateral abdominal veins,** and the **brachial veins** (from the pectoral fins), which unite to form the short **subclavian veins.** On each side the subclavian vein joins with the **posterior cardinal sinus** and the **anterior cardinal vein** to form the short **common cardinal vein** (duct of Cuvier), which empties into the **sinus venosus.** The openings of the anterior cardinal veins can be found by slitting open one of the common cardinal veins and using your probe to locate the aperture of the anterior cardinal vein.

Review of the Vascular System

Review the blood vessels of the shark that you have studied, giving particular attention to the principal divisions of the arterial and venous systems. As an aid in your review and to assist in your understanding of the pattern of circulation, use colored pencils to identify the blood vessels in figures 16.2 and 16.3 according to the type of blood carried. Color vessels carrying oxygenated blood **red** and those with oxygen-depleted blood **blue.**

Internal Structure of Esophagus, Stomach, and Intestine

To review the organization of the digestive system, slit open the digestive tract from the esophagus to the colon. Observe the many, small, fingerlike projections from the lining of the esophagus, the **papillae;** the longitudinal folds (**rugae**) of the inner stomach wall; the sphincter muscles of the **pylorus;** the opening of the bile duct into the duodenum (the opening of the pancreatic duct is in the same region, but is smaller and more difficult to find), and the **spiral valve.**

Cross Sections of the Body

Cut off the head just behind the pectoral fins and save it for the study of the nervous system. Take the posterior portion of the body and cut several transverse sections through the trunk for study of the anatomy of the trunk region.

In the sections of the vertebrae, observe the rounded **centrum** with the **neural arch** above, enclosing the spinal cord (see figure 16.3). In the tail region, find the ventral **hemal arch** that encloses the caudal artery and caudal vein, and note that in the trunk region there is no hemal arch. Find the **transverse processes** on the trunk vertebrae extending from the ventrolateral margin of each centrum to articulate with the rib cartilages of the shark.

Nervous System and Sense Organs

The nervous system of the shark, like that of other vertebrates, consists of two principal components: (1) the **central nervous system** including the brain and the spinal cord and (2) the **peripheral nervous system** including the various nerves that connect the brain and spinal cord with various other parts of the body. The nerve fibers of the peripheral nervous system can be divided anatomically into those that innervate the outer portions of the body, such as the skin and voluntary muscles (the **somatic division**), and those that innervate the internal organs or viscera (the **visceral division**). Both the somatic and visceral divisions of the peripheral nervous system contain numerous **afferent** (sensory) and **efferent** (motor) **nerve fibers**. The **autonomic nervous system** is a specialized and very important part of the visceral division of the peripheral nervous system in higher vertebrates, although it is poorly developed in the shark. In higher vertebrates, the autonomic nervous system plays a vital role in the coordination of the smooth muscles and glands of the body and thus is intimately involved in such processes as digestion, regulation of heart rate, dilation and constriction of blood vessels, and regulation of blood glucose levels.

The basic organization of the vertebrate nervous system can therefore be outlined in the following manner:

> Central Nervous System
> > Brain
> > Spinal cord
>
> Peripheral Nervous System
> > Somatic division
> > > Afferent neurons (sensory)
> > > Efferent neurons (motor)
>
> > Visceral Division
> > > Autonomic Nervous System
> > > (plus a few motor neurons to the branchial musculature)

Your study of the nervous system of the shark will be limited to a brief survey of the brain and cranial nerves since these parts will suffice to introduce you to the pattern of organization of the vertebrate nervous system. Some further aspects of the peripheral nervous system of vertebrate animals will be studied in some of the later exercises in this manual.

The brain and cranial nerves of the shark are illustrated in figures 16.11 and 16.12. Study these figures carefully before undertaking the dissection of the nervous system. The successful dissection and study of the nervous system requires patience, care, and attention to detail. Haste and carelessness most often lead to disappointing results and confusion.

After you have carefully studied figures 16.11 and 16.12, take the shark head saved from your earlier dissection and carefully remove the skin from the dorsal side. When the skin is all removed from the dorsal surface, slice away thin sections of the cartilaginous skull (chondrocranium) until the dorsal and lateral surfaces of the brain are exposed. Be careful not to cut or tear away the delicate nerves that pass through the several small openings (foramena) of the skull.

On either side of the skull, near its posterior margin, are the two fused **otic capsules,** which enclose the **semicircular canals** of the inner ear. Carefully shave away the cartilage from the otic capsule and locate the semicircular canals, which are important organs of equilibrium (figure 16.11).

During the early stages of embryonic development, the brain of vertebrates becomes divided first into **three distinct lobes.** These three lobes are identified as the **forebrain** (prosencephalon), the **midbrain** (mesencephalon), and the **hindbrain** (rhombencephalon). This early embryonic differentiation of the three primary brain divisions (which together form the **brain stem**) reflects the evolutionary history of the vertebrate brain, since the brain of the earliest vertebrates also consisted of three divisions, each associated closely with one of three major sense organs: the nose, the eye, and the ear (plus the lateral line).

The brain of adult sharks and other recent vertebrates, however, consists of five major divisions rather than three. Two of the three basic brain divisions (the forebrain and hindbrain) divide again during later stages of embryonic development. The brain of **adult sharks** therefore consists of **five major divisions:** (1) the **telencephalon,** (2) the **diencephalon,** (3) the **mesencephalon,** (4) the **metencephalon,** and (5) the **myelencephalon.** Locate each of these five divisions of the shark brain on your specimen. These divisions and their relationships are summarized in table 16.1.

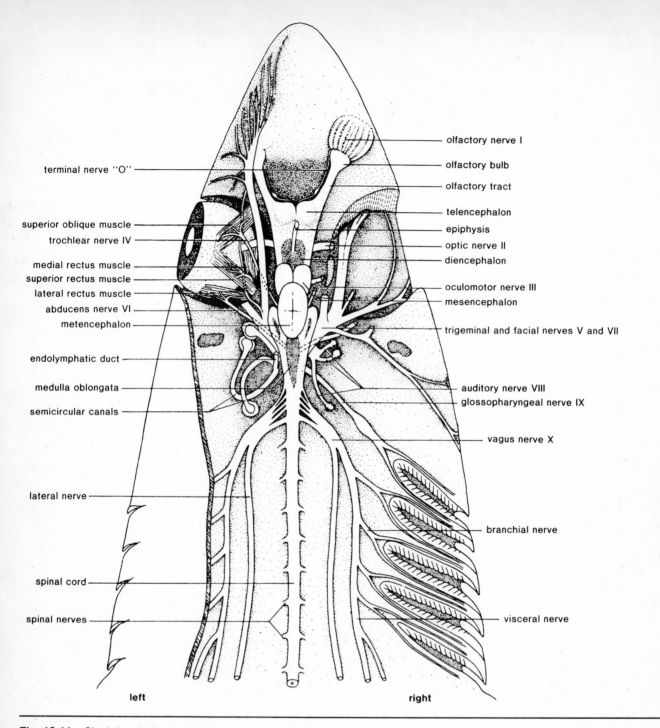

olfactory nerve I
olfactory bulb
olfactory tract
telencephalon
epiphysis
optic nerve II
diencephalon
oculomotor nerve III
mesencephalon
trigeminal and facial nerves V and VII

auditory nerve VIII
glossopharyngeal nerve IX

vagus nerve X

branchial nerve

visceral nerve

terminal nerve "O"

superior oblique muscle
trochlear nerve IV

medial rectus muscle
superior rectus muscle
lateral rectus muscle
abducens nerve VI
metencephalon

endolymphatic duct

medulla oblongata

semicircular canals

lateral nerve

spinal cord

spinal nerves

left

right

Fig. 16.11 Shark head, showing brain, cranial nerves, and eye muscles, dorsal view.

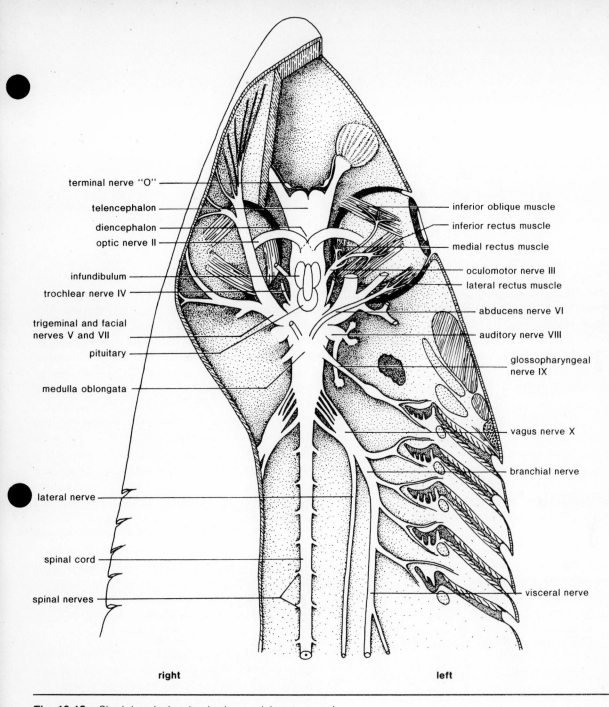

terminal nerve "O"
telencephalon
diencephalon
optic nerve II
infundibulum
trochlear nerve IV
trigeminal and facial nerves V and VII
pituitary
medulla oblongata
lateral nerve
spinal cord
spinal nerves

inferior oblique muscle
inferior rectus muscle
medial rectus muscle
oculomotor nerve III
lateral rectus muscle
abducens nerve VI
auditory nerve VIII
glossopharyngeal nerve IX
vagus nerve X
branchial nerve
visceral nerve

right

left

Fig. 16.12 Shark head, showing brain, cranial nerves, and eye muscles, ventral view.

Table 16.1   Summary of the Organization of the Shark Brain

Embryonic Divisions	Adult Divisions	Component Structures
Prosencephalon (forebrain)	Telencephalon	Olfactory bulbs Olfactory tract Olfactory lobes Cerebral hemispheres
	Diencephalon	Epiphysis (pineal gland) Infundibulum } Pituitary gland Hypophysis
Mesencephalon (midbrain)	Mesencephalon	Optic lobes
Rhombencephalon (hindbrain)	Metencephalon	Cerebellum
	Myelencephalon	Medulla

Survey of the Brain

1. The anteriormost **telencephalon** is represented by the **olfactory sacs** connected with the **olfactory bulbs** via the **olfactory tracts.** Just behind the two olfactory bulbs are the less prominent **cerebral hemispheres.** They can be identified as slight swellings behind the olfactory lobes and are separated from the olfactory lobes by a shallow groove.

2. The **diencephalon** lies behind the telencephalon and appears on the dorsal surface of the brain as a narrow depressed area just anterior to the optic lobes. The diencephalon bears the **epiphysis** or pineal gland— a slender stalk extending anteriorly up through an opening in the roof of the skull. The pineal or "third eye" has long been an enigmatic organ of vertebrates; its function in the shark is not well understood. Ventrally the diencephalon bears the **infundibulum** and the **hypophysis,** which together comprise the **pituitary gland** of the shark. These structures will be studied later.

3. Behind the diencephalon are the two large lateral swellings, the **optic lobes,** important brain centers that play an essential role in vision; here the optic nerves terminate. The optic lobes represent the **mesencephalon.**

4. The **metencephalon** consists mainly of the **cerebellum,** a single large median lobe lying behind the paired optic lobes.

5. The **myelencephalon** consists principally of two parts—the triangular **medulla oblongata,** the most posterior portion of the brain, which tapers into the connecting spinal cord, and the two lateral **auricular lobes.** The latter extend forward from the medulla and can be seen adjacent to the posterior portion of the cerebellum. The auricular lobes serve as important centers of equilibrium.

Cranial Nerves

Sharks exhibit eleven pairs of **cranial nerves.** Each of these nerves attaches to a specific portion of the brain and connects that part of the brain with some specific organ or organs of the body. Some of the cranial nerves consist wholly of **sensory neurons** (carry impulses to the brain), and other cranial nerves are wholly made up of **motor neurons** (carry impulses from the brain), and about half of them contain mixtures of sensory and motor elements.

Cranial nerves were first studied seriously in humans and were given names and numbers (I to XII) based on their location and function in humans. (It is customary to use Roman numerals to designate the cranial nerves.) Subsequent study of other vertebrates, however, has revealed that the human pattern of cranial nerves does not hold for all vertebrates. In sharks, for example, only ten of the cranial nerves correspond to those of man (nerves I through X). The eleventh cranial nerve in the shark is a small anterior nerve, the **terminal nerve,** not present in humans.

Locate each of the cranial nerves on your specimen with the aid of figures 16.11 and 16.12 and the list provided below.

Nerve O. Terminal nerve. A delicate sensory nerve arising from the median surface of the olfactory lobe and extending into the nasal region. It is numbered "O" because it was discovered after the numbering system for cranial nerves became established.

Nerve I. Olfactory nerve. Carries sensory impulses from the olfactory epithelium within the olfactory sac to the olfactory bulb.

Nerve II. Optic nerve. Arises in the retina of the eye and runs to the ventral surface of the diencephalon where it crosses the brain and carries impulses to the optic lobe on the opposite side of the brain. The prominent crossing of the optic nerves on the ventral surface of the diencephalon is called the optic chiasma (to be studied later).

Nerve III. Oculomotor nerve. Originates in the mesencephalon and divides into four branches, which carry motor impulses to the muscles of the eye.

Nerve IV. Trochlear nerve. Originates from the dorsal surface of the mesencephalon and passes through the chondrocranium to carry impulses to the superior oblique muscle of the eye.

Nerve V. Trigeminal nerve. A mixed nerve with both motor and sensory fibers, which arises from the anterior part of the medulla and enters the orbit. It is the largest of the cranial nerves in the shark and divides into four main branches: (1) the superficial ophthalmic nerve to the skin of the head, (2) the deep ophthalmic nerve to the skin of the snout, (3) the infraorbital nerve to the region of the mouth and ventral surface of the snout, and (4) the mandibular nerve to the jaw muscles and skin of the lower jaw.

Nerve VI. Abducens nerve. A motor nerve arising from the ventral surface of the medulla and carrying impulses to the external rectus muscle of the eye.

Nerve VII. Facial nerve. A mixed nerve with both motor and sensory fibers arising with the trigeminal nerve from the medulla. It is made up of three main branches: (1) a branch from the **superficial ophthalmic nerve** described previously, (2) the **buccal nerve** with sensory fibers from the mouth region, and (3) the **hyomandibular nerve** with sensory branches from the tongue, lateral line, and lower jaw.

Nerve VIII. Auditory nerve. Arises from the anterior end of the medulla and innervates the inner ear.

Nerve IX. Glossopharyngeal nerve. A mixed nerve arising from the medulla with branches to the first functional gill slit and to the roof of the mouth.

Nerve X. Vagus nerve. A large mixed nerve, which arises from several roots on the medulla. It divides to form three main branches: (1) the **lateral line trunk** to the lateral line, (2) the **branchial trunk** to the gills (except the first) and (3) the **visceral trunk** to the heart and abdominal organs.

Ventral Surface of the Brain. After you have completed your study of the dorsal parts of the brain and the cranial nerves, study figure 16.12 and carefully dissect away the anterior portion of the roof of the mouth and the cartilage underlying the brain. Study the ventral surface of the brain and locate the **optic chiasma** where the large optic nerves cross, the two lobes of the **infundibulum,** the **hypophysis,** and the origin of the **sixth cranial nerve** (abducens) described above.

Key Terms

Hepatic portal system   portion of the venous circulation in sharks and higher vertebrates that collects blood from the stomach and intestine and returns it to the liver, where many food materials absorbed from the gut are removed for storage.

Heterocercal tail   type of tail found in sharks in which the caudal vertebra are deflected upward into the dorsal lobe. The tail is asymmetrical with the dorsal lobe larger and longer than the ventral lobe.

Lateral line   type of sense organ found in sharks and many other fishes and amphibians. Consists of a series of sensory cells usually found along the two sides of the body. Enables the animals to detect water currents, temperature changes, and electrical currents.

Ostium tubae   opening of the oviduct in the abdominal cavity of the shark and other vertebrates.

Renal portal system   part of the venous circulation in sharks (also other fishes, amphibians, reptiles, and birds; absent in mammals), which returns blood from the tail, hind limbs, and posterior portion of the body to the kidneys.

Notes and Sketches

17
Perch Anatomy

Objectives

After you have completed the laboratory work for this chapter, you should be able to perform the following tasks:

1. Locate and name the main external features of a perch.
2. Locate the axial skeleton, appendicular skeleton, and visceral skeleton on a prepared perch skeleton and explain the function of each. Describe and point out the main parts of a trunk vertebra.
3. Describe and locate the main divisions of the perch musculature.
4. Locate the gills, show the main parts of a gill, and explain the function of each part.
5. Identify the parts of the digestive system.
6. Describe the heart of a perch and locate its main parts.
7. Describe the basic pattern of circulation in a perch.
8. List the five major divisions of the perch brain and demonstrate them on a specimen.
9. Discuss the principal morphological similarities and differences between the perch and the shark.

The Yellow Perch: *Perca flavescens*

The yellow perch, *Perca flavescens,* is a common bony fish found in lakes and streams throughout most of the United States; it is native to the central U.S. and southern Canada, but is widely stocked elsewhere. A similar species, *Perca fluviatilis,* the European perch, is common in Europe. Although less commonly studied in zoology and comparative anatomy laboratories than the shark, the perch is more typical of modern fishes.

The yellow perch is a member of the Class Osteichthyes, the bony fishes, the largest class of living vertebrates. Over 20,000 species of bony fishes have been described from the lakes, streams, rivers, and oceans of the world. Among the principal distinguishing features of the Osteichthyes are a **bony skeleton, terminal mouth, dermal scales, homocercal tail, paired nostrils,** and ears with **three semicircular canals.**

Mature specimens range from 6 to about 12 inches in length. Preserved or freshly killed specimens may be dissected, but preserved, latex-injected specimens are best suited for study of the circulatory system.

You should also study a goldfish or other small bony fish in an aquarium to observe swimming movements and other aspects of fish behavior.

Materials List

Living specimens
  Small perch or goldfish
Preserved specimens
  Perch
  Perch, double or triple injected to show details of circulatory system
Prepared microscope slides
  Ctenoid scale, whole mount
  Fish gill, cross section
  Fish skin, cross section to show origin of scales
  Freshwater fish, longitudinal section of head to show gills
Plastic mounts
  Fish heart
  Fish skeleton
  Perch skull
  Perch, dissected
Miscellaneous
  Aquarium
  Model of dissected perch

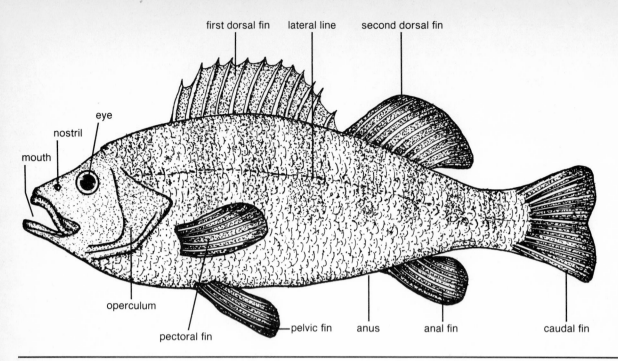

first dorsal fin   lateral line   second dorsal fin

eye

nostril

mouth

operculum

pectoral fin   pelvic fin   anus   anal fin   caudal fin

Fig. 17.1   Perch, external anatomy, lateral view.

External Anatomy

Obtain a preserved perch and study the principal features of its external anatomy (figure 17.1). Note the streamlined **fusiform** (spindle-shaped) **body,** which is thickest about one-third of the distance from the mouth to the tail and tapers in both directions. Pick up the fish and look directly at the mouth from the front; observe the ovoid cross section of the fish. How would this shape facilitate movement of the fish through the water? Numerous mucous glands in the skin further aid in reducing resistance during movement through the water.

Identify the three regions of the body: the anterior **head,** which extends to the rear of the bony operculum covering the gills; the **trunk,** extending from the operculum to the anus; and the **tail,** extending from the anus posteriorly.

On the head find the two double **nostrils,** two large **eyes** (no eyelids), and the large **mouth** equipped with **teeth.** Where are the teeth found? A bony **operculum** covers the gills on each side of the head; under each operculum are four **gills.** The operculum is attached at the front and on the dorsal side, but is open behind and ventrally for the release of water.

On the ventral surface, just anterior to the tail, locate the **anus** and the **urogenital opening**(s). Female perch have a single urogenital opening anterior to the anus; males have a separate genital pore and also a urinary pore located on a small urinary papilla.

Observe the several fins attached to the body: four unpaired **median fins** (two dorsal fins, one anal fin, and one caudal fin) and two sets of **paired fins** (two pectoral fins and two pelvic fins). All the fins are membranous extensions of the skin supported by numerous **fin rays.** The fins

are important aids in swimming; they aid in stabilizing the fish and in directing its movements through the water.

On each side of the fish, extending from the operculum behind the eye to the base of the tail, is a **lateral line.** The lateral lines are specialized sense organs which serve to detect vibrations and current directions in the water. They appear to aid fishes in orientation, in avoiding obstacles in the water, and in escaping predators.

The exterior surface of the perch is covered by a tough **skin** which contains many **mucous glands** and which produces the **scales.** The scales protect the surface of the body and are arranged in a well-ordered pattern of longitudinal and diagonal rows. Note how the posterior portion of each scale overlaps the anterior portion of the next scale. Each scale is formed in an epidermal pocket and extends posteriorly from the pocket. The scales grow continuously during the life of the fish and are not regenerated if lost. Seasonal differences in the growth of the fish are revealed in the deposition of new material around the margin of the scales. Thus, a scientist can determine the age of a fish by microscopic examination of its scales. This is an important technique frequently used in research on the biology of fishes.

Remove a scale from your specimen and make a wet mount on a microscope slide and observe it under low power. Note the numerous concentric ridges (**annuli**) on the scale and the many fine **teeth** on the posterior portion of the scale. This type of scale is called a **ctenoid scale** because of the presence of these teeth.

Draw a ctenoid scale in figure 17.2.

Fig. 17.2 Drawing of a ctenoid scale.

Internal Anatomy

Skeletal system

The scales, fin rays, and some of the bones of the skull of the perch represent elements of a **dermal exoskeleton,** but the chief supporting structure of the body consists of a **bony endoskeleton.**

Observe a prepared skeleton of a perch or other bony fish on demonstration. Locate the **axial skeleton** consisting of the skull, the vertebral column, the ribs, and the medial fins. The **appendicular skeleton** is made up of the pectoral girdle and the pectoral fins and a small pelvic girdle which supports the pelvic fins.

The **vertebral column** is made up of many individual **vertebrae.** The trunk vertebrae have a large cylindrical **centrum** with a dorsal **neural arch** through which the dorsal nerve cord passes, and a single **neural spine.** Lateral processes on each side of the trunk vertebrae articulate with the ribs.

Draw an anterior view of a trunk vertebra in figure 17.3. Label each part.

The caudal vertebrae also have a ventral **haemal arch,** through which the caudal artery passes, and a **haemal spine.**

There is also a **visceral skeleton,** first formed of cartilage and later replaced by bone, which supports the gills. There are **seven paired arches** which correspond to similar structures in the skeleton of the shark. The upper part of

Fig. 17.3 Drawing of a trunk vertebra.

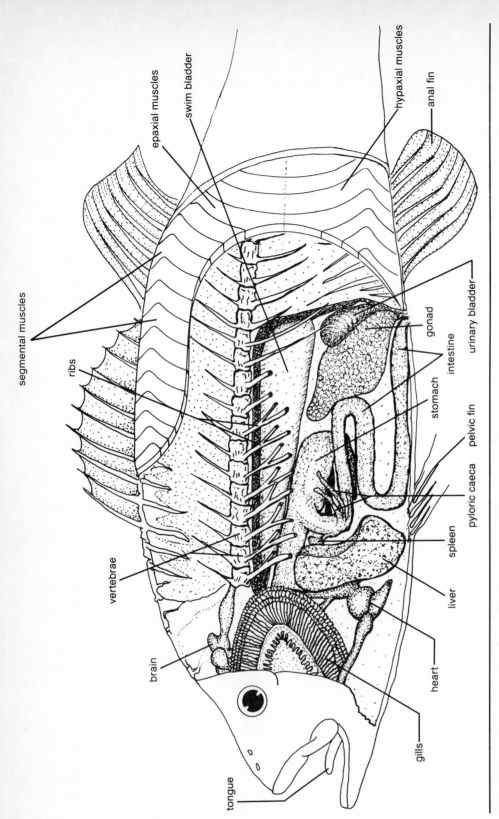

epaxial muscles

swim bladder

hypaxial muscles

anal fin

segmental muscles

ribs

gonad

intestine

urinary bladder

vertebrae

stomach

pelvic fin

brain

spleen

pyloric caeca

liver

heart

gills

tongue

Fig. 17.4 Perch, partly dissected to show muscles and internal organs, lateral view.

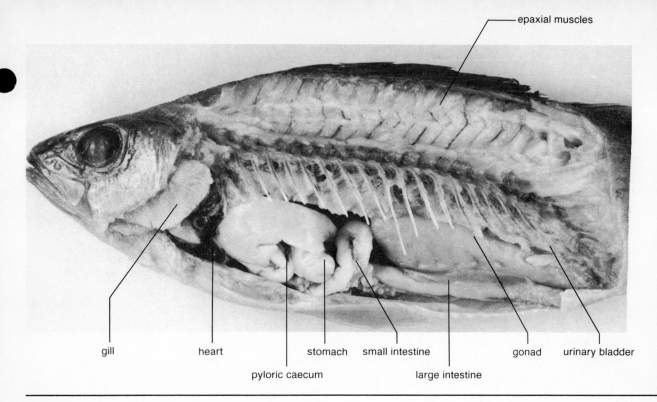

epaxial muscles

gill   heart   stomach   small intestine   gonad   urinary bladder

pyloric caecum   large intestine

Fig. 17.5 Perch, dissected to show visceral organs, lateral view. (Photograph by Carol Majors.)

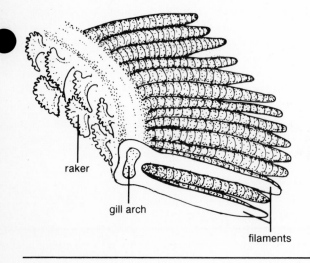

raker

gill arch

filaments

Fig. 17.6 Perch, portion of a gill.

the first arch connects to the skull; the second arch (hyoid arch) supports the tongue; the four gill arches each support a gill; and the last arch has no gill.

We shall observe the gill arches later when we study the gills.

Muscular System

The muscular system of the perch is relatively simple compared to that of terrestrial vertebrates. Most of the body musculature consists of **segmental muscles** (myotomes) (see figure 17.4). Contractions of these myotomes result in flexing of the body, which aids in swimming. Adjacent myotomes are separated by a **myoseptum** of connective tissue. The myotomes are also separated into dorsal and ventral portions by a **transverse septum.** The muscle segments dorsal to the transverse septum are called **epaxial muscles,** and the muscle segments ventral to the transverse septum are called **hypaxial muscles.**

More specialized muscles in the head region serve to move the lateral fins, mouth parts, jaws, gill opercula, gill arches, and associated parts.

Respiratory System

Carefully cut away the bony operculum from one side of the perch to expose the **gills** (figure 17.5). Locate the four gills within the gill chamber. Observe the numerous fingerlike **gill filaments** extending posteriorly from each gill (figure 17.6). The large surface area of these filaments facilitates gas exchange with the **capillary beds** within each filament.

Remove one gill and locate the bony **gill arch** which supports the gill and the hard, fingerlike projections, the **gill rakers,** which protect the gills and prevent passage of coarse material through the gills. Each filament consists of many thin **lamellae** which contain the capillaries and provide a large surface area for gas exchange.

Perch Anatomy

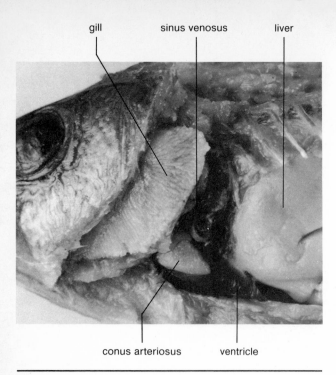

Fig. 17.7 Perch heart, dissected. (Photograph by Carol Majors.)

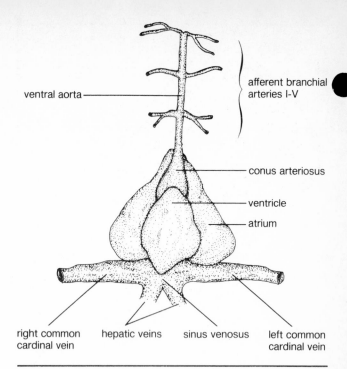

Fig. 17.8 Perch heart and associated blood vessels, ventral view.

Coelom and Visceral Organs

The **coelom** of the perch consists of a large **peritoneal cavity** and a small **pericardial cavity.** The peritoneal cavity contains the stomach, liver, and other digestive organs, swim bladder, and other visceral organs. The pericardial cavity is located anterior to the peritoneal cavity and encloses the heart. The location of the principal internal organs is illustrated in figure 17.4.

To study the internal organs, make a longitudinal cut along the ventral abdominal wall with your scalpel. Start your incision just anterior to the anus and carefully cut anteriorly to the level of the pelvic girdle. Take care not to cut too deeply or you may damage internal organs to be studied later.

After you have completed the midventral incision, make a second incision from the posterior end of the first incision and cut dorsally to the level of the lateral line. Make a similar incision from the anterior end of the midventral incision. Raise this portion of body wall to locate the visceral organs in the peritoneal cavity. The outer lining of the cavity is the parietal peritoneum. Cut through the peritoneum if you have not already done so in opening up the body wall and observe the large **liver.** Beneath the liver, locate the short **esophagus,** the **stomach,** and the **small intestine.** Extending from the anterior part of the small intestine are three short sacs, the **pyloric caeca.** Posteriorly, the small intestine empties into the **large intestine,** which terminates at the anus.

Dorsal to the digestive tract, find the large **swim bladder** or air bladder. (It is deflated in preserved specimens.) The swim bladder is a hollow, gas- or air-filled sac which serves as a buoyancy organ. Alterations of the volume of gas within the swim bladder assist the perch in compensating for the differences in the specific gravity between its body and that of the surrounding water while swimming at various depths.

Above the swim bladder are two long, dark **kidneys.** Other organs in the peritoneal cavity include the **spleen,** an elongate organ lying along the posterior surface of the stomach; the **pancreas,** on the ventral surface of the intestine (often difficult to find); the **gonads,** posterior to the stomach and dorsal to the intestine; and the **urinary bladder,** found posterior to the gonads.

Circulatory System

The **heart** is located in the pericardial cavity, which lies ventral to the gills and anterior to the pelvic fins (see figure 17.7). Carefully cut through the pectoral girdle and the muscles anterior to the girdle and cut away part of the lateral body wall to expose the heart and major blood vessels.

Like other fishes, the perch has a **two-chambered heart** (figure 17.8). Find the thin-walled **atrium** and the thick-walled, muscular **ventricle.** Blood passes from the **sinus venosus** to the **atrium** and from the atrium to the muscular **ventricle.** Contraction of the ventricle forces the blood into the short **conus arteriosus** and out through the short **ventral aorta.** From the ventral aorta, the blood passes to the gills via four different pairs of **branchial arteries.**

The **afferent branchial arteries** lead to extensive capillary beds in the lamellae of the gills, where the blood

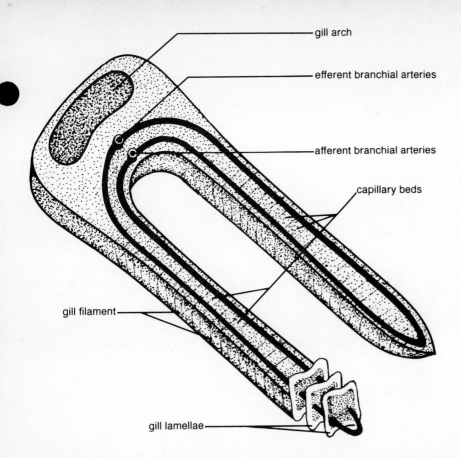

gill arch

efferent branchial arteries

afferent branchial arteries

capillary beds

gill filament

gill lamellae

Fig. 17.9 Perch gill, pattern of circulation within filaments.

is oxygenated (figure 17.9). The oxygenated blood is collected by the **efferent branchial arteries,** which empty into the **dorsal aorta.** From the dorsal aorta, arteries carry oxygenated blood to the organs and tissue of the head, trunk, and caudal regions. Some of the principal blood vessels are shown in figure 17.10.

The principal veins include a pair of **anterior cardinals** that collect blood returning from the head region and a pair of **posterior cardinals** that collect blood from the posterior region of the body. The perch also has a **hepatic portal system** through which blood from the stomach, intestine, and other visceral organs is carried to capillary beds in the liver. From the liver, blood is collected by the hepatic vein and is carried to the **sinus venosus** and thence into the heart for recirculation. The perch, like most freshwater bony fishes, lacks a well-developed renal portal system.

Urogenital System

The **kidneys** are two long, slender organs lying dorsal to the swim bladder. They filter nitrogenous wastes from the blood and empty posteriorly through the **archinephric,** (or Wolffian) **ducts,** which lead to the **urinary bladder.** From the bladder, urine passes into the **urogenital sinus** and out through the **urogenital pore** (see figure 17.11). In the male,

the urinary pore and the genital openings are separate. In the female, there is a common urogenital pore through which both systems empty.

The reproductive system of the **male** includes two long lobed **testes** lying posterior to the stomach and ventral to the swim bladder. Two **vas deferens** carry the sperm to a common **genital sinus,** which opens via the **genital pore.**

In the **female** is a single, fused **ovary,** a large, saclike structure which releases the eggs through a short **oviduct** to the **urogenital pore.**

Nervous System

Like the nervous system of the shark, that of the perch consists of two main divisions, the **central nervous system** (brain and spinal cord) and the **peripheral nervous system** (nerves connecting the brain and spinal cord with other parts of the body).

We shall confine our brief study of the nervous system of the perch largely to the **brain** (figures 17.12 and 17.13). The brain of the adult perch consists of five major divisions: (1) the **telencephalon,** (2) the **diencephalon,** (3) the **mesencephalon,** (4) the **metencephalon,** and (5) the **myelencephalon.** The specific parts of the brain making up these five divisions are summarized in table 17.1.

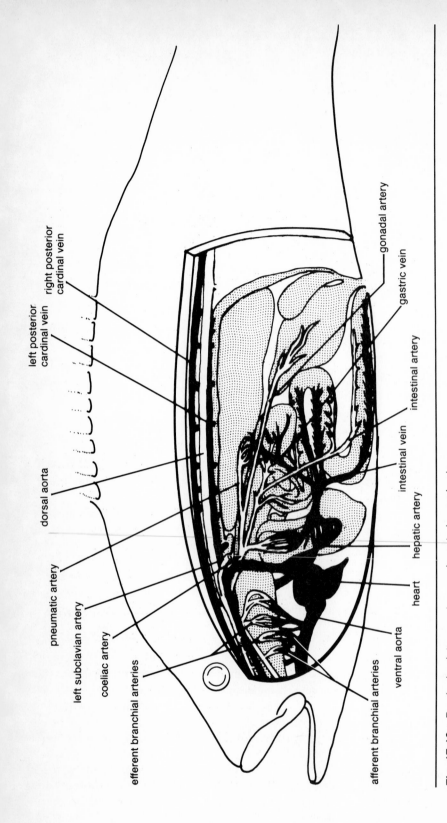

right posterior
cardinal vein

left posterior
cardinal vein

dorsal aorta

pneumatic artery

left subclavian artery

coeliac artery

efferent branchial arteries

gonadal artery

gastric vein

intestinal artery

intestinal vein

hepatic artery

heart

ventral aorta

afferent branchial arteries

Fig. 17.10 Perch circulatory system, major arteries and veins.

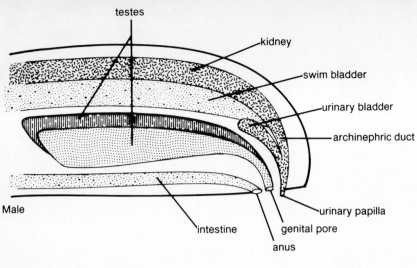

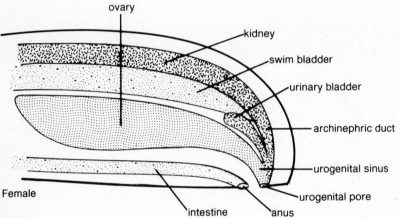

Fig. 17.11 Perch urogenital system, diagrammatic.

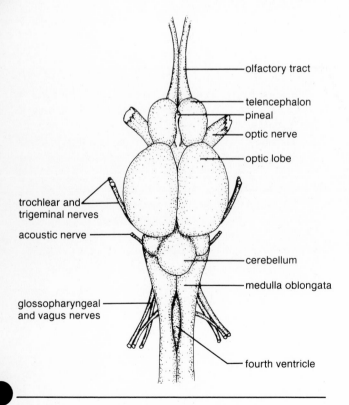

Fig. 17.12 Perch brain, dorsal view.

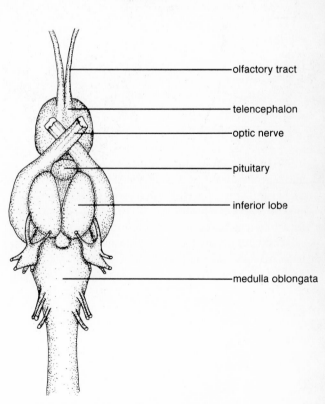

Fig. 17.13 Perch brain, ventral view.

Perch Anatomy

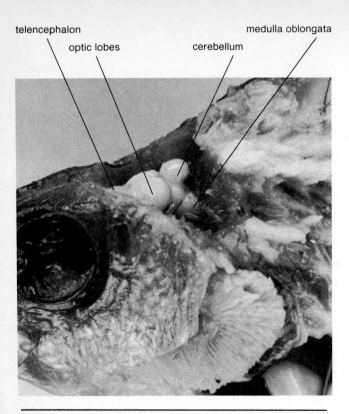

telencephalon   optic lobes   cerebellum   medulla oblongata

Fig. 17.14   Perch brain, dissected, dorsolateral view. (Photograph by Carol Majors.)

Table 17.1   Components of the Perch Brain

Division	Chief Structures
Telencephalon	Olfactory lobes, cerebral hemispheres
Diencephalon	Thalamus, hypothalamus, pineal, pituitary
Mesencephalon	Optic lobes
Metencephalon	Cerebellum
Myelencephalon	Medulla oblongata

To expose the brain for study, you must remove the skin from the dorsal surface of the skull behind the eyes (see figure 17.14). Carefully shave away the bony roof of the skull above the brain, taking care not to damage the delicate tissues beneath. You will find the brain enclosed in a gelatinous mass which must be removed to expose the brain. Also around the brain is a pigmented membrane. Carefully remove the membrane and identify the principal structures of the brain with the aid of figure 17.12.

Associated with the brain of the perch and other bony fishes, as in the shark, are **ten pairs of cranial nerves** (table 17.2).

Table 17.2   Cranial Nerves of the Perch

Nerve	Function
I. Olfactory	Olfaction (smell)
II. Optic	Vision
III. Oculomotor	Eye movements
IV. Trochlear	Superior oblique muscle
V. Trigeminal	Jaw muscles, touch
VI. Abducens	Lateral rectus muscle
VII. Facial	Taste, lateral line, skin of head
VIII. Acoustic	Inner ear and lateral line
IX. Glossopharyngeal	Gill muscles and lateral line
X. Vagus	Gills, heart, anterior part of digestive tract, lateral line

The spinal cord leads posteriorly from the brain to the tail and passes through the neural arches of the vertebrae. One pair of **spinal nerves** arises from the spinal cord in each segment.

Demonstrations

1. Microscope slide of ctenoid scale.
2. Microscope slide of fish skin showing origin of scale.
3. Mounted or plastic-embedded fish skeleton.
4. Model of dissected perch.
5. Fish heart, plastic mount.
6. Microscope slide of fish gill, sectioned through filaments.
7. Living fish in aquarium.

Key Terms

Archinephric duct   a primitive type of kidney duct which collects urine from a series of segmentally arranged nephrons; also called a Wolffian duct.

Ctenoid scale   type of dermal scale made of a thin sheet of bonelike material, circular or oval in shape, and bearing spines on the posterior margin.

Homocercal tail   type of tail with upper and lower parts of the tail symmetrical and with the vertebral column ending at the center of the base; found in most bony fishes.

Swim bladder   a gas- or air-filled sac found in the abdominal cavity of most bony fishes; serves mainly as a hydrostatic organ in modern fishes.

Notes and Sketches

18
Frog Anatomy

Objectives

After completing the laboratory exercise for this chapter, you should be able to complete the following tasks:

1. Identify the main morphological features within the oral cavity of the frog and explain the function of each.
2. Describe the external features (secondary sexual characteristics) that distinguish mature male and female frogs.
3. Describe the major divisions of the frog skeleton and explain the chief functions of each division.
4. Describe the principal parts of the digestive system of the frog and explain their function.
5. Describe the structure of the frog heart and identify its major parts.
6. Describe the basic pattern of blood circulation in the frog and identify the principal blood vessels involved.
7. Identify the important parts of the urogenital system of both male and female frogs and explain their functions.
8. Identify the five main regions of the frog brain and the components of each. Briefly explain the main function of each part.
9. Locate the ten pairs of cranial nerves on a frog, name each pair, and explain its main function.

Rana pipiens or *Rana catesbeiana*

Frogs are the most commonly studied representatives of the Class Amphibia, Subphylum Vertebrata, Phylum Chordata. Although there is no "typical vertebrate" any more than there is a "typical person," a study of frog anatomy does serve to illustrate effectively the basic body organization of a vertebrate animal.

It is important to remember that the amphibians are transitional animals that typically live a portion of their lives on land and another portion in the water. Thus they exhibit a peculiar mixture of characteristics—some of which represent adaptations for terrestrial life, and some represent adaptations for life in the water. Mating nearly always occurs in the water, since the eggs lack the protective outer coverings that permit the terrestrial existence of birds and reptiles. Amphibian eggs hatch into **tadpole larvae** with gills and a muscular tail for swimming. Later the tadpoles metamorphose into four-legged adults, which are usually semiterrestrial. Some amphibians, however, spend all of their lives in the water, and a very few spend all their lives on land, having developed special mechanisms to protect their eggs from desiccation.

The skin of adult amphibians is smooth and usually moist. Generally the adults live in wet or moist environments, since they are very susceptible to water loss through the skin. Toads are rather exceptional amphibians that have developed a tough, horny skin that reduces water loss.

Since the amphibians represent an evolutionary transition between the fishes and the terrestrial animals, they also exhibit numerous morphological advances over the fishes. The skull is broad, flat, and lighter in weight; they have jointed tetrapod limbs rather than fins for locomotion; adult amphibians often develop lungs for breathing; a three-chambered heart is formed; and the circulatory system exhibits two distinct circuits, a **systemic division** to supply the body organs and a **pulmonary division** to carry blood to and from the lungs.

Frogs exhibit most of the typical amphibian features, but they also show some features peculiar to their own mode of life. The salamanders more nearly represent the typical features of the Amphibia. Some of the specialized features of the adult frog are: (1) the absence of a tail, (2) the loss of certain skull bones, (3) the posterior attachment of the tongue, (4) the absence of ribs, (5) the lack of a distinct neck, and (6) the powerful and highly developed hind limbs.

Materials List

Living Specimens
   Rana pipiens (or *R. catesbiana*)
Preserved Specimens
   Rana pipiens (or *R. catesbiana*)
   Frog skeleton
   Representative vertebrate skeletons

External Anatomy and Behavior

Study a living frog and note the smooth, moist, and pliable skin. Observe the pattern of coloration of the dorsal and ventral surfaces of the body. How does it differ on the two surfaces? Compare the color pattern of your specimen with those of others in the laboratory. How much variation in color patterns do you observe? The specific pattern of spots has been shown to be genetically determined.

Note the broad, flat head with the large mouth, the nostrils or **external nares,** two conspicuous **eyes,** and the circular **tympanic membranes** located behind the eyes. Bordering the eye is a fleshy lower eyelid and a less-prominent upper eyelid. A third transparent inner eyelid, the nictitating membrane, helps keep the eye moist while the frog is on land and also helps protect the eye under water.

Observe demonstrations of living frogs in an aquarium and in a terrarium. Study a frog in the aquarium and observe its swimming behavior. How does the frog propel itself through the water? Which limbs are most prominent in swimming? What adaptations of the limbs are most important for swimming? What is the posture of the frog while it is floating? Compare the behavior of frogs in an aquarium filled with water at room temperature and the behavior of frogs in an aquarium filled with cold water. Which frogs are more active? How can you explain the difference in their activity? Which frogs remain submerged for longer periods of time, those in the cold aquarium or those in the warm (room temperature) aquarium? Frogs are **poikilothermal** or "cold-blooded" animals. This means that their body temperature approximates that of their environment. How does this fact help you to explain the difference in behavior between the frogs in the cold aquarium and those in the warm aquarium?

Observe also the behavior of frogs in a terrarium. When exposed to the air, frogs breathe by drawing air through the external nares into the oral cavity. Later air is forced into the lungs by closing the internal nares and raising the floor of the mouth. Some gas exchange with the blood also occurs across the moist epithelium of the oral cavity and through the skin covering the exterior of the body.

Compare the locomotion of frogs in the water with frogs in a terrarium or on a laboratory table. How are the limbs used in locomotion on land?

To observe the feeding behavior of a frog, place some living house flies, fruit flies, or other small insects in a small aquarium with a frog. How does the frog react to the food organisms? How are the insects captured?

Oral Cavity

Place the frog on its back and make a small cut with your scissors at each corner of the mouth so the jaws can be opened widely. Rinse out the oral cavity with cold water if there is an accumulation of mucus. Consult figure 18.1 and locate the following structures in the oral cavity of your specimen: the **maxillary teeth** on the margin of the upper jaw, the **vomerine teeth** on the palate (the teeth are used for holding food rather than for chewing since the food is swallowed whole); the **internal nares** (probe through them from the **external nares**); and the openings into the **eustachian tubes.** Observe also the **tongue** and its attachment. What advantage does this peculiar attachment of the tongue have for the frog? Posterior to the oral cavity (behind the tongue) is the **pharynx.** At the rear of the pharynx is the opening into the esophagus, and on the ventral surface find the **glottis,** a slitlike opening into the **laryngotracheal chamber.** A pair of short **bronchi** connect the laryngotracheal chamber to the two lungs.

The males of many species of frog have openings into two **vocal sacs** located at the posterior corners of the mouth. Inflation of the vocal sacs serves to amplify the croaking sounds. Vocal sacs are absent in female frogs.

Internal Anatomy

Review the general instructions for dissection provided in chapter 9. When dissecting freshly killed or anesthetized specimens, take care to minimize the cutting of blood vessels to prevent the blood from obscuring your view of the internal organs. Keep your specimen covered with cold water during dissection. Fasten the frog, ventral side up, to the bottom of the dissection pan with pins through the tip of the jaw and all the limbs.

Study figures 18.1 and 18.2 to make certain that you understand the approximate location of the internal organs before you begin your dissection. Proceed carefully in your dissection to avoid unnecessary damage to body parts that you may need to study later.

Lift the skin from the body with your forceps and make a **longitudinal cut** slightly to one side of the mid-ventral line and forward from the pelvis to the tip of the lower jaw. Observe the subcutaneous lymph spaces and the attachment of the skin to the body as you free the skin from the ventral body surface. Make short transverse cuts in the skin at the anterior and posterior ends of the trunk and pin back the flaps of skin on both sides.

Consult figure 18.7 and locate the following three large muscles before continuing with your dissection: the **pectoralis,** the **rectus abdominis,** and the **external oblique.**

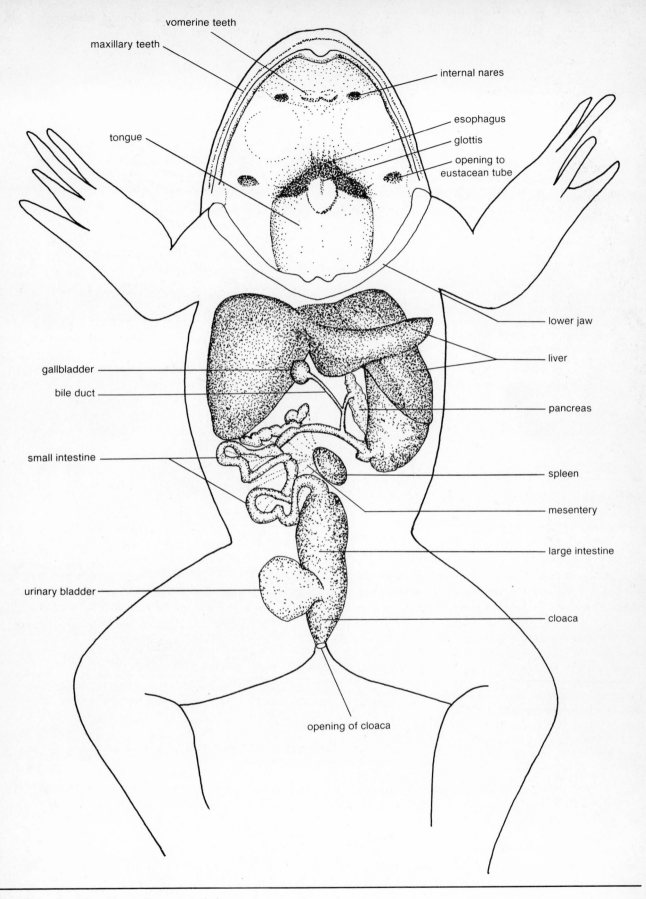

Fig. 18.1 Frog digestive system, ventral view.

vomerine teeth

maxillary teeth

internal nares

esophagus

glottis

opening to eustacean tube

tongue

lower jaw

liver

gallbladder

bile duct

pancreas

small intestine

spleen

mesentery

large intestine

urinary bladder

cloaca

opening of cloaca

Frog Anatomy

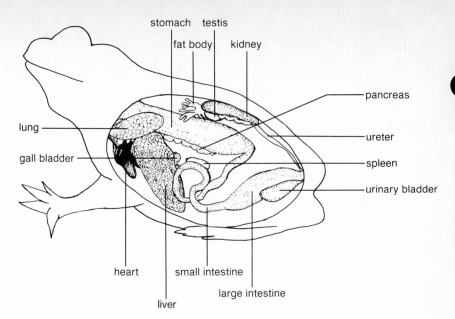

Fig. 18.2 Frog, side view, showing location of major internal organs.

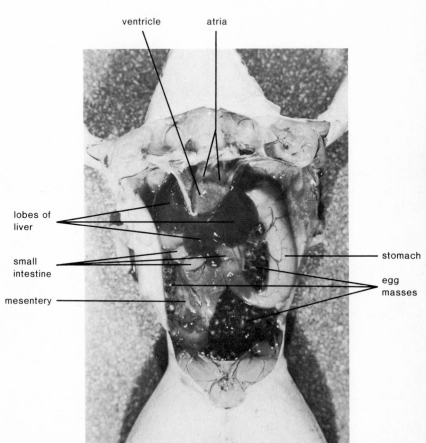

Fig. 18.3 Frog, dissected, ventral view, showing digestive system and other internal organs. (Photograph by John Vercoe.)

Note also the **linea alba** along the midventral line. The ventral abdominal vein lies beneath the muscular body wall along the midventral line (see figure 18.10).

Now lift the muscles of the abdomen with your forceps and make a **longitudinal incision** through the body wall with your scissors a little to one side of the linea alba. Cut through the bony **sternum** which supports the forelimbs (see figure 18.5), and up to the posterior end of the

lower jaw, taking care not to cut the ventral abdominal vein and to avoid damage to the internal organs within the coelom. Trace the ventral abdominal vein to the liver and make two short transverse incisions through the body wall just in front of this vein so the muscles can be pinned back to expose the internal organs. Remove about a half-inch section from the middle portion of the sternum and pin the forelimbs back to provide access to the internal organs in this region as shown in figure 18.3.

Chapter 18

Survey of Internal Organs

Now that you have completed the preliminary phase of your dissection, make a brief survey of the internal anatomy, referring to the figures cited in the following description. Note the position, size, shape, color, and texture of each organ. Consider also the relationship of each organ to other organs and attempt to determine the function(s) of each organ as you work. If you have a female specimen and the ovaries and oviducts are filled with eggs and greatly enlarged, consult figure 18.11 and carefully remove the ovary and oviduct from one side of the body to facilitate your study of the other parts.

The location of the principal internal organs is illustrated in figures 18.1, 18.2, and 18.3. The large, three-lobed **liver,** located just posterior to the pectoral girdle, is one of the most prominent internal organs. The liver is dark in color in both preserved and living specimens. Adjacent to the liver on the right side of the frog (viewed from the ventral surface as in a dissecting pan) is the tubular **stomach.**

The **heart** is enclosed in a membranous sac, the **pericardial cavity,** which lies anterior to the liver and partly beneath the pectoral girdle. You will open the pericardial cavity later to study the heart and its relationship with its attached blood vessels. The two **lungs** are located on the sides of the visceral cavity posterior and lateral to the heart and dorsal to the liver. In a living or freshly killed frog the lungs will be semitransparent and inflated. In preserved specimens they will often be deflated.

If your specimen is a mature female, much of the visceral cavity will be filled with many black and white **eggs** enclosed in the large, membranous **ovaries.**

After you have located these major interior organs, you should proceed with a more detailed study of specific organ systems as directed by your laboratory instructor and as described in the following sections.

Skeletal System

The **endoskeleton** of the frog (figures 18.4 and 18.5) consists chiefly of bone and cartilage. It supports the various parts of the body, protects delicate organs such as the brain and spinal cord, and provides points of attachment for the skeletal muscles. Study a prepared skeleton of a bullfrog (plastic-embedded specimens are excellent for this purpose) and compare the general organization of the frog skeleton to that of other vertebrates on demonstration.

The skeleton of **vertebrates** consists of the **somatic skeleton** (skeleton of the body wall and appendages) and the **visceral skeleton** (skeleton of the pharyngeal wall—prominent in fishes as support for the gills and as a part of the jaws but much reduced in higher vertebrates). In the frog, the visceral skeleton is represented principally by the **hyoid apparatus,** a small bone and cartilage structure that helps support the floor of the mouth, the base of the tongue, and parts of the jaws and larynx. In prepared skeletons, you will find the hyoid apparatus embedded or mounted separately from the rest of the skeleton.

The major components of the frog skeleton and some of their relationships are shown in the following outline.

> Visceral Skeleton
> > Hyoid apparatus
>
> Somatic Skeleton
>
> > Axial skeleton
> > > Vertebral column
> > > Sternum
> > > Skull
> >
> > Appendicular skeleton
> > > Pectoral girdle
> > > Forelimbs
> > > > Humerus
> > > > Radio-ulna
> > > > Carpals
> > > > Metacarpals
> > > > Phalanges
> > >
> > > Pelvic girdle
> > > Hindlimbs
> > > > Femur
> > > > Tibio-fibula
> > > > Tarsals
> > > > Metatarsals
> > > > Phalanges

The **skull** consists of three main parts (figure 18.4): (1) the narrow braincase or **cranium;** (2) the paired **sensory capsules** of the ears, nose, and the large eye sockets, or orbits, for the eyes, and (3) the **visceral skeleton** (consisting of parts of the jaws, hyoid apparatus, and the laryngeal cartilages). Study the figures and locate the various bones of the skull on your specimen.

The **vertebral column** is made up of ten vertebrae. The first vertebra, the **atlas,** articulates with the base of the skull (see figure 18.4). It has no transverse processes and is the only **cervical** (neck) **vertebra** in the frog. The next seven vertebrae are the **abdominal vertebrae** (figure 18.5). Behind the abdominal vertebrae is a large **sacrum** with two strong transverse processes, which join with the **ileum.** The last vertebra is the long **urostyle** (figure 18.5). Study the parts of an abdominal vertebra with the aid of figure 18.5.

Muscular System

The muscles of the frog are illustrated in figures 18.6 and 18.7. The skeletal muscles of the adult are well adapted for swimming and for locomotion on land. Note the highly developed muscles associated with the hindlimbs.

A skeletal muscle typically consists of a fixed end, called the **origin,** and a movable end, called the **insertion.** The fleshy middle portion is called the **belly.** Most of the skeletal muscles taper at their ends into tough white cords of connective tissue, the **tendons,** which serve to attach them to bones or to other muscles.

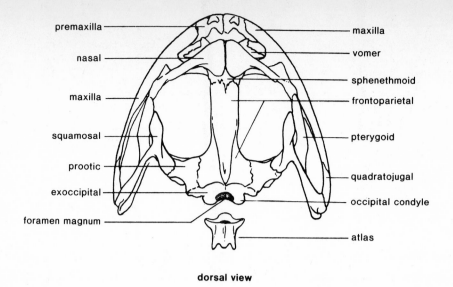

dorsal view

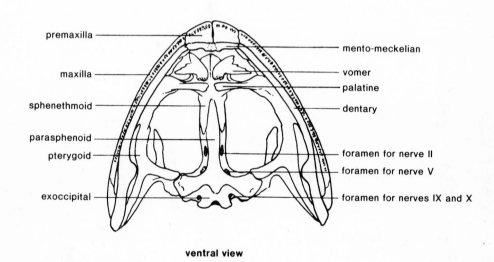

ventral view

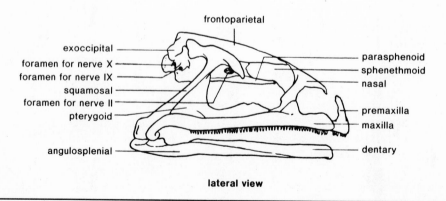

lateral view

Fig. 18.4 Frog skull.

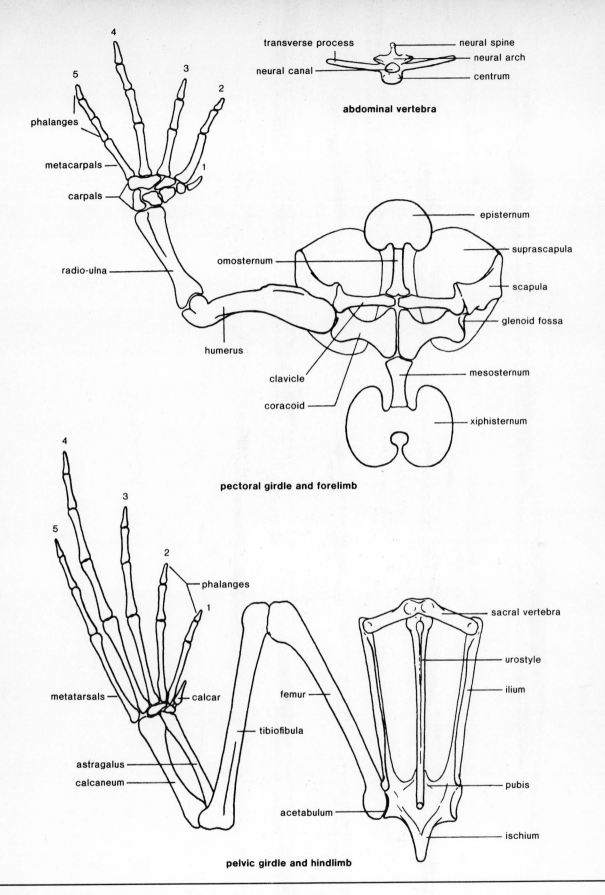

Fig. 18.5 Frog skeletal structures.

Frog Anatomy

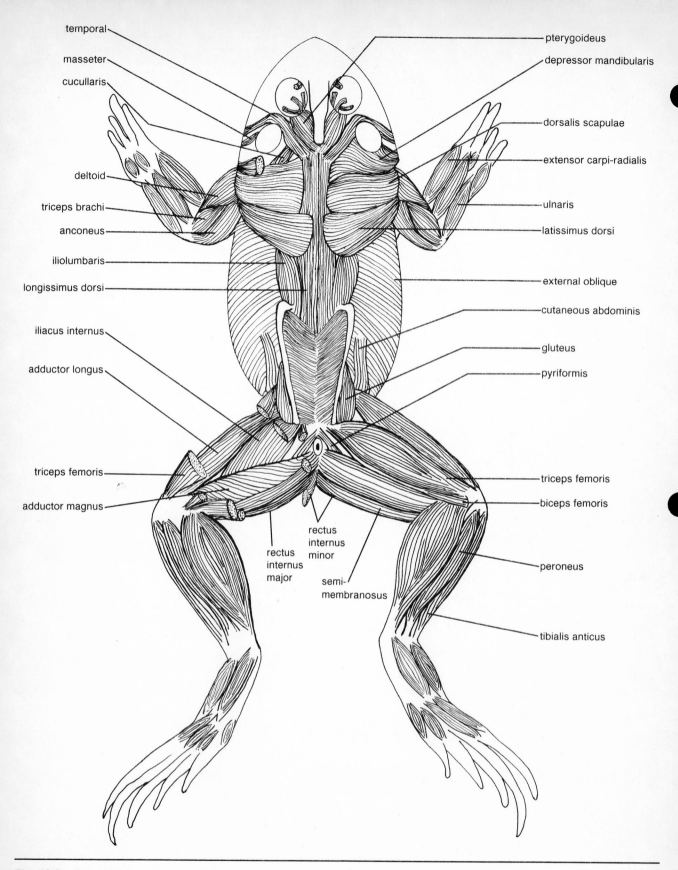

Fig. 18.6   Frog muscles, dorsal view.

Chapter 18

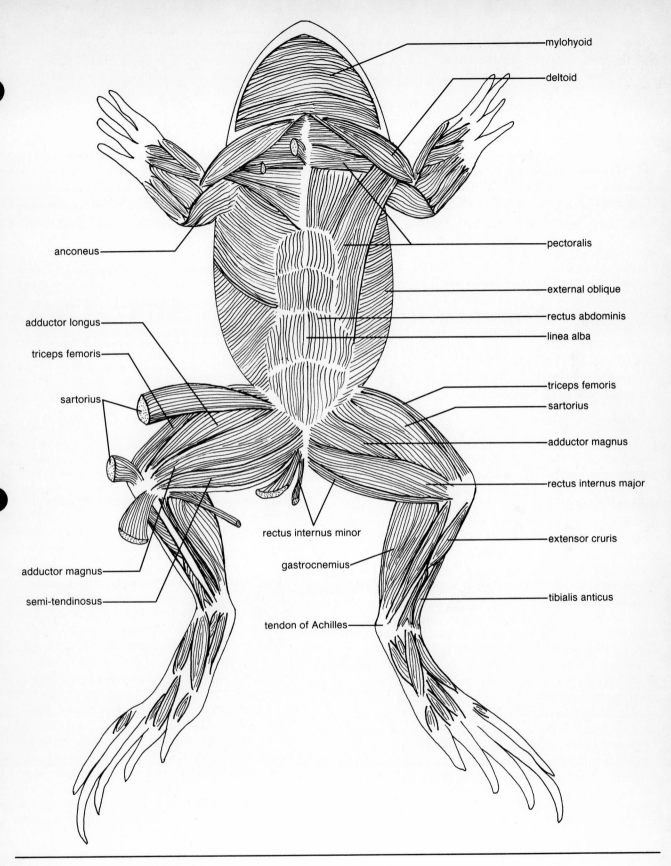

Fig. 18.7 Frog muscles, ventral view.

Labels (clockwise from top right):
mylohyoid
deltoid
pectoralis
external oblique
rectus abdominis
linea alba
triceps femoris
sartorius
adductor magnus
rectus internus major
extensor cruris
tibialis anticus
tendon of Achilles
gastrocnemius
rectus internus minor
semi-tendinosus
adductor magnus
sartorius
triceps femoris
adductor longus
anconeus

The movement or effect caused by a muscle is its **action.** Thus, each skeletal muscle has a characteristic origin, insertion, and action. Most muscles are arranged in pairs or groups that are **antagonistic;** they have opposite actions.

Muscles are classified according to their actions. Some general classes of skeletal muscles are: **extensors,** muscles that straighten or extend a part; **flexors,** muscles that bend one part toward another part; **adductors,** muscles that pull a part back toward the axis of the body; and **abductors,** muscles that draw a part away from the axis of the body.

For a detailed study of the muscles of the frog, you must first remove the skin from a specimen. Specimens especially preserved in alcohol are best for this purpose, although formalin-preserved specimens may also be used. Carefully free the skin from the underlying muscles and study figures 18.6 and 18.7. Identify first the surface muscles and then cut the three muscles on the ventral side of the right hind legs, as illustrated in figure 18.7, to expose the deep-lying muscles for study. Carefully free and separate the muscles with a blunt (not sharp) instrument as you work. Complete table 18.1 showing the origin, insertion, and action of the various muscles of the hindlimb.

Digestive System

Trace the pathway of the digestive system starting with the mouth. The **mouth** opens into the **oral cavity,** as seen previously. Behind the **tongue** locate the **pharynx,** the posterior portion of the oral cavity that opens into the **esophagus.** The esophagus is a short, cylindrical tube that passes food material to the **stomach,** where the food is stored temporarily and digestion begins. The muscular stomach also provides a thorough mixing of the food mass by its contractions. At the posterior end of the stomach find the **pyloric valve,** a sphincter that controls the release of the stomach contents into the small intestine.

The anterior segment of the small intestine is the **duodenum,** which receives secretions from the liver and pancreas through the common bile duct. Behind the duodenum is the convoluted **ileum,** the posterior section of the small intestine where digestion is completed and where most absorption of nutrients into the bloodstream occurs. The ileum empties into the **large intestine,** where most water as well as certain vitamins and ions are absorbed. The large intestine is also where the undigested residue is temporarily stored as fecal material. Posteriorly the large intestine narrows and empties into the **cloaca.** The cloaca is a common chamber that collects materials from the digestive, excretory, and reproductive systems prior to their discharge through the cloacal opening.

Circulatory System

In many respects the circulatory system of an amphibian is similar to the sharks or cartilaginous fishes. There are, however, some important differences. A shark has a two-chambered heart that receives only venous blood. This blood is pumped through the respiratory organs (gills) and then passes directly into the systemic blood vessels.

Amphibians have a **three-chambered heart** with two **atria** and a **ventricle.** In the adult frog, part of the venous blood moves from the heart to the respiratory organs (lungs) and then returns to the heart before it is pumped out to the various organs of the body. Thus there are two major divisions of the circulatory system: (1) the **pulmocutaneous circulation** and (2) the **systemic circulation.** Preserved frogs that have been injected with latex are most satisfactory for detailed study of the circulatory system. Injected specimens may be singly injected (arteries only) or doubly injected (both arteries and veins injected with latex).

Heart. Carefully remove the enveloping membrane, the **pericardium,** from around the heart. Consult figures 18.8, 18.9, and 18.10 and identify the thin-walled right and left **atria,** the thick-walled **ventricle,** and the **truncus arteriosus,** which divides into right and left aortic arches. Lift the ventricle and find the dark-colored, thin-walled, triangular **sinus venosus** on its dorsal side (figure 18.10), the **posterior vena cava** entering it at the posterior end, and the right and left **anterior vena cavae** entering it at the front.

Arterial System. Study the arterial system with the aid of figure 18.9. Note the two **aortic arches,** each giving rise to three large arteries.

1. The **common carotid artery.** This divides into: (a) the **external carotid** or lingual artery leading to the ventral part of the head and to the tongue; (b) the **internal carotid** to the dorsal part of the head. (Note the carotid gland.)
2. The **systemic arch.** The two aortic arches bend dorsally around the pharynx and unite to form the large **dorsal aorta.** Note the arteries leading from the arch: (a) the short **occipito-vertebral** dividing into the **occipital** leading to the skull and the **vertebral** leading to the vertebral column; and (b) the **subclavian,** giving off branches to the shoulder region and extending into the arm where it becomes the brachial; (c) the **dorsal aorta;** (d) the **coeliacomesenteric,** which divides into the **coeliac** and the **mesenteric;** (e) the **right** and **left gastric** to the stomach, the **pancreatic** to the pancreas, and the **hepatic artery** to the liver (these latter arteries are all branches of the coeliac); (f) the **mesenteric** to the intestine, with branches to the spleen (the **splenic artery**) and to the rectum and large intestine (**posterior mesenteric artery**); (g) the **urogenital arteries** to the kidneys, gonads, and fat bodies (**renal** to kidneys and **genital** to gonads). The urogenital arteries vary considerably in number and arrangement in different individuals (how many are there in your specimen?); (h) the **lumbar arteries** to the dorsal body wall (several pairs

Table 18.1   Muscles of the Hindlimb

Name	Origin	Insertion	Function
Sartorius	Pelvis	Head of tibia	Flexes the shank and adducts the entire leg.

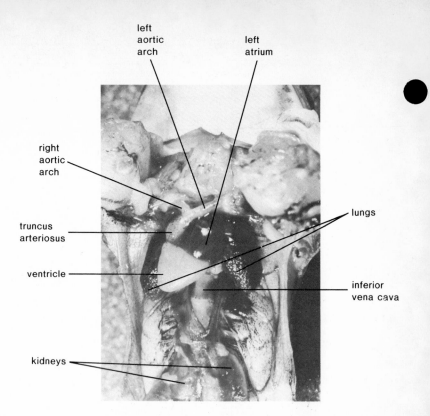

left aortic arch

left atrium

right aortic arch

truncus arteriosus

ventricle

lungs

inferior vena cava

kidneys

Fig. 18.8 Frog, dissected, showing heart and other internal organs. (Photograph by John Vercoe.)

of small arteries arising from the dorsal surface of the aorta—not shown in the figure); (i) the **common iliac arteries,** divisions of the aorta; (j) the **epigastric** to the bladder and body wall of that region; (k) the **femoral** in the thigh; (l) the **sciatic,** a continuation of the iliac furnishing branches to most of the muscles of the leg; and (m) the **peroneal** and **tibial arteries,** divisions of the sciatic to lower parts of the leg as shown in the figure.

3. The **pulmocutaneous artery.** This divides into: (a) the **pulmonary artery** going to the lung and (b) the **cutaneous artery** to the skin. Note that this artery and its branches carry nonoxygenated blood.

Venous System. Veins are blood vessels that carry blood **toward** the heart. Study the venous system of your specimen with the aid of figure 18.10 and locate the following veins.

1. The two (right and left) **anterior venae cavae** into which enter: (a) the **external jugular** with its branches, the **lingual** from the tongue and floor of the mouth and **maxillary** or mandibular from the jaw; (b) the **innominate vein** into which flows the **internal jugular** from the deeper parts of the head and the **subscapular** from the shoulder; (c) the **subclavian vein** formed by the union of the **musculocutaneous** from the muscles and skin of the side and back, and the **brachial** from the forelimb.

2. The **posterior vena cava** into which enter: (a) the **hepatic veins** from the liver; (b) the **renal veins** from the kidneys; (c) the **genital veins** from the gonads.

3. The **hepatic portal system** consists of: (a) the **abdominal vein,** which enters the liver and is formed by the union of the two **pelvic veins;** (b) the **hepatic portal vein,** which carries blood from the stomach (**gastric vein**), intestine (**mesenteric vein**), and the spleen (**splenic vein**).

4. The **renal portal system.** The **renal portal veins** carry the blood to the kidneys. They receive the blood from (a) the **dorso-lumbar vein;** (b) a branch of the **femoral vein** from the hindlimb; (c) the **sciatic vein** from the thigh. Note that the blood from the femoral vein may go to the kidney by way of the renal portal vein or to the liver by way of the pelvic and abdominal veins.

5. The **pulmonary veins.** These veins carry oxygenated blood from the lungs and unite to enter the left auricle.

The Urogenital System

The excretory and reproductive organs are closely associated, and together comprise the **urogenital system.** If your specimen is a female with large ovaries concealing the other visceral organs, consult figures 18.11 and 18.12 and carefully remove one of the ovaries. The excretory structures are similar in the two sexes of the frog. With the help of figure 18.11, again note the two **kidneys,** each with the adrenal gland on its ventral surface; the **ureter** leading from the posterior border of the kidney to the cloaca; the **urinary bladder,** which empties into the cloaca on the ventral side; and the **fat bodies.**

The kidneys are the major excretory organs of the frog. They are responsible for the removal of most of the

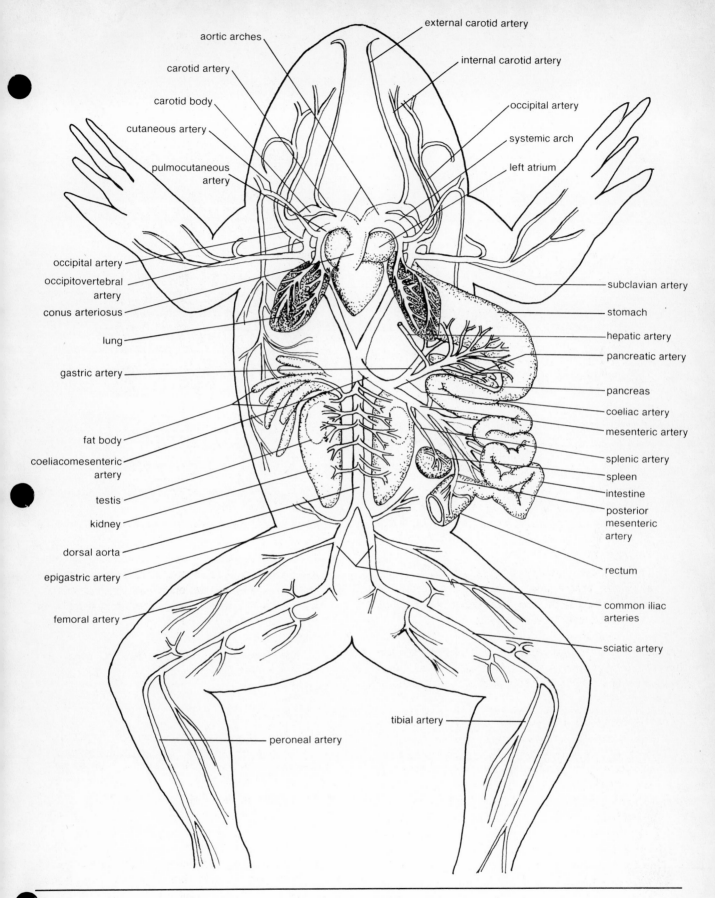

aortic arches

external carotid artery

carotid artery

internal carotid artery

carotid body

occipital artery

cutaneous artery

systemic arch

pulmocutaneous
artery

left atrium

occipital artery

subclavian artery

occipitovertebral
artery

stomach

conus arteriosus

hepatic artery

lung

pancreatic artery

gastric artery

pancreas

coeliac artery

mesenteric artery

fat body

splenic artery

coeliacomesenteric
artery

spleen

intestine

testis

posterior
mesenteric
artery

kidney

dorsal aorta

rectum

epigastric artery

common iliac
arteries

femoral artery

sciatic artery

tibial artery

peroneal artery

Fig. 18.9 Frog arteries, ventral view.

Frog Anatomy

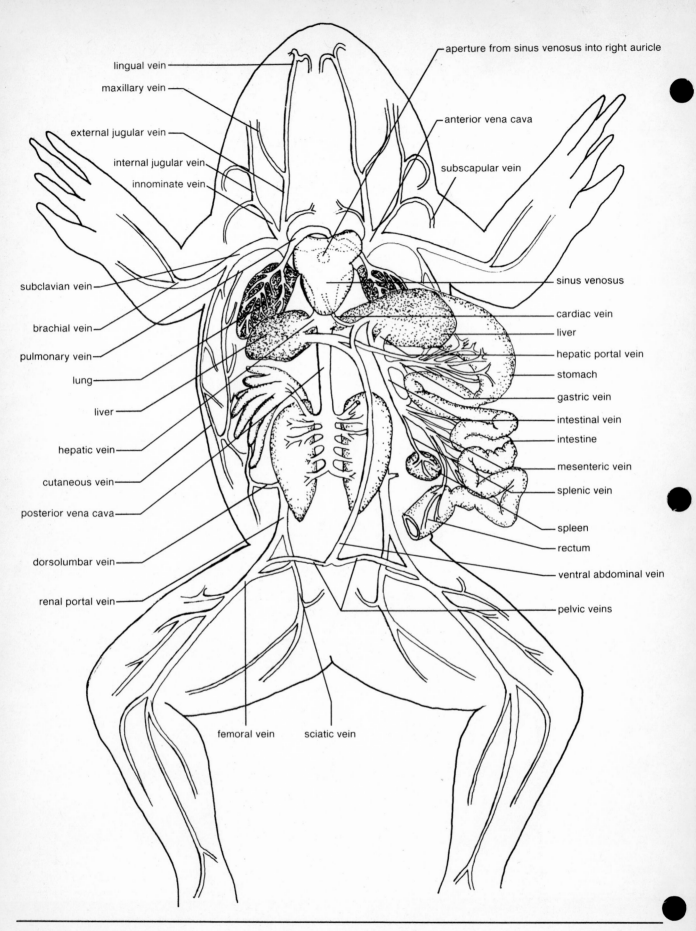

Fig. 18.10   Frog veins, ventral view.

Chapter 18

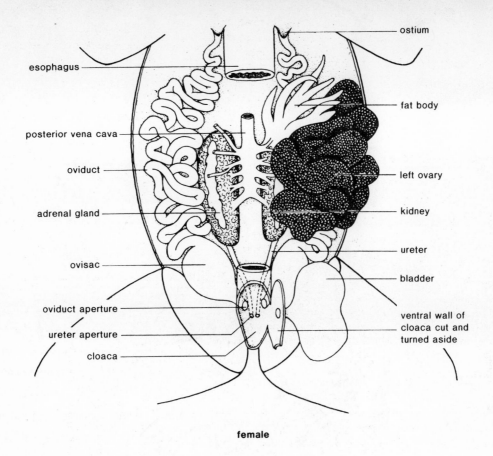

female

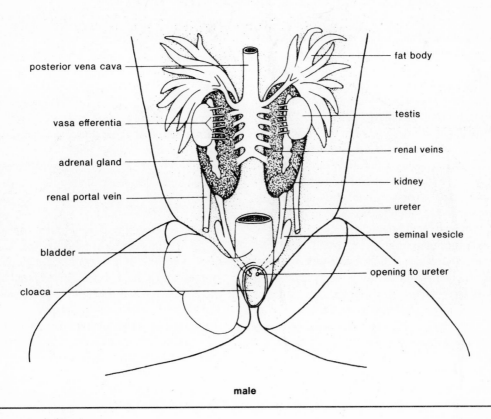

male

Fig. 18.11 Frog urogenital system.

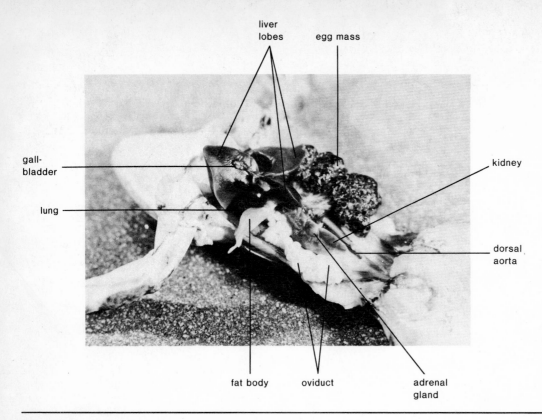

liver lobes — egg mass

gall-bladder —

kidney —

lung —

dorsal aorta —

fat body — oviduct — adrenal gland

Fig. 18.12 Frog, dissected, female reproductive system.
(Photograph by John Vercoe.)

wastes from the body, although some wastes are also lost through the skin. The kidneys remove most of the nitrogenous wastes from the blood, which are excreted in soluble form as urea and ammonia. These important organs also play a major role in **homeostasis,** the maintenance of a constant internal environment by removing excess ions and other substances from the blood and by conserving those substances of limited availability.

Specimens should be distributed in the class so that both male and female specimens are available at each table. After you have completed the study of the urogenital system on your own specimen, find a specimen of the opposite sex and make a comparative study so that you will be familiar with the organization of the urogenital system in both the male and female.

In a female specimen (figures 18.11 and 18.12), locate on one side the **ovary** attached dorsally and suspended into the coelom by a **mesentery;** the coiled **oviduct** with a funnel-shaped opening, the **ostium,** at its anterior end; and a **uterus,** an enlargement near the cloaca. Cut open the cloaca and use your probe to find the openings of the oviducts, the ureters, and the urinary bladder.

Near the time of maturation of the eggs, the eggs are released from the ovaries into the coelom and pass through the ostia into the oviducts. As the eggs pass down the oviduct, they are covered with several layers of jellylike material secreted by glands in the walls of the oviducts. The eggs collect in the ovisacs before they are

discharged through the cloaca and into the water. Fertilization is external and in mating the male mounts the female and discharges sperm on the eggs as they are discharged.

In a male specimen (figure 18.11), locate the two **testes,** each suspended from the dorsal abdominal wall by a mesentery, and the several **vasa efferentia,** the small ducts that carry sperm from the testes to the kidneys. The sperm are carried from the kidneys to the cloaca via the **ureters,** which serve as genital ducts in the male. Male leopard frogs also frequently have vestigial oviducts (mesonephric ducts) alongside the kidneys.

Nervous System

The nervous system of vertebrates consists of three main parts: (1) the **central nervous system,** which includes the brain and the spinal cord; (2) the **peripheral nervous system,** which includes the nerves extending from the central nervous system; and (3) the **autonomic nervous system,** a specialized portion of the peripheral nervous system that regulates the glands of the body and the visceral organs.

Successful study of the nervous system of the frog requires careful dissection. Do not rush the dissection, because you may damage important structures necessary for your study. Carefully remove the skin from the dorsal surface of the head between the eyes and along the vertebral column. Clear away the muscles and connective tissue underlying the skin to expose the skull and vertebral column

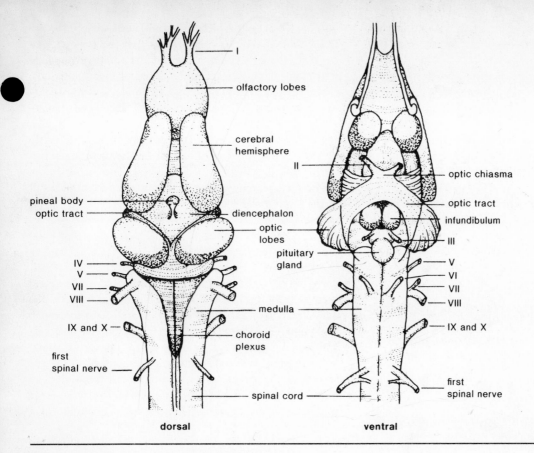

Fig. 18.13 Frog brain and cranial nerves.

With your scalpel carefully shave thin sections of bone from the skull until you expose the brain. Then carefully pick away additional small pieces of bone with your forceps until the brain is completely exposed. Continue the same procedure to expose the vertebral column and the spinal cord, but leave the nervous system in place.

The Brain. Consult figure 18.13 and identify the five main regions of the brain: the **telencephalon,** the **diencephalon,** the **mesencephalon,** the **metencephalon,** and the **myelencephalon.** The anterior-most telencephalon bears four distinct lobes: two **olfactory lobes** and, posterior to them, two **cerebral hemispheres.** Extending anteriorly from the olfactory lobes are the **olfactory nerves,** which carry impulses from the olfactory epithelium to the brain.

Posterior to the telencephalon is the diencephalon, a diamond-shaped depressed area directly behind and somewhat between the posterior portion of the cerebral hemispheres. The **pineal gland** (epiphysis) is a small, inconspicuous body attached to the dorsal wall of the diencephalon, which probably will be removed in your dissection of the skull. The stalk by which it was attached, however, may still be visible. The lateral walls of the diencephalon constitute the **thalamus.** Below the diencephalon lies the **pituitary gland,** which you will study later. The principal function of the telencephalon and diencephalon is the integration of olfactory signals. Chemical perception is of great importance to the frog in feeding and in defense.

The mesencephalon, immediately posterior to the diencephalon, bears two large **optic lobes** that serve to integrate nerve impulses from the eyes. The optic lobes in the frog also provide an important overall coordination of sensory information, a function carried out mainly by the cerebral hemispheres in higher vertebrates.

Behind the mesencephalon find the metencephalon, which is represented by the **cerebellum,** a narrow transverse portion of the brain lying immediately posterior to the optic lobes.

The most posterior part of the brain is the myelencephalon, consisting of the elongated **medulla oblongata,** which tapers gradually into the **spinal cord.** The anterior portion of the medulla oblongata has a thin roof that is frequently removed during dissection. The cerebellum and medulla receive and integrate signals from the ears and from the skeletal muscles. They serve to monitor the general state of activity in the body and also coordinate respiration, feeding, and muscular activity.

When you have completed your study of the dorsal side of the brain, **carefully** detach the anterior portion of the brain and lift it gently from the floor of the skull. Continue detaching the brain and spinal cord from its ventral connections and gradually free the entire nervous system.

Place the isolated nervous system ventral side up in your dissecting pan and study the structures visible from this aspect.

Note the crossed optic nerves, called the **optic chiasma,** on the ventral side of the diencephalon. Observe how the nerves from the left eye carry their impulses across the brain to the right optic lobe. The optic nerves are the **second pair** of **cranial nerves** extending from the brain to various parts of the body. The olfactory nerves extending forward from the olfactory lobes are the **first pair** of **cranial nerves.**

There is a total of ten pairs of cranial nerves in the frog, extending from the brain to various parts of the body. (Fishes and amphibians generally have ten distinct pairs of cranial nerves; reptiles, birds, and mammals have twelve pairs.) Refer to figures 18.13 and 18.14 and locate the following cranial nerves.

Nerve I. Olfactory, extends from the nose to the olfactory lobe.

Nerve II. Optic, extends from the eye to the ventral side of the diencephalon. Note the optic chiasma; see figure 18.13.

Nerve III. Oculomotor, extends from the ventral side of the mesencephalon to the muscles of the eye.

Nerve IV. Trochlear, extends from the dorsal side of the mesencephalon to the superior oblique muscles of the eye.

Nerve V. Trigeminal, closely associated with Nerve VII, the facial, from the side of the medulla to the skin of the face, muscles of the jaw, and tongue.

Nerve VI. Abducens, extends from the ventral surface of the medulla to the external rectus muscle of the eye.

Nerve VII. Facial, mentioned in connection with Nerve V above.

Nerve VIII. Auditory (statoacoustic), extends from the side of the medulla to the ear.

Nerve IX. Glossopharyngeal, extends from the side of the medulla to the muscles and membranes of the tongue and pharynx.

Nerve X. Vagus, closely associated with Nerve IX, extends from the side of the medulla to the heart, lungs, and digestive organs.

Next, observe the ventral outgrowth from the diencephalon, the **infundibulum,** which corresponds to the posterior lobe of the pituitary gland in birds and mammals. Located behind and close to the infundibulum is a dorsal outgrowth originating from the roof of the mouth, the **anterior lobe of the pituitary** (also called the hypophysis). Together these two lobes constitute the **pituitary gland,** an important gland of internal secretion (endocrine gland). Note that the adult pituitary gland develops from two different sources—a ventral outgrowth from the diencephalon (the infundibulum) and a dorsal outgrowth from the mouth (the hypophysis). The hypophysis frequently adheres to the floor of the skull and may be detached during your removal of the brain. If you do not find it on the isolated nervous system, search on the floor of the skull in the appropriate location.

Spinal Cord and Spinal Nerves. Observe now the spinal cord. Note that the terminal portion of the spinal cord in the adult frog is threadlike and is called the **filum terminale.** Ten pairs of **spinal nerves** are attached to the spinal cord (see figure 18.14). Each nerve has two roots, the **dorsal root** and the **ventral root.** Be careful not to disturb the sympathetic nerves while tracing the spinal nerves with which they unite. Note the large size of the **second spinal nerve.** Find the **brachial plexus** and the **sciatic plexus** where several of the spinal nerves leading to the forelimbs and hindlimbs first unite and then rebranch. Trace the large **sciatic nerve** and its principal branches in the hindlimb.

Autonomic Nervous System. Examine figure 18.15, which shows the autonomic system in solid black. Note the two **sympathetic nerve trunks,** which originate anteriorly as the Gasserian ganglia and extend backward along the systemic arches and dorsal aorta. Find the **sympathetic ganglia,** the **sympathetic nerves** connecting with the spinal nerves, the **cardiac plexus,** the **solar plexus,** the large peripheral **splanchnic nerves** with branches to the stomach, intestines, and other organs, and the delicate **peripheral nerves,** which pass to the gonads, kidneys, spleen, adrenal glands, and other organs.

Demonstrations

1. Bullfrogs—injected and dissected to show the various systems.
2. Various vertebrate skeletons for comparison with the bullfrog.
3. Beating of the frog's heart and circulation of the blood through capillaries of the web of the frog's foot.

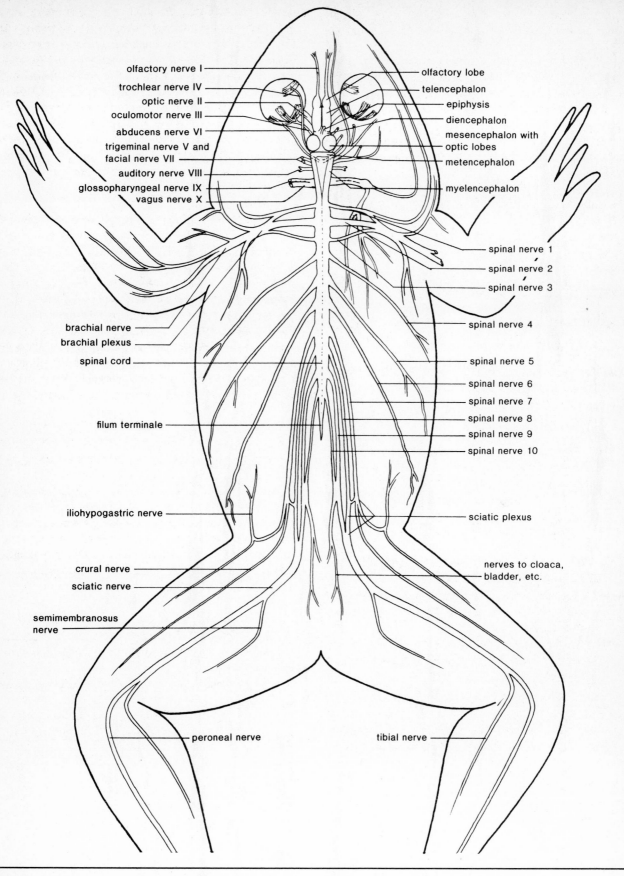

olfactory nerve I

trochlear nerve IV

optic nerve II

oculomotor nerve III

abducens nerve VI

trigeminal nerve V and
facial nerve VII

auditory nerve VIII

glossopharyngeal nerve IX
vagus nerve X

olfactory lobe

telencephalon

epiphysis

diencephalon

mesencephalon with
optic lobes

metencephalon

myelencephalon

spinal nerve 1

spinal nerve 2

spinal nerve 3

brachial nerve

brachial plexus

spinal cord

spinal nerve 4

spinal nerve 5

spinal nerve 6

spinal nerve 7

spinal nerve 8

filum terminale

spinal nerve 9

spinal nerve 10

iliohypogastric nerve

sciatic plexus

crural nerve

nerves to cloaca,
bladder, etc.

sciatic nerve

semimembranosus
nerve

peroneal nerve

tibial nerve

Fig. 18.14 Frog nervous system, dorsal view.

Frog Anatomy

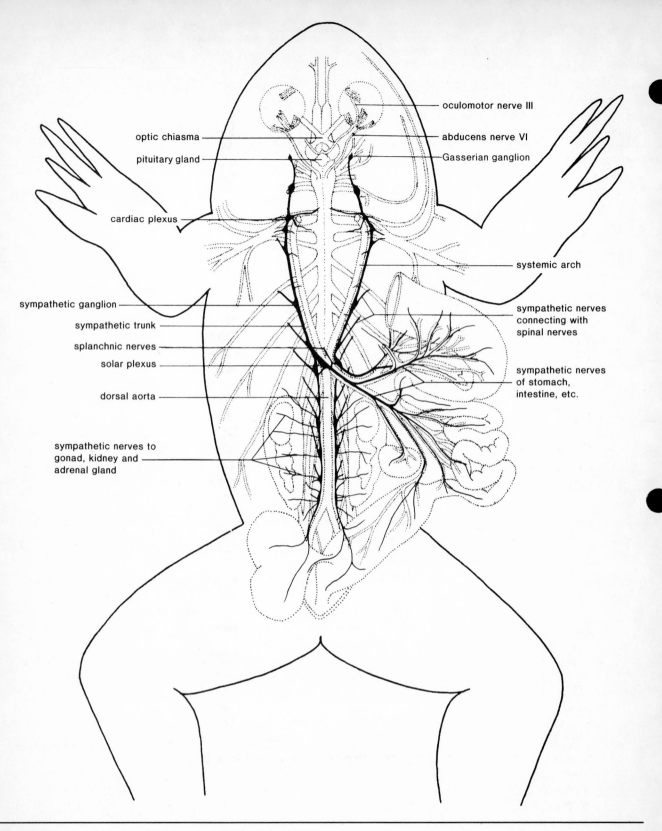

Fig. 18.15 Frog nervous system, ventral view. Autonomic
nervous system shown in solid black.

The labels on the figure are:

optic chiasma

pituitary gland

cardiac plexus

sympathetic ganglion

sympathetic trunk

splanchnic nerves

solar plexus

dorsal aorta

sympathetic nerves to
gonad, kidney and
adrenal gland

oculomotor nerve III

abducens nerve VI

Gasserian ganglion

systemic arch

sympathetic nerves
connecting with
spinal nerves

sympathetic nerves
of stomach,
intestine, etc.

Key Terms

Abductor muscle a skeletal muscle that pulls an appendage or body part away from the central axis of the body.

Adductor muscle a skeletal muscle that pulls an appendage or body part toward the central axis of the body.

Autonomic nervous system division of the vertebrate nervous system that controls involuntary functions, such as breathing or heart rate. Consists of branches from several cranial and spinal nerves leading to various internal organs.

Central nervous system portion of the vertebrate nervous system consisting of the brain and the spinal cord.

Cranium portion of the skull that encloses the brain; the braincase.

Extensor muscle a skeletal muscle that extends or straightens an appendage.

Flexor muscle a skeletal muscle that bends an appendage.

Peripheral nervous system portion of the nervous system outside the central nervous system made up of the nerves connecting the organs and tissues with the central nervous system.

Poikilothermal refers to an animal whose body temperature roughly parallels that of his environment; cold-blooded.

Tadpole larva larval form typical of amphibians with an ovoid body and a thin muscular tail.

Urogenital system organ system of vertebrates specialized for excretion and reproduction.

Vertebra unit of the spinal (vertebral) column or "backbone" of the frog and other vertebrates.

Visceral skeleton skeletal parts associated with the support of the gills in fishes and other primitive vertebrates; vestiges remain in the frog and higher vertebrates mainly in the hyoid apparatus that helps support the tongue.

19
Fetal Pig Anatomy

Objectives

After completing the laboratory exercises for this chapter, you should be able to perform the following tasks:

1. Locate and identify the principal external features of a fetal pig.

2. Identify the chief components of the axial and appendicular skeleton of the fetal pig.

3. Locate five major skeletal muscles of the fetal pig and explain their origins, insertions, and actions. Distinguish between extension, flexion, adduction, and abduction.

4. Demonstrate on a specimen the divisions of the coelom and the five principal mesenteries that support the abdominal organs.

5. Locate and identify the organs of the digestive system of the fetal pig.

6. Find and explain the function of the parts of the urogenital system of the fetal pig. Demonstrate the differences between male and female specimens.

7. Trace the main circulatory pathway on a specimen. Explain the main changes in circulation which occur at birth.

8. Point out the similarities and differences in the heart of a fish, a frog, and a fetal pig or other mammal. Discuss the significance of the differences.

9. Locate the main arteries and veins on a dissected specimen.

10. Identify the five divisions of the mammalian brain on a pig or sheep brain and point out the brain parts that make up these divisions.

The Fetal Pig: *Sus scrofa*

The fetal pig is often chosen for study as a representative mammal because it contains the organs and organ systems typical of most mammals. The principal distinguishing characteristics of the Order Mammalia include: body surface covered with hair; an integument (skin) with several types of glands; skull with two occipital condyles; seven cervical (neck) vertebrae; teeth borne on bony jaws; movable eyelids and fleshy external ears (pinnae); a four-chambered heart; persistent left aorta; and a muscular diaphragm separating the thoracic and abdominal cavities. Mammals also are warm-blooded. The young develop within the uterus of the female, have a placental attachment for nourishment, and are enveloped by special fetal membranes (amnion, chorion, and allantois). Milk to nourish the young after birth is produced by mammary glands. Other common representatives of this group include moles, bats, whales, mice, deer, monkeys, horses, cattle, and humans.

The fetal pig serves well as an example of mammalian anatomy because of its convenient size, ready availability, and relatively low cost. Fetal pigs are removed from the uteri of pregnant sows sold to meat packers. The gestation period in swine is sixteen to seventeen weeks; full-term fetal pigs measure about twelve inches in length and weigh two to three pounds. At fourteen weeks, pig embryos average about nine inches in length. The number of pigs in a litter averages about seven to twelve, although occasionally litters may number as high as eighteen.

Many of the anatomical features of the fetal pig are typical of mammals, although some special features related to its embryonic condition will also be apparent

body
regions: head   neck   trunk   tail
vertebrae: cervical   thoracic   lumbar   sacral
                                                                        caudal

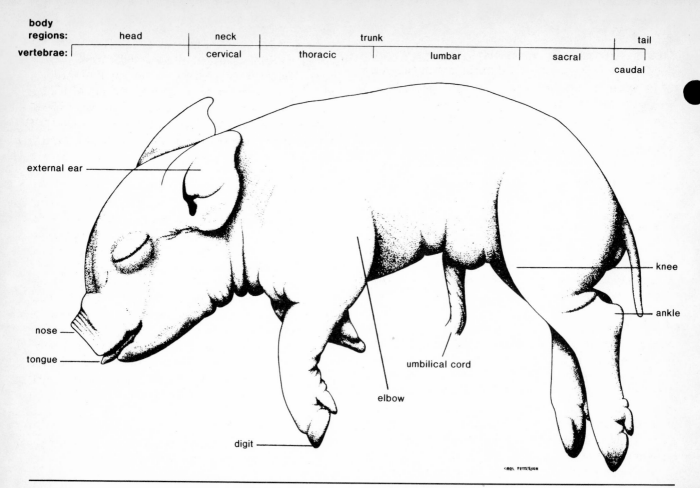

external ear

nose

tongue

digit

elbow

umbilical cord

knee

ankle

CARL PETTERSON

Fig. 19.1 Fetal pig external features.

during your study. Among these special embryonic features are the rudimentary development of the reproductive system, the low degree of ossification in many of the bones, and the soft texture of and indistinct separations between the skeletal muscles. There are also certain peculiarities of the circulatory system associated with the intrauterine development of placental mammals.

The pig embryo develops within the uterus of the mother and obtains its nourishment and oxygen supply through the umbilical cord, which attaches to the placenta. The placenta is a specialized structure made up partly of uterine tissues and partly of embryonic tissues and provides a route for the supply of nutrients and oxygen and for the removal of metabolic wastes from the embryo. Some further details of the fetal circulation and the changes in the circulatory system will be studied during this exercise.

Materials List

Living Specimens
  Mammalian spermatozoa (Demonstration)
Preserved Specimens
  Fetal pig
  Cat skeleton (Demonstration)
  Mammalian placenta (Demonstration)
  Mammalian lung (Demonstration)
  Fetal pig, injected, dissections to illustrate circulatory system (Demonstration)
  Mammalian heart, dissected (Demonstration)
  Sheep brain (Demonstration)
  Fetal pig, dissected to show cranial and spinal nerves (Demonstration)
Prepared Microscope Slides
  Pig testis, cross section (Demonstration)
  Mammalian spermatozoa (Demonstration)
  Mammalian lung, cross section (Demonstration)
  Mammalian ovary, cross section (Demonstration)
  Mammalian spinal cord, cross section (Demonstration)

External Anatomy

Obtain a preserved fetal pig from your instructor and study first the general organization of the mammalian body and the principal features of external anatomy (figure 19.1). Note that the body of the fetal pig is divided into three major regions—the **head, neck,** and **trunk.** At the posterior end of the trunk is the **tail.** The trunk consists of two principal subdivisions, the anterior **thorax** and the posterior **abdomen,** separated internally by the muscular diaphragm. Observe the two pairs of appendages, one pair of **forelimbs** and one pair of **hindlimbs.** Note that the limbs of the pig are directed ventrally rather than laterally as in the case of the frog studied earlier. This difference in orientation of the appendages represents an important advance of higher vertebrates and is closely linked with the evolution of the rapid and efficient locomotion characteristic of most mammals.

Head

Principal external features of the head include the large **mouth,** the **eyes,** and the **ears.** Note the elongated **snout** with a pair of nostrils at the anterior end. On the snout, find the stiff sensory hairs or **vibrissae.** Observe also the strong lower jaws below the mouth and the hard, rounded dorsal cap between the ears. This bony "cap" represents a portion of the **braincase** or cranium, which protects the soft brain.

Neck

Connecting the head with the trunk is a short, stout **neck.** The neck is equipped with powerful dorsal muscles. Squeeze the dorsal portion of the neck with your fingers to feel these muscles. What is the significance of these muscles in the characteristic rooting of pigs? Locate these neck muscles in figure 19.6.

Thorax

The thoracic region comprises the anterior half of the trunk and includes the **pectoral girdle, forelimbs,** and the **rib cage,** which encloses the lungs. Observe carefully the forelimbs and the feet. Note that the feet consist of **four digits.** The pentadactyl ("five-fingered") structure of the primitive vertebrate limb has been lost. The third and fourth digits are large and make up the two halves of the "cloven hoof" of the pig. The second and fifth digits are present but are reduced in size and are posterior in location. Feel the ribs and sternum in the thoracic region. These important skeletal parts provide protection for the lungs, heart, and major blood vessels within the pleural cavity.

Abdomen

The abdominal region lies posterior to the thorax and contains the **peritoneal cavity** into which the abdominal organs are suspended. Observe the ventral **umbilical cord.**

The point of attachment of the umbilical cord to the abdominal wall is called the **umbilicus;** after detachment, the scar marking the former site of the umbilicus becomes the **navel.** Observe the severed end of the umbilical cord and note that there are **four large tubes** or channels within the tube. These tubes include an **umbilical vein,** two **umbilical arteries,** and an **allantoic duct.** The former three channels provide important routes for the exchange of food, oxygen, and other material between the mother and embryo; the latter represents an important channel for the disposal of metabolic waste products produced by the embryo.

At the posterior end of the abdomen are the **pelvic girdle** and the **hindlimbs.** Note the basic similarity in the organization of the forelimbs and hindlimbs. Find the several pairs of nipples or **mammae** on the ventral abdominal wall. Also locate the anus at the base of the tail.

You can determine the sex of your specimen by observing certain external features in the abdominal region. Male specimens exhibit a **urogenital opening** just posterior to the umbilical cord and two scrotal sacs at the posterior end of the body, just below the anus. In older specimens, the **testes** descend into these scrotal sacs from the peritoneal cavity. Female specimens exhibit a urogenital opening, the **vulva,** just ventral to the anus.

Tail

The **tail** is short, flexible, and usually curled. Its skeletal axis is a continuation of the vertebral column.

Skeletal System

The general organization of the mammalian body can best be understood after study of the skeleton, the underlying supporting framework for the body. Unfortunately, the skeleton of the fetal pig is not very suitable for such study because in fetal pigs the skeleton is incompletely ossified and, consequently, some of the bones are not distinct. Therefore, you should refer to a mounted skeleton of a cat or similar vertebrate on demonstration for the purpose of general orientation. Figure 19.2 is provided for your assistance in identifying various parts of the cat skeleton; compare it with figure 19.3, which illustrates the skeleton of a late-term pig embryo.

The skeleton of the pig and other mammals is made up of two divisions, the **axial skeleton** (skeleton of the body axis including the skull) and the **appendicular skeleton.** The appendicular skeleton provides articulation for the four limbs with the axial skeleton. The skull is a complex structure consisting of many fused bones, which enclose and protect the brain and other organs of the head. The cat skull is more convenient to study than the skull of the pig and is more typical of the adult condition. Both skulls are made up of the same bones, although their shape varies slightly. Figures 19.4a, b, and c will help you identify the principal bones of the skull.

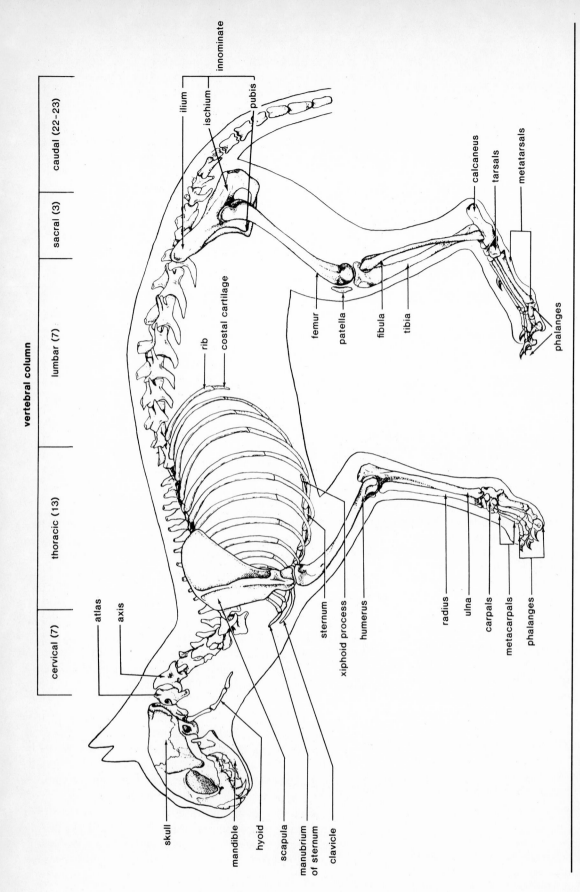

Fig. 19.2 Cat skeleton.

vertebral column

cervical (7) | thoracic (13) | lumbar (7) | sacral (3) | caudal (22–23)

ilium
ischium
innominate
pubis

atlas
axis

rib
costal cartilage

calcaneus
tarsals
metatarsals

femur
patella
fibula
tibia
phalanges

skull
mandible
hyoid
scapula
manubrium of sternum
clavicle

sternum
xiphoid process
humerus

radius
ulna
carpals
metacarpals
phalanges

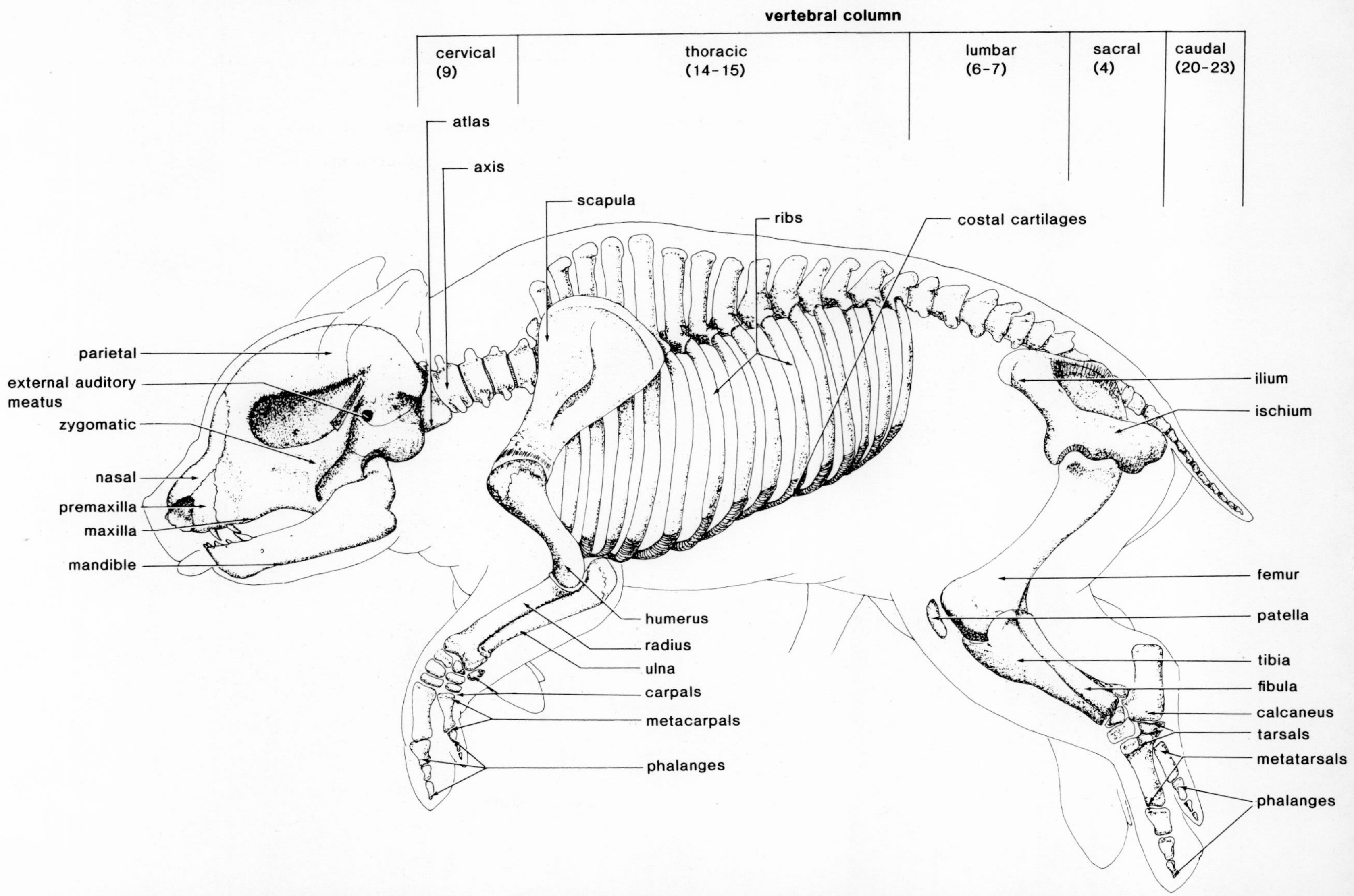

Fig. 19.3 Fetal pig skeleton.

281

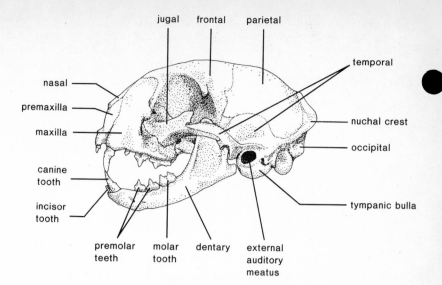

Fig. 19.4a Cat skull, lateral view.

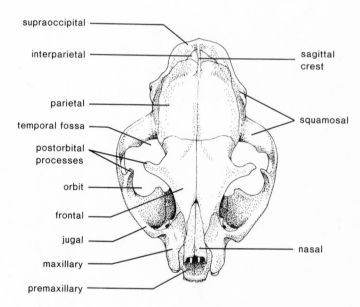

Fig. 19.4b Cat skull, dorsal view.

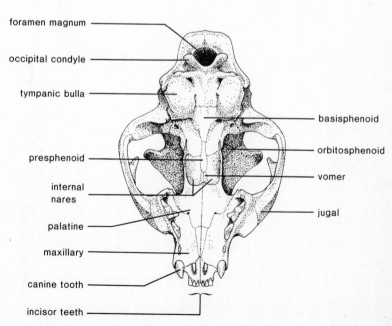

Fig. 19.4c Cat skull, ventral view.

The principal bones making up the skeleton of the pig, cat, and other mammals are listed in the following outline.

Components of the Skeleton

A. **Axial skeleton**
1. Skull
2. Mandible (lower jaw)
3. Hyoid apparatus (several small bones which support the tongue and larynx; remnant of the visceral skeleton)
4. Vertebral column
   Cervical vertebrae (9)
   Thoracic vertebrae (14–15)
   Lumbar vertebrae (6–7)
   Sacral vertebrae (4)
   Caudal vertebrae (20–23)
5. Ribs
6. Sternum

B. **Appendicular skeleton**
1. Pectoral girdle
   Scapula
2. Forelimb
   Humerus
   Radius
   Ulna
   Carpals
   Metacarpals
   Phalanges
3. Pelvic girdle
   Ilium
   Ischium   } fused
   Pubis
4. Hind limb
   Femur
   Patella
   Tibia
   Fibula
   Tarsals (The calcaneus, or heel bone, is the largest of the tarsals.)
   Metatarsals
   Phalanges

Muscular System

The muscles of the fetal pig are soft, incompletely developed, and easily torn. Special care is therefore necessary if you are to be successful in your study of the muscular system.

A skeletal muscle typically consists of a fixed end, called the **origin,** and a movable end, called the **insertion.** The fleshy middle portion of the muscle is called the **belly.** The whitish connective tissue covering the muscles is the **deep fascia,** and skeletal muscles are connected to cartilage, bone, ligaments, or skin by this fascia, by tendons, or by aponeuroses (singular: aponeurosis, thin flat sheets of tough connective tissue).

The movement or effect caused by a muscle is its **action.** Thus, each skeletal muscle has a characteristic origin, insertion, and action. Most muscles are arranged in pairs or groups that are **antagonistic;** they have opposite actions. Several actions of muscles have special names. Some common actions are:

Adduction—moving the distal end of a bone closer to the ventral median line of the body
Abduction—moving the distal end of a bone farther away from the ventral median line of the body
Flexion—bending a limb at a joint
Extension—straightening a limb

The pig must be skinned before you can study its muscles. Make a midventral incision as indicated in figure 19.5. Extend the cut forward to the lower jaw and backward to the hindlimbs. Take care to cut only through the skin; do not cut into the underlying musculature. Next make an incision through the skin on the medial (inner) surface of the right forelimb to the hoof. Make a similar incision on the right hindlimb. Remove the skin from the right side of the body and from the right forelimb and hindlimb.

After you have removed the skin, you will observe a thin whitish layer of connective tissue. This is the **superficial fascia,** which must be removed to expose the underlying muscles. Carefully remove the superficial fascia and study first the muscles of the dorsal side (figure 19.6). Then study the muscles of the ventral surface with the aid of figure 19.7. Tables 19.1 and 19.2 give the origin, insertion, and action of the most prominent muscles of the pig.

General Internal Anatomy

Place your specimen ventral side up in a dissecting pan and tie a stout cord to one of the forelimbs. Pass the cord under the pan and attach the cord to the other leg in order to secure the specimen in place, to spread the legs apart, and to provide access to the ventral body surface. Repeat the tying process with the hindlimbs. Study figure 19.8 to obtain a general idea of the thickness of the body wall and the approximate location of the internal organs before starting your dissection.

Begin your dissection by making an incision through the skin and muscles of the ventral body wall overlying the sternum. Locate the sternum between the forelimbs and make a longitudinal incision as shown in figure 19.5. Continue the incision anteriorly to the level of the lower jaw and posteriorly to a point just anterior to the umbilical cord. Use a sharp scalpel for your dissection and cut cleanly through the skin and muscles; take care not to damage any underlying organs.

Make a second incision from the region of the umbilical cord toward the side of the body and then connect the first and second incisions by cutting around the umbilicus to leave intact a portion of the ventral body wall

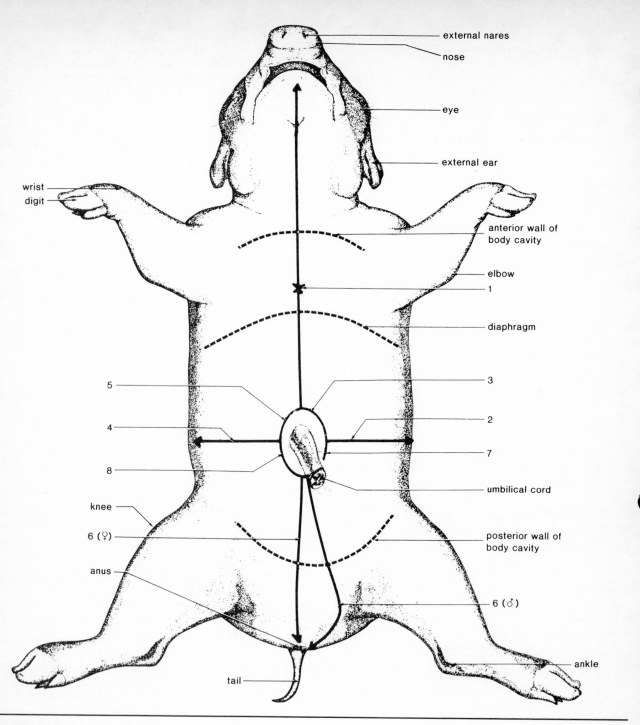

Fig. 19.5 Fetal pig, ventral view with lines to indicate incisions for dissection. Numbers show sequence of recommended incisions.

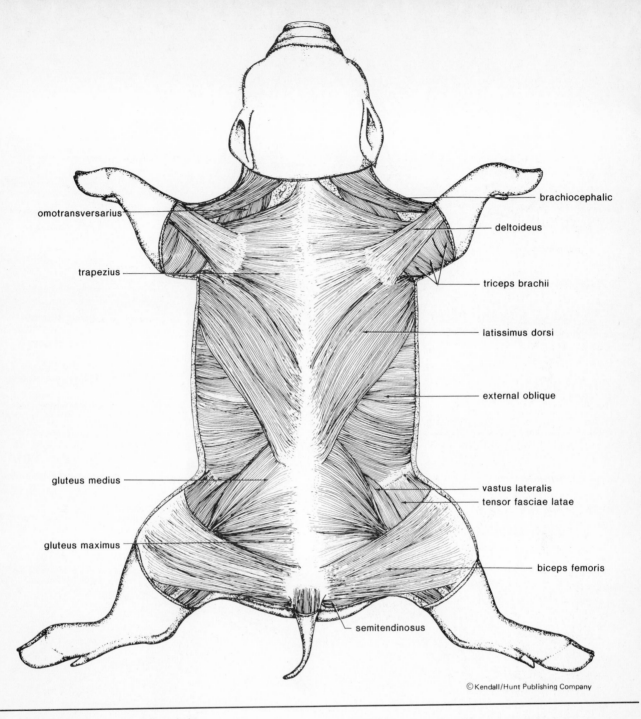

omotransversarius

brachiocephalic

deltoideus

trapezius

triceps brachii

latissimus dorsi

external oblique

gluteus medius

vastus lateralis

tensor fasciae latae

gluteus maximus

biceps femoris

semitendinosus

© Kendall/Hunt Publishing Company

Fig. 19.6 Fetal pig muscles, dorsal view.

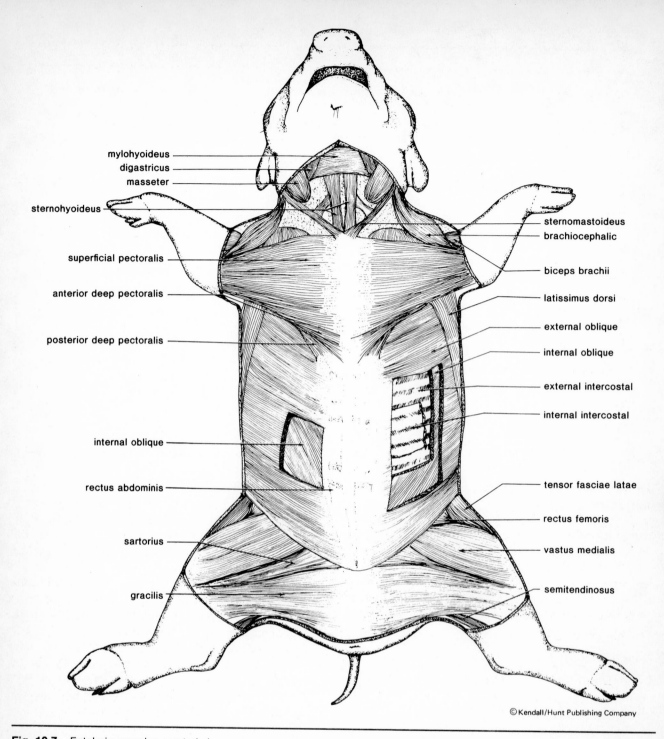

mylohyoideus
digastricus
masseter
sternohyoideus
superficial pectoralis
anterior deep pectoralis
posterior deep pectoralis
internal oblique
rectus abdominis
sartorius
gracilis

sternomastoideus
brachiocephalic
biceps brachii
latissimus dorsi
external oblique
internal oblique
external intercostal
internal intercostal
tensor fasciae latae
rectus femoris
vastus medialis
semitendinosus

© Kendall/Hunt Publishing Company

Fig. 19.7   Fetal pig muscles, ventral view.

Table 19.1. Muscles of Dorsal Side

Muscle	Origin	Insertion	Action
Back and Side			
Latissimus dorsi	Lumbar and thoracic vertebrae	Humerus	Draws humerus upward and backward
Trapezius	Skull; cervical and thoracic vertebrae	Scapula	Elevates shoulder
External oblique	Last 9 or 10 ribs, lumbodorsal fascia	Linea alba, ilium	Flexes trunk, constricts abdomen
Forelimb			
Splenius	Thoracic vertebrae	Skull, cervical vertebrae	Elevates head and neck
Deltoideus	Scapula	Humerus	Raises humerus
Brachiocephalic	Skull	Humerus, shoulder muscles	Raises head
Triceps brachii	Scapula, humerus	Ulna	Extends forearm
Pelvic Region and Hindlimb			
Gluteus medius	Longissimus dorsi muscle, ilium, sacroiliac	Femur	Abducts thigh
Vastus lateralis	Femur	Patella and tibia	Extends shank
Tensor fasciae latae	Ilium	Patella and tibia	Flexes hip joint, extends knee joint
Gluteus maximus	Sacral and caudal vertebrae	Fascia lata	Abducts thigh
Biceps femoris	Ischium, sacrum	Patella, leg, thigh	Abducts and extends limb
Semitendinosus	Caudal vertebrae, ischium	Tibia, calcaneus	Extends hip, flexes knee joint

immediately surrounding the umbilical cord. The fourth incision should be made laterally from the umbilical region toward the side of the pig. Continue your incisions in the sequence indicated in figure 19.5 to provide good access to the internal organs of the pig.

Remove the skin from the ventral surface of the neck, the lower jaw, and the left side of the head up to the base of the ear. Also remove the muscles on the ventral side of the **larynx** and **trachea.** Take care not to damage any of the larger blood vessels.

Neck Region

Observe the large **thymus gland** that lies along the trachea and extends forward to the larynx (see figure 19.8). Find also the small, bilobed **thyroid gland** located just posterior to the larynx. The thyroid gland is darker and located dorsal to (beneath) the larger thymus gland. Just ventral to the ear, see the large, light-colored parotid gland (see figure 19.17). Ventral and slightly anterior to the parotid gland, find the smaller **submaxillary gland.** Also locate the flat, narrow **sublingual gland** anterior to the submaxillary gland. The sublingual gland lies beneath the mylohyoid and geniohyoid muscles and surrounds the anterior portion of the submaxillary duct. The ducts from both the submaxillary and sublingual glands empty in the floor of the mouth cavity.

The Coelom and Its Divisions

The coelom of the pig is divided into two major regions by a muscular partition, the **diaphragm.** Anterior to the diaphragm is the **thoracic cavity,** which is subdivided into **two pleural cavities** containing the lungs and the **pericardial cavity** (mediastinum), which contains the heart and the large blood vessels connected with the heart. The thin epithelial lining of the pleural cavity is the **pleura;** that of the pericardial cavity is the **pericardium.** Posterior to the diaphragm is the large **peritoneal** (abdominal) **cavity** within which the abdominal organs are suspended. Each of the abdominal organs is enclosed in a thin layer of mesodermal epithelium (**visceral peritoneum**); a similar layer also covers the inner surface of the body wall (**parietal peritoneum**). The double sheets of peritoneum that support the abdominal organs are called **mesenteries.**

Several mesenteries that support specific organs can be identified: (1) the **falciform ligament,** which connects the liver to the ventral body wall; (2) the **coronary ligament,** which attaches the anterior surface of the liver to the diaphragm; (3) the **greater omentum,** which attaches the greater curvature of the stomach to the colon and most of the small intestine; (4) the **lesser omentum,** which attaches the lesser curvature of the liver with the anterior portion of the duodenum; and (5) the **mesentery proper,** which suspends the small intestine from the dorsal midline. Other smaller mesenteries support the colon, the rectum, and the gonads from the dorsal body wall.

Table 19.2 Muscles of Ventral Side

Muscle	Origin	Insertion	Action
Head and Neck Region			
Mylohyoideus	Mandibles	Hyoid bone	Raises floor of mouth, tongue, and hyoid bone
Digastricus	Mastoid process of skull	Mandible	Depresses mandible
Masseter	Zygomatic arch of skull	Mandible	Elevates jaw, closes mouth
Sternohyoideus	Sternum	Hyoid bone	Retracts and depresses hyoid and base of tongue
Forelimb and Pectoral Girdle			
Brachiocephalic	Nuchal crest, mastoid process of skull	Humerus and shoulder	Inclines or extends head, draws forelimb forward
Biceps brachii	Scapula	Radius, ulna	Flexes forearm
Superficial pectoralis	Sternum	Humerus	Adducts humerus
Anterior deep pectoralis	Sternum	Scapula, supraspinatus muscle	Adducts and retracts forelimb
Posterior deep pectoralis	Sternum, 4th to 9th ribs	Humerus	Retracts and adducts forelimb
Side and Belly			
Latissimus dorsi	Lumbar and thoracic vertebrae	Humerus	Draws humerus upward and backward
External oblique	Last 9 or 10 ribs	Linea alba, ilium	Constricts abdomen, flexes trunk
Internal oblique	Lumbodorsal fascia	Linea alba	Constricts abdomen, flexes trunk
External intercostal	Ribs	Adjacent rib	Pulls ribs together
Internal intercostal	Ribs	Adjacent rib	Pulls ribs together
Rectus abdominis	Pubic symphysis	Sternum	Constricts abdomen
Pelvic Girdle and Hindlimb			
Tensor fasciae latae	Ilium	Patella and tibia	Flexes hip joint, extends knee joint
Rectus femoris	Femur	Patella and tibia	Extends shank
Vastus medialis	Femur	Patella and tibia	Extends shank
Semitendinosus	Caudal vertebrae and ischium	Tibia and calcaneus	Extends hip, flexes limb
Sartorius	Iliac fascia and tendon of psoas minor muscle	Patella and tibia	Abducts hindlimb, flexes hip joint
Gracilis	Pubis	Patella and tibia	Adducts hindlimb

Thoracic Cavity

Within the thoracic cavity identify the **diaphragm,** the two **pleural cavities** enclosing the lungs, and the **pericardial cavity** around the heart.

The heart of the pig, as that of all mammals, consists of four chambers: **two atria** (auricles) and **two ventricles.** How does the heart of mammals compare in this regard with that of the shark?

Locate the four lobes of the right lung. How many lobes of the left lung can you locate? Observe the demonstration materials showing the internal structure of the mammalian lung.

Peritoneal Cavity

Within the peritoneal cavity (figures 19.8, 19.9, 19.10, and 19.11) identify the four-lobed **liver,** the **gall bladder,** the common **bile duct,** and the **stomach.** (In addition to the four main lobes of the liver, there is also a small caudate lobe attached to the posterior portion of the right lateral lobe.) Connecting with the stomach, find the **esophagus.** The upper part of the stomach, into which the esophagus empties, is called the **cardiac portion** and the lower part is the **pyloric portion.** The constriction between the stomach and the small intestine is the **pylorus.** A specialized muscle, the **pyloric sphincter,** acts as a valve to regulate passage of material from the stomach into the small intestine.

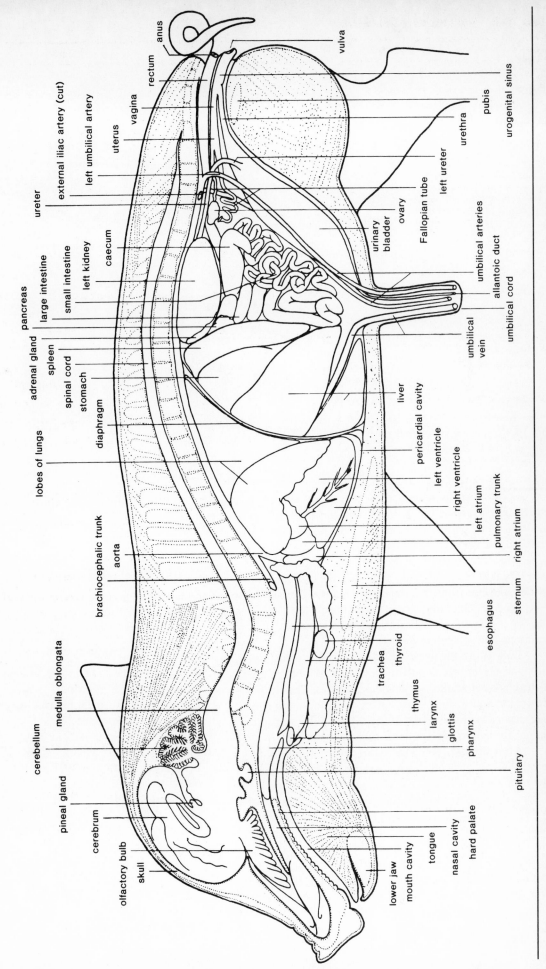

Fig. 19.8 Fetal pig, internal anatomy of female.

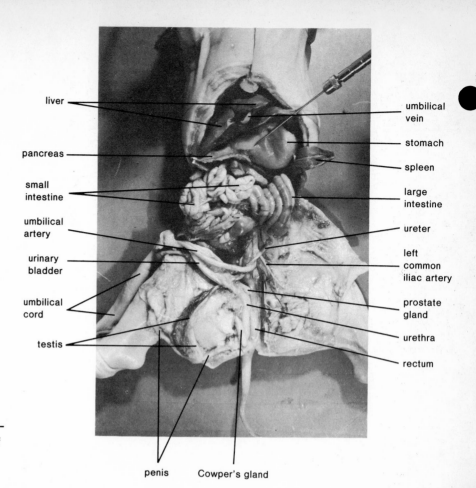

liver

pancreas

small
intestine

umbilical
artery

urinary
bladder

umbilical
cord

testis

umbilical
vein

stomach

spleen

large
intestine

ureter

left
common
iliac artery

prostate
gland

urethra

rectum

penis   Cowper's gland

Fig. 19.9 Fetal pig, internal organs of abdomen, male. (Photograph by John Vercoe.)

Locate the elongated, dark-red **spleen** along the greater curvature of the stomach and the light-colored **pancreas** found posterior to the stomach. Observe also the loosely coiled **small intestine** followed by the tightly coiled **large intestine**. Near the junction of the small intestine with the large intestine, find a blind sac, the **caecum**. Posteriorly, the large intestine is modified to form the **rectum**.

Find the chief mesenteries that support the abdominal organs: **falciform ligament, greater omentum, lesser omentum,** and **mesentery proper.**

Locate also the **urinary bladder,** which extends dorsally and posteriorly from the umbilicus. On either side of the bladder find the two **umbilical arteries.** The outline drawing in figure 19.12 is provided for you to show a general view of your dissection of the fetal pig. Draw in the major internal organs and label each of them.

After you have completed your general survey of internal anatomy and completed figure 19.12, sever the small intestine about an inch behind the pylorus and cut the mesentery close to the intestine. Remove the small and large intestines from the body cavity after you have cut through the rectum about two inches from its posterior end. Observe the large blood vessels that supply the intestines (see figures 19.16 and 19.17) as you remove this portion of the digestive tract. Examine the interior of the large and small intestines and of the caecum. How do their interior walls differ in structure? Notice also the numerous, light-colored lymph nodes found between the two layers of the mesentery.

The Urogenital System

The excretory and reproductive (genital) systems are best studied together because they are closely related in development and because of their use of common ducts.

The Excretory (Urinary) System

Refer to figures 19.8, 19.13, 19.14, and 19.15 and locate the kidneys on your specimen. A **ureter** leads from each kidney to the large **urinary bladder** (an enlargement of the allantois). Find the **allantoic duct,** which leads from the bladder through the umbilical cord. Note also the **renal artery** (figure 19.16), which enters each of the kidneys, and the large renal veins (figure 19.17) through which blood leaves the kidneys. Locate also the two large umbilical arteries (figures 19.14 and 19.16), which enter the umbilical cord. Along the medial border of each kidney find the long, narrow, whitish **adrenal glands.** The adrenals are important endocrine glands.

The specimens provided for each laboratory section should consist of both male and female pigs. Study the genital (reproductive) system of your specimen according

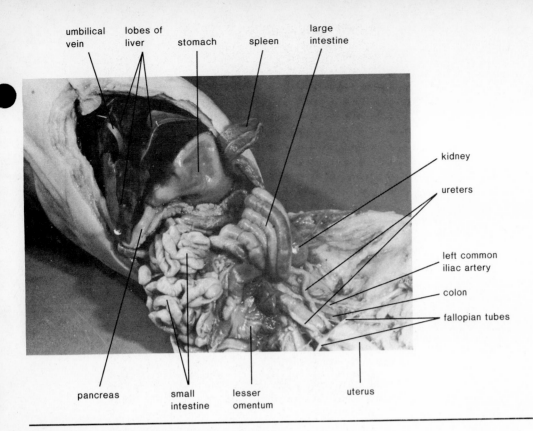

Fig. 19.10   Fetal pig abdominal organs, female.
(Photograph by John Vercoe.)

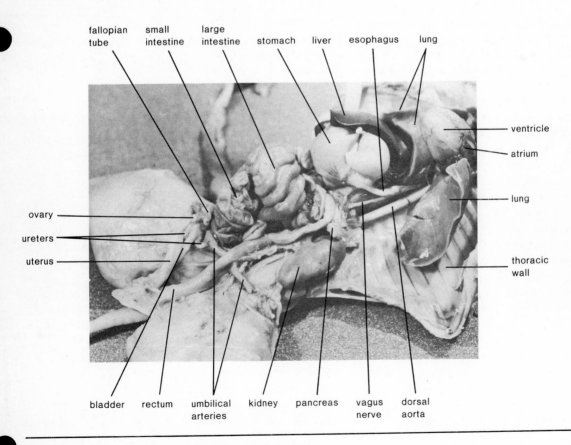

Fig. 19.11   Fetal pig, organs of thoracic and abdominal
regions, female. (Photograph by John Vercoe.)

Fetal Pig Anatomy

Fig. 19.12 Drawing of internal organs of fetal pig.

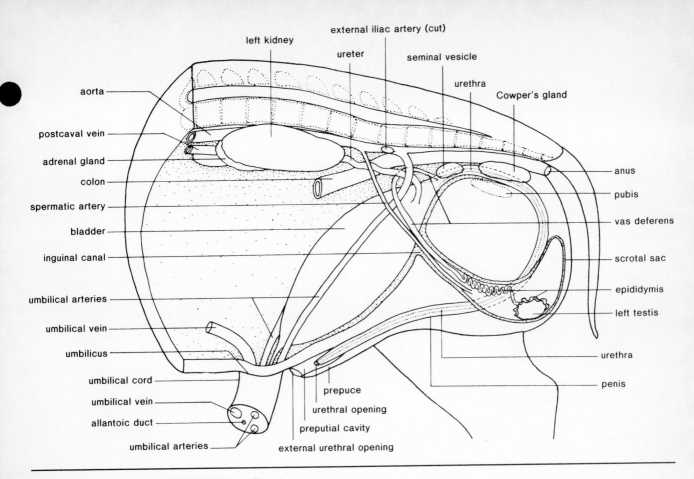

Fig. 19.13   Fetal pig urogenital system, male.

to the following directions and then study also a specimen of the opposite sex on a specimen assigned to one of your classmates.

The Female Genital System

Study figure 19.8 and locate the principal female reproductive organs. Then cut through the skin, muscles, and pelvis along the midventral line of your specimen and pull the limbs wide apart to locate these parts on your specimen. Find the two small, bean-shaped **ovaries** near the posterior end of the peritoneal cavity. Attached to the dorsal surface of each ovary is a small convoluted duct, the **Fallopian tube** (oviduct). Locate the wide, ciliated funnel, called the **infundibulum,** at the terminus of each Fallopian tube. The opening into the end of the Fallopian tube is the **abdominal ostium.** Trace the Fallopian tubes to the larger **horns of the uterus,** which connect with a single median **body of the uterus.** The uterus is continued posteriorly by a thick, muscular tube, the **vagina.** The vagina and the urethra both empty into a common chamber, the **urogenital sinus** or vestibule. Carefully cut open the urogenital sinus by making an incision along one side, and use a blunt probe to locate the openings of the urethra and the vagina. Also search on the ventral floor of the urogenital sinus for the **clitoris,** a small, rounded papilla. **Do not** confuse the clitoris (homologous with the penis of the

male) with the larger **genital papilla.** The exterior opening of the urogenital sinus is the **vulva,** a slitlike opening located immediately below the anus. The genital papilla is an external ventral projection of the vulva.

Study also the demonstrations of microscopic sections of a mammalian ovary to see germ cells in various stages of development and of pig embryos at various stages.

The Male Genital System

Study figures 19.9, 19.13, and 19.14 and note the relative position of the principal male reproductive organs. Cut through the scrotal sac and down through the pubis about one-eighth inch to one side of the midventral line, pull the hindlimbs apart, and identify the principal male organs.

Locate first the two **testes.** In larger and older specimens, the testes will be located within two **scrotal sacs** (collectively referred to as the scrotum), but in younger specimens the testes may not yet have descended from the peritoneal (abdominal) cavity where they initially grow and differentiate. As the male pig grows larger and the testes develop, the testes descend from the peritoneal cavity through the **inguinal canal** into the scrotal sacs. In younger specimens the testes may be found anywhere along this descending path.

Locate also the **epididymis,** a mass of coiled tubules lying along one side of the testis. The epididymis connects

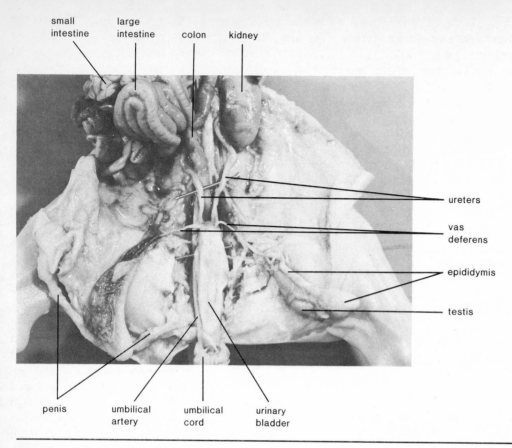

small intestine large intestine colon kidney

ureters

vas deferens

epididymis

testis

penis umbilical artery umbilical cord urinary bladder

Fig. 19.14 Fetal pig urogenital system, dissected, male. (Photograph by John Vercoe.)

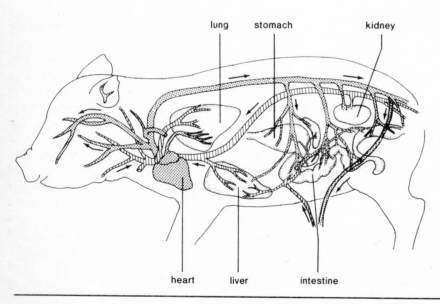

lung stomach kidney

heart liver intestine

Fig. 19.15 Adult pig, pattern of circulation. The heart, systemic arteries, and pulmonary veins are identified by the dotted areas; the pulmonary arteries and the venous circulation are shaded with vertical lines.

with the **vas deferens,** which passes through the inguinal canal along the **spermatic artery** and **vein** and crosses over the **ureter** to enter the **urethra** (see figure 19.13).

Locate the **penis,** a long, muscular cylinder lying beneath the skin just posterior to the umbilical cord. Cut through the skin overlying the penis to free the penis and trace it posteriorly to its junction with the urethra. Dorsal to the urethra, near the entrance of the **vasa deferentia,** find the two small **seminal vesicles.** Adjacent to the seminal vesicles is the **prostate gland** (poorly developed in younger specimens). Note also the two **Cowper's glands** (bulbourethral glands) lying alongside the urethra near its junction with the penis.

The tissue overlying the penis on the ventral abdominal wall is called the **prepuce,** and the cavity enclosed within the prepuce is the **preputial cavity.** Opening from the preputial cavity to the exterior is the **external urethral orifice.**

Observe the demonstration of the pig testis showing **germ cells** in various stages of maturation. See also the demonstrations of mammalian **spermatozoa.**

Circulatory System

To assure your understanding of circulation in mammals, you should review the basic organization of the mammalian circulatory system prior to undertaking a dissection and study of the circulatory system of the pig. Consult figure 19.15 for the basic pattern of mammalian circulation. The basic features of mammalian circulation should already be familiar to you from previous high school courses in health and biology. Among these essential features are:

1. The presence of a four-chambered heart, which provides an efficient separation of the systemic and pulmonary divisions of the circulatory system.

2. Oxygenated blood from the lungs is carried to the heart by the pulmonary veins, which empty into the left atrium. From the left atrium, the blood passes to the left ventricle, which pumps it out through the aorta and its branches to all parts of the body **except the lungs.**

3. Blood from the organs of the body (except the lungs) is returned to the right atrium by the large precaval and postcaval veins. From the right atrium, it goes to the right ventricle and then back via the **pulmonary arteries** to the lungs to be oxygenated.

4. Mammals have a **hepatic portal system,** but the **renal portal system,** characteristic of many lower vertebrates such as the frog and shark, has been lost during the evolution of the mammals.

Fetal Circulation

As noted previously, the circulatory system of the fetal pig differs in some important respects from the typical pattern of adult circulation in a mammal. These differences are due to the fact that the fetus is dependent upon the placenta for its supply of oxygen and nutrients and for the removal of carbon dioxide and nitrogenous wastes. At birth, certain important changes occur in the pattern of circulation of the pig and other placental mammals, and a change to the adult pattern of circulation is effected.

Consult your textbook for further details on the pattern of fetal circulation in mammals and the important circulatory changes at birth. Also observe the demonstration materials in the laboratory, which further illustrate the nature of the placenta and its relationship to the fetus in mammals.

Arterial System

Observe the external surface of the heart and locate its four chambers using figures 19.16 and 19.17 as guides. Identify the large pulmonary trunk and the aorta, which carry blood away from the heart. Study figure 19.16 carefully and locate the major systemic arteries described in the following paragraphs.

The **brachiocephalic trunk** is the first large branch from the arch of the aorta and leads toward the head; it soon branches into the **right subclavian artery,** which supplies the right forelimb, the right anterior body wall, and the right side of the neck. Shortly anterior to its origin from the aorta, the right subclavian branches into the left and right **common carotid arteries,** which supply the skull, brain, tongue, cheek, and face. The right subclavian artery arises from the aorta independently and slightly to the left of the brachiocephalic trunk. Branching from the right subclavian artery, locate the large **brachial artery** leading into the forelimb, the **thyro-cervical artery** leading to the thyroid and parotid glands, and the **sternal artery** leading to the muscles of the thoracic and abdominal walls.

In the abdominal region locate the large **dorsal aorta** and the numerous **intercostal arteries,** which supply muscles between the ribs. Trace the large **coeliac artery** from its origin on the dorsal aorta just posterior to the diaphragm. Three branches of the coeliac artery supply the stomach (**gastric artery**); the spleen, stomach, and pancreas (**splenic artery**); the stomach, liver, pancreas, and duodenum (**gastrohepatic artery**). Locate also the **anterior mesenteric artery** leading from the dorsal aorta to the pancreas, small intestine, and large intestine. It originates just posterior to the coeliac artery. Two **renal arteries** lead to the kidneys and two **genital arteries** lead to the gonads. Posterior to the origin of the two genital arteries, locate the **posterior mesenteric artery** leading ventrally to the colon and rectum.

Near the posterior end of the abdominal cavity, the dorsal aorta divides into two large **external iliac arteries,** which supply the hind limbs (see figure 19.18). Posterior to the origin of the two external iliac arteries, locate the two large **umbilical arteries** (figures 19.18 and 19.19) leading into the umbilical cord and the single **caudal artery** continuing into the tail region. The umbilical arteries are fetal accessories and convey blood to the embryonic

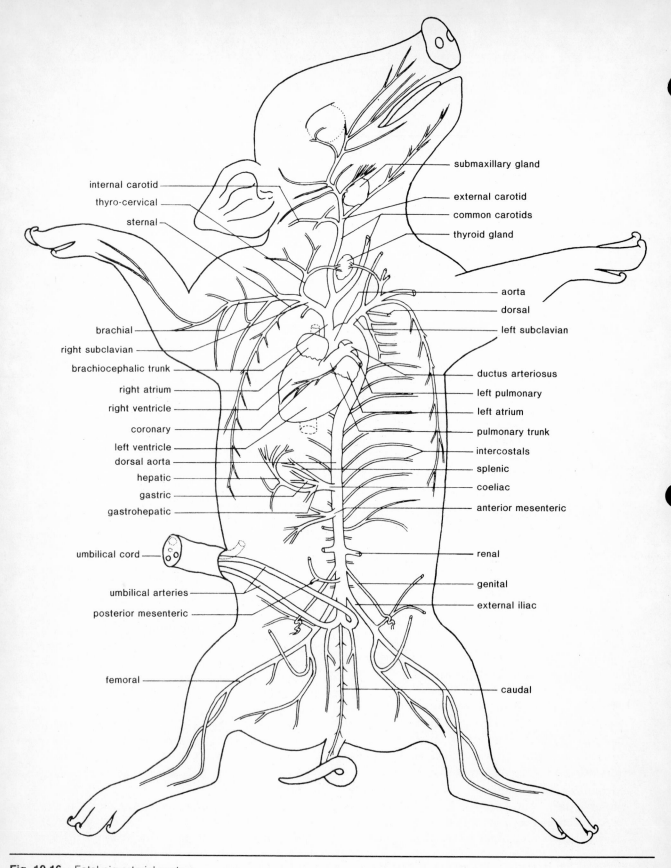

internal carotid

thyro-cervical

sternal

brachial

right subclavian

brachiocephalic trunk

right atrium

right ventricle

coronary

left ventricle

dorsal aorta

hepatic

gastric

gastrohepatic

umbilical cord

umbilical arteries

posterior mesenteric

femoral

submaxillary gland

external carotid

common carotids

thyroid gland

aorta

dorsal

left subclavian

ductus arteriosus

left pulmonary

left atrium

pulmonary trunk

intercostals

splenic

coeliac

anterior mesenteric

renal

genital

external iliac

caudal

Fig. 19.16   Fetal pig arterial system.

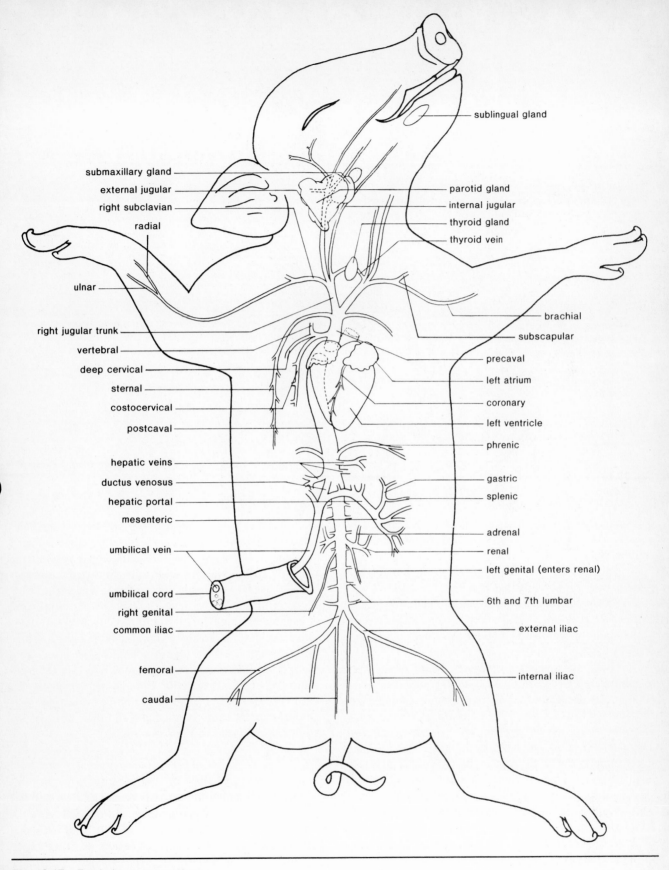

Fig. 19.17 Fetal pig venous system.

sublingual gland

submaxillary gland
external jugular
right subclavian
radial

parotid gland
internal jugular
thyroid gland
thyroid vein

ulnar

brachial
subscapular

right jugular trunk
vertebral
deep cervical
sternal
costocervical
postcaval

precaval
left atrium
coronary
left ventricle
phrenic

hepatic veins
ductus venosus
hepatic portal
mesenteric

gastric
splenic

adrenal
renal
left genital (enters renal)
6th and 7th lumbar

umbilical vein

umbilical cord
right genital
common iliac

external iliac

femoral
caudal

internal iliac

Fetal Pig Anatomy

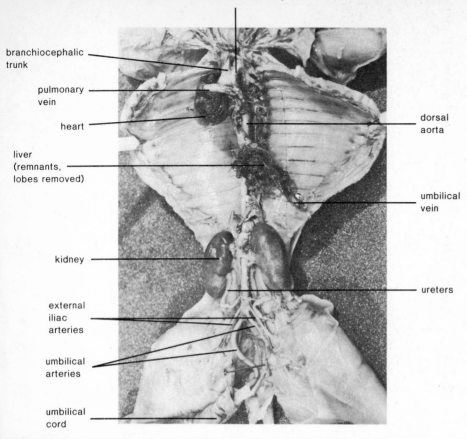

left subclavian artery

branchiocephalic
trunk

pulmonary
vein

heart

liver
(remnants,
lobes removed)

kidney

external
iliac
arteries

umbilical
arteries

umbilical
cord

dorsal
aorta

umbilical
vein

ureters

Fig. 19.18 Fetal pig, principal blood vessels of the thoracic and abdominal areas. (Photograph by John Vercoe.)

portion of the placenta. They degenerate after birth, and only those branches of the umbilical artery supplying the bladder persist in the adult pig.

Trace each of the arteries from its origin along the route to the chief organs that it supplies with blood and observe its main branches. When you complete your study of the arterial system, you should be able to identify each of the main arteries and to describe the chief organs supplied by each.

After you have completed your study of the systemic arteries, you should next study the pulmonary circulation. Locate again the pulmonary trunk leading from the right ventricle (figure 19.16) and trace it to the point of its branching into the **right** and **left pulmonary arteries.** These arteries carry the blood to the lungs and are relatively small in the fetus, but they become much larger at birth when the lungs become functional. Note also the short duct connecting the pulmonary trunk with the aorta. This duct is called the **ductus arteriosus** and is a **special fetal structure.** The duct ceases to function after birth and gradually degenerates to form a fibrous cord, the **ligamentum arteriosum.** In the adult pig, all of the blood in the pulmonary circulation passes to the lungs for oxygenation.

Another special feature of the fetal circulation is also related to the mixture of blood from the pulmonary and systemic divisions. Within the heart, blood passes from the right atrium into the left atrium through a temporary opening, the **foramen ovale,** in the median wall of the heart. Blood returning to the heart through the postcaval veins is shunted through this opening directly back into systemic circulation, thus bypassing the lungs. The foramen ovale becomes sealed at birth and all of the blood from the right atrium thereafter goes to the lungs via the right ventricle, pulmonary trunk, and the remainder of the pathway. The former site of the foramen ovale appears as an oval depression in the medial wall of the heart of the adult pig and is called the **fossa ovalis.**

Venous System

Study figures 19.15 and 19.17 and identify the major veins of the pig, including the **precaval veins,** the **right** and **left jugular trunks,** the **subclavian veins,** the **brachial veins,** the **external** and **internal jugular veins,** the **postcaval veins,** and the **hepatic veins.** Locate also the **hepatic portal vein** with its three main branches; the **gastric vein** carrying blood from the stomach, the **splenic vein** carrying blood from the spleen; and the **mesenteric vein** carrying blood chiefly from

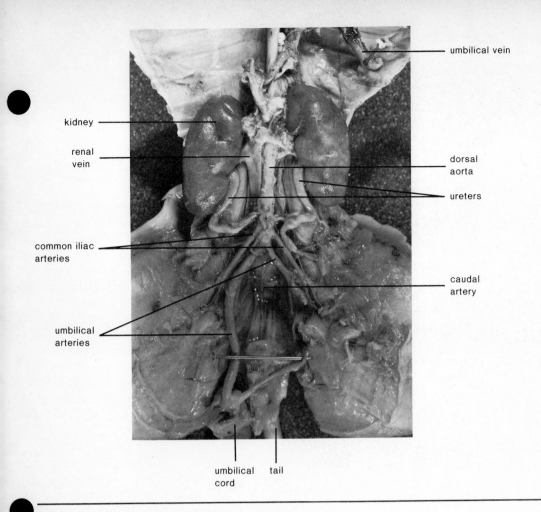

kidney

renal
vein

common iliac
arteries

umbilical
arteries

umbilical vein

dorsal
aorta

ureters

caudal
artery

umbilical   tail
cord

Fig. 19.19   Fetal pig, principal blood vessels of the
abdominal and pelvic areas. (Photograph by John Vercoe.)

the small intestine. Find the **renal veins** leading from the
kidneys, and from the hindlimbs find the **common iliac
veins** into which blood flows from the **internal** and **external
iliac veins.**

Locate the **umbilical vein** in the umbilical cord once
again and trace it forward. Note the various branches that
carry blood to the **hepatic portal vein** and into the liver.
Note also that part of the blood from the **umbilical vein** is
shunted directly to the **postcaval vein** through the short
ductus venosus. Shortly after birth, the umbilical vein and
the ductus venosus cease to function and degenerate. Like
the umbilical arteries, the ductus arteriosus, and the foramen ovale, the umbilical vein and the ductus venosus
are functional only in the fetus.

Compare the blood of the umbilical vein with that
of the umbilical arteries in regard to the content of:
(1) oxygen, (2) nitrogenous wastes, (3) carbon dioxide,
and (4) nutrients.

Observe the several **pulmonary veins** arising from the
lungs and uniting to form two main **pulmonary trunks,**
which enter the left atrium of the heart.

Heart Anatomy

Observe the demonstration of dissected adult pig and/or
beef hearts. Locate the **four chambers,** the valves between
the atria and ventricles, and the relationship of the major
blood vessels attached to the four chambers of the heart.

Remove the heart of your fetal pig by severing the
large blood vessels about one-fourth to one-half inch from
the heart and by freeing the heart and the blood vessels
from the surrounding tissues. Make a clean cut across the
heart to expose the interior of the two ventricles. Wash
out the clots of blood with cold water and note the interior
structure of the heart.

Insert the tips of your scissors into the cavity within
the right ventricle and cut through the ventricular wall to
expose the cavity within the right atrium. Note the
opening, or orifice, between the right ventricle and right
atrium guarded by the **tricuspid valve.** Make a similar incision on the left side of the heart and observe the orifice
between the two left chambers of the heart guarded by
the **bicuspid** (mitral) **valve.** How does the bicuspid valve
differ in structure from the tricuspid valve? Next locate
the opening from the left ventricle into the aorta guarded
by the **semilunar valves.** Note that these semilunar valves
consist of three pouchlike structures.

Review the general plan of circulation in the fetal pig and note especially those features of **fetal circulation** that distinguish it from the pattern of **adult circulation.**

Summary of changes in fetal circulation occurring at birth:

Arterial system
  1. Ductus arteriosus closes
  2. Umbilical arteries degenerate

Venous system
  1. Ductus venosus closes
  2. Umbilical vein degenerates

Heart
  1. Foramen ovale closes

Nervous System

Review figure 19.20 and observe the location and relative position of the principal components of the central nervous system of the fetal pig. Note the large anterior **brain** enclosed in the bony **braincase** of the skull and the **spinal cord** within the **neural canal** of the vertebral column. Cut away the skin, muscles, and dorsal half of the skull to expose the brain. Remove the covering tissues carefully to avoid damage to the brain. Note the three membranes or **meninges** surrounding the brain. The tough outer layer that adheres to the skull is the **dura mater;** underneath is the delicate **arachnoid layer,** and the thin layer that dips into the sulci (crevices) of the brain is the **pia mater.** The same three meninges also cover the spinal cord but are more difficult to observe than on the surface of the brain.

Cut away the left side of the skull and remove the dura mater. Locate and study the five main regions of the pig brain listed in table 19.3 (see also figure 19.21).

It is often more convenient and satisfactory to study the anatomy of the mammalian brain with a specimen that has been specially preserved and prepared for this purpose. The brain of the cat and the sheep are most commonly used for this purpose. The principal parts of the sheep brain are illustrated in figure 19.22a, b, and c.

You should also review your previous studies of the brains of the shark and frog and compare the brain of the pig with that of these two lower vertebrates. What are the principal differences in brain structure among these animals? Which parts of the brain are more highly developed in the pig? Which structures are less well developed or absent in the pig? Can you relate these structural differences to differences in behavior of the three species?

Dissect away the skin, connective tissue, and muscles to expose the anterior portion of the vertebral column. Next remove the neural arches of several of the cervical (neck) vertebrae to observe the spinal cord within the neural canal. Note in figure 19.20 the **enlargements** in the spinal cord in the **brachial** and in the **lumbosacral regions** and the **nerve plexes** (networks) associated with these enlargements. Locate the **cervical** and **lumbosacral nerve**

Table 19.3   Principal Regions of the Mammalian Brain

Region	Principal structure(s)
Telencephalon	Cerebral hemispheres
Diencephalon	Pituitary gland
Mesencephalon	Corpora quadrigemina
Metencephalon	Cerebellum
Myelencephalon	Medulla oblongata

plexes on your specimen by carefully dissecting away the tissues in these two regions of the vertebral column. After you have removed the muscles and connective tissues from the brachial region of the vertebral column, you will find an interconnected network of tough, whitish nerves connecting with the spinal cord. This is the **brachial plexus;** it is made up of branches from several of the spinal nerves of the pig.

There are thirty-three pairs of spinal nerves in the pig; other mammals may have more or fewer spinal nerves. The cat, for example, has thirty-eight pairs of spinal nerves. In the pig, the spinal nerves consist of eight pairs in the cervical region, fourteen pairs in the thoracic region, seven pairs in the lumbar region, and four pairs in the sacral region.

Remove the skin, connective tissue, and muscles in the lumbosacral region to locate the **lumbosacral plexus.**

Use your scalpel or scissors to cut out a section of the spinal cord about an inch in length from the cervical region. Sever also the nerves attached to the spinal cord in this region leaving sufficient nerve attached to the cord to study the attachment of the nerves to the spinal cord. Observe that each spinal nerve is formed by the union of **two roots,** one **dorsal** and one **ventral.** The former carries principally **sensory fibers,** and the latter carries principally **motor fibers.** Observe also the microscopic demonstration illustrating a cross section of a mammalian spinal cord.

The pig has twelve pairs of cranial nerves, the same number found in man. Nerves I-X correspond with those of the frog and the shark studied in previous exercises. Posterior to these are found Nerve XI, the **spinal accessory nerve,** and Nerve XII, the **hypoglossal nerve** (see figure 19.21). The spinal accessory nerve (XI) arises from several roots on the lateral surface of the spinal cord and medulla and is made up of several motor and sensory fibers connected with the shoulder muscles. The hypoglossal nerve arises from several roots on the ventral surface of the medulla and contains sensory and motor fibers connected with the tongue.

Carefully free the brain, starting from the severed spinal cord in the cervical region and proceeding anteriorly, cutting nerves close to the skull on each side and gently freeing the brain from the skull. As you work, identify as many of the cranial nerves as possible, using figures 19.21 and 19.22a, b, and c as guides.

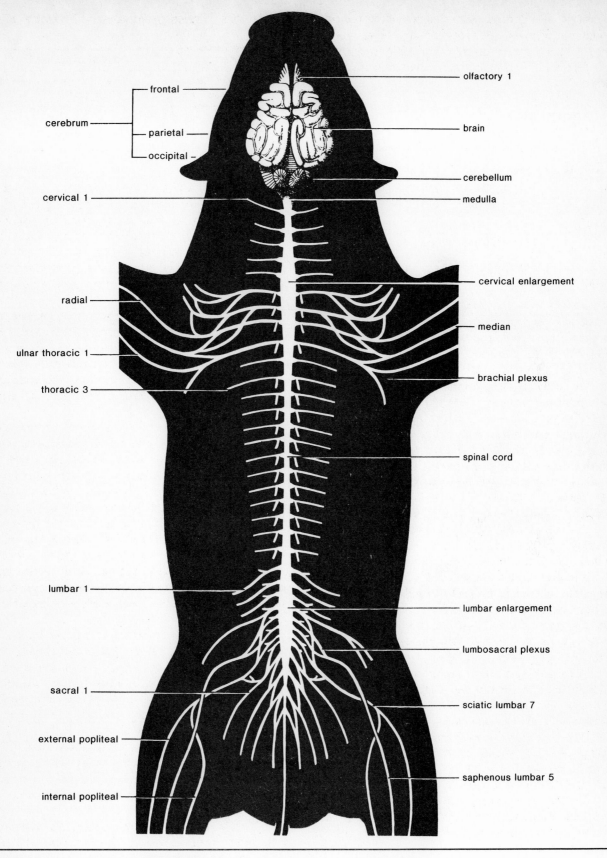

Fig. 19.20 Fetal pig, central nervous system.

olfactory 1

frontal

cerebrum

parietal

occipital

brain

cerebellum

cervical 1

medulla

cervical enlargement

radial

median

ulnar thoracic 1

brachial plexus

thoracic 3

spinal cord

lumbar 1

lumbar enlargement

lumbosacral plexus

sacral 1

sciatic lumbar 7

external popliteal

saphenous lumbar 5

internal popliteal

Fetal Pig Anatomy

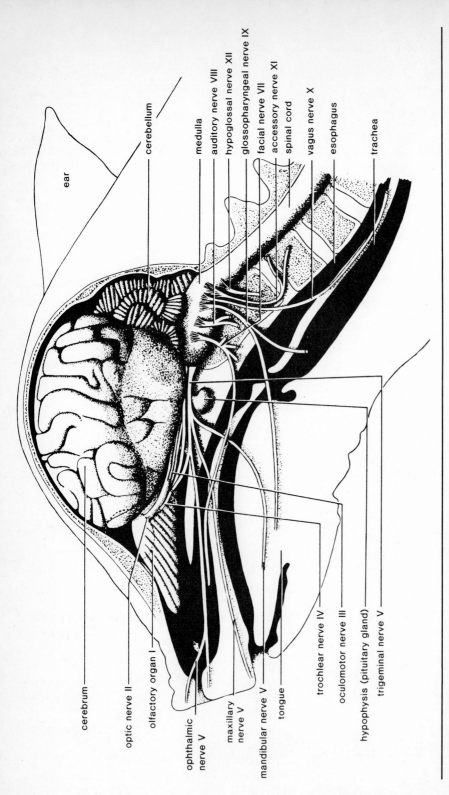

ear

cerebellum

medulla
auditory nerve VIII
hypoglossal nerve XII
glossopharyngeal nerve IX
facial nerve VII
accessory nerve XI
spinal cord
vagus nerve X
esophagus
trachea

cerebrum

optic nerve II

olfactory organ I

ophthalmic
nerve V

maxillary
nerve V

mandibular nerve V

tongue

trochlear nerve IV

oculomotor nerve III

hypophysis (pituitary gland)

trigeminal nerve V

Fig. 19.21 Fetal pig, lateral view of head with brain and cranial nerves.

Chapter 19

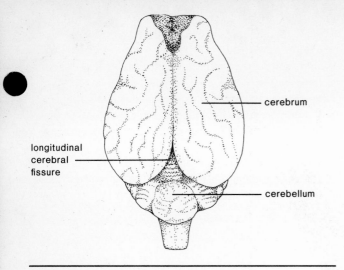

Fig. 19.22a   Sheep brain, dorsal view.

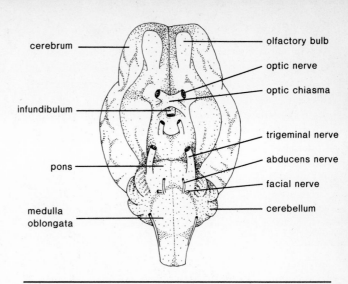

Fig. 19.22b   Sheep brain, ventral view.

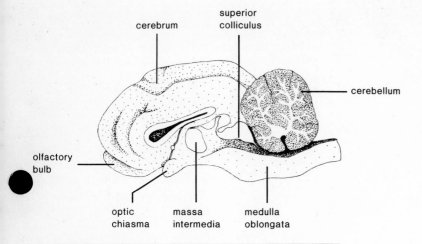

Fig. 19.22c   Sheep brain, sagittal section.

Find the **pituitary gland** and **optic chiasma** on the ventral surface, and with a *sharp* scalpel carefully bisect the brain from front to rear into equal right and left halves. Locate and study the various brain structures on your specimen with the aid of figures 19.21 and 19.22a, b, and c.

Demonstrations

1. Placentae of pig and other mammals.
2. Skeleton of fetal pig.
3. Mounted cat skeleton.
4. Fresh or dried specimens of mammalian lung.
5. Microscope slides of mammalian lung tissue.
6. Microscope slide showing germ cells in mammalian ovary.
7. Microscope slides showing mammalian testis tissue and developing spermatozoa.
8. Living mammalian sperm.
9. Microscopic demonstration of stained mammalian sperm.
10. Dissection of fetal pigs to illustrate injected arterial and venous systems.
11. Dissected pig, sheep, or beef hearts to illustrate interior chambers and valves.
12. Preserved sheep brains, whole and sections.
13. Dissected fetal pigs to show cranial and spinal nerves.
14. Microscope slide to show cross section of mammalian spinal cord.

Key Terms

Appendicular skeleton portion of the skeleton of the pig and other vertebrates that includes the bones of the appendages and the girdles, which provide their articulation with the axial skeleton.

Axial skeleton portion of the skeleton that provides support for the main axis of the body; includes the skull, the vertebral column, etc.

Mesentery a double sheet of mesodermal epithelium that serves to support various internal organs in the coelom of vertebrates.

Pericardial cavity portion of the coelom that encloses the heart; lined by the pericardium.

Peritoneal cavity (Abdominal cavity) portion of the coelom lying posterior to the diaphragm; contains the abdominal organs.

Pleural cavity portion of the coelom that encloses a lung; lined by the pleura.

Thoracic cavity portion of the coelom lying anterior to the diaphragm; subdivided in the pig and other mammals into the pericardial cavity and two pleural cavities.

Umbilical cord the cord that attaches the fetus to the placenta of the mother; contains the umbilical arteries, the umbilical vein, and the allantoic duct.

Notes and Sketches

Notes and Sketches

20
Animal Behavior

Objectives

After completing the laboratory work for this chapter, you should be able to perform the following tasks:

1. Describe an experiment to examine the effects of light on the swimming behavior of *Daphnia* and discuss the effects of light on this behavior.

2. Explain the importance of controls in experiments on animal behavior and give examples of the types of controls needed in the experiment described for Objective No. 1.

3. Describe how you would test the effects of gravity on some selected invertebrate animal. Give examples of the type of responses you would expect.

4. Tell how you would set up an experiment to demonstrate territorialism in crickets or a similar animal.

5. Define dominance hierarchy and design an experiment to demonstrate dominance relationships in crickets or a similar animal.

6. Discuss the importance of careful observation and recording in experiments on animal behavior.

Introduction

Behavior plays a central role in the existence of every animal; it is the animal's direct response to the challenges of its environment. The study of behavior is just as important to the understanding of an animal's adaptations to its environment as is the study of anatomy or physiology. The possession of long limbs and strong muscles is of little use for escaping predators unless an animal also has the appropriate behavioral response to flee from danger.

The study of behavior is especially difficult because it requires a lot of time, careful attention to the environment of the animals to be observed, and other special considerations. In this exercise we shall study two simple types of behavior: (1) the locomotory responses of animals to physical changes in their environment, such as light intensity, temperature, and humidity, and (2) social behavior, in which animals respond to other members of their own species.

Materials List

(Listed with individual experiments.)

Locomotory responses to environmental changes or stimuli are divided into two categories: **kineses** (singular: kinesis) and **taxes** (singular: taxis). In a taxis, movement is either toward (positive taxis) or away from (negative taxis) the source of the stimulus. In a kinesis, movement is random and not specifically oriented either toward or away from the source of stimulus. For example, a bright light may inhibit locomotion and thus effectively prevent movement away from a certain area. In contrast, a dim light may increase the locomotion of an animal and thus increase the probability that individuals will move away from that area. Both of these examples involve essentially random processes and thus are kineses rather than taxes.

Kineses and taxes in response to specific kinds of stimuli are given special names to indicate the nature of the eliciting stimulus: for example, phototaxis (response to light), thermotaxis (response to temperature), hygrotaxis (response to humidity), and geotaxis (response to gravity).

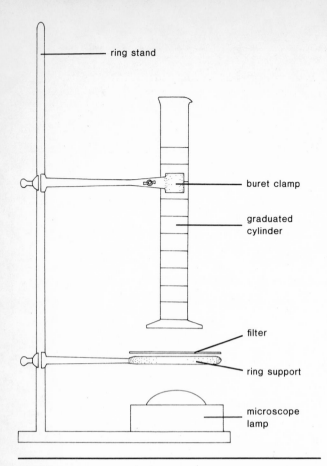

ring stand

buret clamp

graduated
cylinder

filter

ring support

microscope
lamp

Fig. 20.1 Apparatus to study light responses of *Daphnia*.

Social behavior is even more complex and may vary even among closely related species. Types of social behavior are classified according to their apparent function: for example, territorial behavior, courtship behavior, mating behavior, and aggressive behavior. Because social behavior is more complex, it is more difficult to study than taxes and kineses, but it is also more interesting. Many types of social behavior are closely associated with special morphological and physiological adaptations of animals. Consider, for example, the large antlers of a male deer, which are used for defense and in courtship, or the phosphorescent lantern of fireflies used for communication and courtship.

In the experiments that follow, we shall first study some simple taxes and kineses and then an example of social behavior.

Experiments with Taxes and Kineses

Responses of *Daphnia* to Light

The swimming behavior of the water flea *Daphnia* (Phylum Arthropoda, Subphylum Crustacea) is influenced by light. Both the intensity of light and the wavelength (color) of light have an effect on this and related aquatic crustacea. Experimental studies have indicated

that these responses to light are adaptations that aid *Daphnia* and other crustacea in feeding. It is possible to study some aspects of this behavior of *Daphnia* with relatively simple apparatus.

Materials needed:

Filtered pond water
100 ml glass graduated cylinder
Daphnia culture
Ring stand
Buret clamp
Small ring for ring stand
Thermometer
Microscope lamp
Set of colored filters
Photographic exposure meter (light meter)

Procedure

1. Attach the graduated cylinder to the ring stand using the buret clamp. Tighten the buret clamp so that the cylinder is held securely in place (see figure 20.1). Place the cylinder high enough on the ring stand that the ring and the microscope lamp can be situated beneath the cylinder.
2. Fill the graduated cylinder with filtered pond water and add about 100 *Daphnia*. Review information about *Daphnia* in chapter 13.
3. Place the ring and the microscope lamp (light off) below the cylinder as shown in figure 20.1.
4. Measure the temperature of the water with your thermometer and record the temperature.
5. Allow the animals to adjust to the new conditions for 5–10 minutes.
6. Count or estimate the number of *Daphnia* in each 10 ml section of the graduated cylinder and record the numbers in table 20.1.
7. Repeat your observations after five minutes and again after ten minutes to provide three repeat measurements. Record the temperature prior to each trial and allow the *Daphnia* to recover from the disturbance of the immersion of the thermometer prior to making your estimates of their abundance in each of the 10 ml sections.
8. Repeat the series of three observations and numerical estimates with the light at low intensity, medium intensity, and high intensity if you have a variable light source. Make a table similar to table 20.1 and record your observations carefully at each light intensity. If your lamp does not have multiple steps or variable intensity, you can achieve the same effect by varying the distance between the light source and the bottom of the cylinder. Remember that light intensity decreases with the square of the distance.
9. Repeat the same series of observations with filters of different colors placed on the ring attached to the

Table 20.1. Swimming Response of *Daphnia* to Light

Location	No. of Animals			
	Trial 1 (start)	Trial 2 (after 5 min.)	Trial 3 (after 10 min.)	Average
1–10 ml				
10–20 ml				
20–30 ml				
30–40 ml				
40–50 ml				
50–60 ml				
60–70 ml				
70–80 ml				
80–90 ml				
90–100 ml				

ring stand below the graduated cylinder. This will enable you to compare the effects of light of different parts of the spectrum (different wavelengths) on the swimming behavior of *Daphnia*. Make additional tables like table 20.1 as needed to record your data for each wavelength.

A major problem encountered in such studies of the effects of light of different wavelengths is that different filters often allow differing amounts of light energy to pass. Thus a yellow filter may produce a greater light intensity than a green filter. Be sure to measure the light intensity for each filter with a light meter. You can compensate for some minor differences in light intensity by increasing the distance between the lamp and the cylinder and checking intensity with a light meter. The best solution, however, is to obtain special (expensive!) filters, which are compensated for intensity and which pass equivalent amounts of light energy. Remember also that the energy of light rays is inversely proportional to wavelength; for example, blue light has higher energy than red light of the same intensity. How could this affect your results?

Results

After you have made several series of observations and recorded your data for *Daphnia* exposed to several conditions of illumination, study your data and answer the following questions.

1. Did the animals tend to move toward or away from the light?
2. Could you tell from watching particular individuals whether they moved toward or away from the light? Did the movement of specific individuals appear to be random or did they seem to be directed toward or away from the light? Would you classify the movements of *Daphnia* as a taxis or a kinesis?
3. To what extent did the population of *Daphnia* tend to aggregate before the light was turned on? In dim light? In bright light? Where did they tend to aggregate?
4. If you do not have filters compensated to provide equal light intensity, prepare a table with your observations of the *Daphnia* distribution with different filters arranged in order of increasing intensity (from

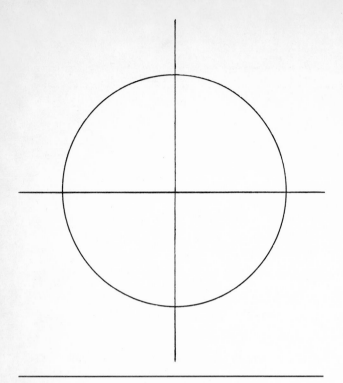

Fig. 20.2 Pattern to study movements of snails in response to gravity.

low intensity to high intensity). Viewing your data in this way, can you distinguish between the effect of light intensity and color of light?

5. Is gravity a factor in this experiment? Explain.
6. Compare your observations with those of other members of your class. How well do your results agree? Can you account for any differences in your observations with others in the class?

Responses of Some Invertebrates to Gravity

The movements of many species of animals are influenced by gravity. Some simple experiments on this type of behavior can be performed with terrestrial snails, isopods, and other slow-moving invertebrates.

Materials needed:

Drawing board
Several large sheets of graph paper
Thumbtacks
Ruler
Bricks (three or four)
Protractor
Compass
Stopwatch or watch with second hand
Several terrestrial snails or other slow-moving invertebrates

Procedure

1. Cut a piece of graph paper slightly smaller than the drawing board and attach it to the board with thumbtacks.
2. Draw a circle eight inches or more in diameter in the center of the paper with the compass. Divide the circle into four equal sections as in figure 20.2.
3. Tilt the drawing board at a 20° angle from horizontal using two bricks as supports. Measure the angle of the board with your bench top with the protractor and adjust the bricks as necessary to bring this angle to 20°.
4. Place a snail at the intersection of the lines on the graph paper in the center of the circle. Time its movements with a stopwatch or with a watch with a second hand. Observe the path of the snail and the slime trail it leaves on the graph paper as it moves out of the circle. Record the time required by the snail to reach the edge of the circle.
5. Draw a pencil line from the center of the circle where the snail started to the perimeter of the circle approximating the mean (average) path of the snail. The line of travel of the snail will almost always be irregular and wavy. Your "mean path" line should divide this irregular path so that about half the total area between the line and the actual path of the snail should be above the line and half below the line. This is really a simple form of curve fitting.
6. Measure the angle of the mean path from a horizontal line drawn through the origin and record this angle in table 20.2.
7. Replace the graph paper with a new sheet, make a new circle, and readjust the angle of the board to 40° from horizontal and repeat the experiment using the same snail. Record your results on table 20.2.
8. Replace the graph paper again, readjust the angle of the drawing board to 60°, repeat the experiment with the same snail, and record your results.
9. Repeat with an 80° angle and the same snail and record your results.
10. Repeat the experiment with a new snail and using angles of 20°, 40°, 60°, and 80°. Record your data in table 20.2 as before.

When you have collected all your data, prepare a graph with the angle of movement plotted as the ordinate (Y axis) against the angle of the board as the abscissa (X axis) on figure 20.3. Plot the data for each snail separately. Make a similar graph of the time required to reach the perimeter as the ordinate against the angle of the board as the abscissa on figure 20.4.

Table 20.2.   Response of Animals to Gravity

Animal No. 1

Angle of Board	Mean Angle of Movement	Time to Escape Circle
20°		
40°		
60°		
80°		

Animal No. 2

Angle of Board	Mean Angle of Movement	Time to Escape Circle
20°		
40°		
60°		
80°		

Questions

1. Did your snails move toward or away from the pull of gravity?
2. Does the angle of the board influence the angle of movement? How?
3. Does the angle of the board influence the amount of time required to reach the perimeter of the circle? In what way?
4. Was the behavior of the two snails the same? How did they differ? Why? If they differed, were their responses basically similar or basically different?
5. Does your data indicate that the snails you tested are positively geotactic or negatively geotactic? Why or why not? From this experiment, what can you conclude about the responses of other species of snails to gravity?

Social Behavior in Crickets

Several important aspects of social behavior can be investigated with crickets (Phylum Arthropoda, Subphylum Uniramia). In this experiment you will study the social interactions between male crickets, between male and female crickets, and between adult males and juvenile crickets. Crickets are members of the Class Insecta, Order Orthoptera, and are closely related to grasshoppers. Review the material on the grasshopper *Romalea* in chapter 13, including the description of sound production, or stridulation.

Two important aspects of social behavior in crickets are the establishment of a social hierarchy among males and territorialism. A **dominance hierarchy** is established through aggressive encounters among competing males. An aggressive encounter between males is progressive and

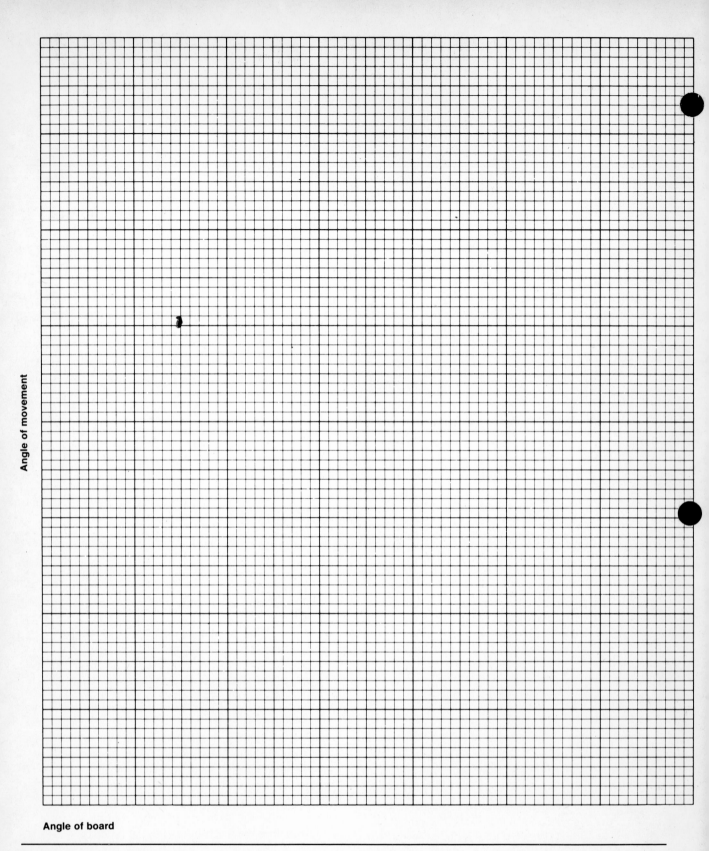

Angle of movement

Angle of board

Fig. 20.3 Diagram showing effect of board angle on direction of movement.